English	Metric

LENGTH

English	Metric
1 foot = 12 inches	1 millimeter (mm) = 0.001 meter
1 yard = 3 feet	1 centimeter (cm) = 0.01 meter
= 36 inches	1 decimeter (dm) = 0.1 meter
1 rod = 16½ feet	1 meter (m) = 1 meter
= 5½ yards	1 dekameter (dam) = 10 meters
1 statute mile = 5,280 feet	1 hektometer (hm) = 100 meters
= 1,760 yards	1 kilometer (km) = 1,000 meters
= 320 rods	

AREA

English	Metric
1 square foot = 144 square inches	1 square millimeter (mm^2)
1 square yard = 9 square feet	= 0.000001 square meter
1 square rod = 30.25 square yards	1 square centimeter (cm^2)
1 acre = 160 square rods	= 0.0001 square meter
= 4,840 square yards	1 square decimeter (dm^2)
= 43,560 square feet	= 0.01 square meter
1 square mile = 640 acres	1 square meter (m^2)
	= 1 square meter
	1 square dekameter (dam^2)
	= 100 square meters
	(common name—are)
	1 square hectometer (hm^2)
	= 10,000 square meters
	(common name—hectare)
	1 square kilometer (km^2)
	= 1,000,000 square meters

VOLUME

English	Metric
1 cubic foot = 1,728 cubic inches	1 cubic millimeter (mm^3)
1 cubic yard = 27 cubic feet	= 0.000000001 cubic meter (m^3)
	1 cubic centimeter (cm^3)
	= 0.000001 cubic meter (m^3)
	1 cubic decimeter (dm^3)
	= 0.001 cubic meter (m^3)

LIQUID

English	Metric
1 pint = 16 ounces	1 liter = 1 cubic decimeter
= 4 gills	= 1 kilogram
1 quart = 2 pints	1 milliliter = 0.001 liter
1 gallon = 4 quarts	
1 imperial gallon = 1.2 US gallons	

WEIGHT

English	Metric
1 pound = 16 ounces	1 milligram (mg) = 0.001 gram (g)
1 short ton = 2,000 pounds	1 centigram (cg) = 0.01 gram (g)
1 long ton = 2,240 pounds	1 decigram (dg) = 0.1 gram (g)
	1 kilogram (kg) = 1,000 grams (g)
	1 metric ton = 1,000 kilograms

Mathematics
for
Technical
and
Vocational
Students

Eighth Edition

Mathematics for Technical and Vocational Students

John G. Boyce
Former Chairman, Mathematics
    Department
William Tennent High School
Warminster, Pennsylvania

Consultant in Mathematics to the
Bucks County Technical School
Fairless Hills, Pennsylvania

Louis Margolis
late, Chairman, Mathematics Department
Alexander Hamilton High School
Brooklyn, New York

Samuel Slade
late, Department of Mathematics
Manhattan Vocational-Technical
    High School
New York, New York

WILEY

JOHN WILEY & SONS

New York   Chichester   Brisbane   Toronto   Singapore

Art director: Karin Gerdes Kincheloe
Text design: Katrine Stevens
Cover design: Lee Goldstein
Cover illustration: Carolyn Joseph
Text illustrations: John Balbalis with the assistance
of the Wiley Illustration Department
Production editor: Marcia Samuels
Production supervisor: Sheila Anderson

Library of Congress Cataloging-in-Publication Data:

Boyce, John G.
    Mathematics for technical and vocational students / John G. Boyce,
Louis Margolis, Samuel Slade. — 8th ed.
        p.   cm.

    Bibliography: p.
    Includes index.
    ISBN 0-471-88828-1
    1. Shop mathematics.   I. Margolis, Louis.   II. Slade, Samuel.
III. Title.

TJ1165.B66 1988
513′.14—dc19   88-22341
                                                                   CIP

Printed in the United States of America

10 9 8 7 6 5 4 3 2 1

Preface

Mathematics for Technical and Vocational Students is a textbook of practical mathematics applied to technical and trade work. Many students choose to keep the book as a reference tool. Theoretical discussions and the derivation of formulas have been reduced to their simplest forms, and there are many illustrative examples and problems that will be of great value for self-instruction and for classwork. The tables in the back of the book have been chosen to assist the student in solving problems. These tables are more comprehensive than the ones usually available to students.

The eighth edition features a review of whole numbers in Chapter 1, which provides practice in adding, subtracting, multiplying, and dividing whole numbers. A review of this material and some practice problems will probably benefit many students. We have included practice problems with large numbers in this section, as well as word problems that use large numbers (numbers in the billions, etc.).

Students are encouraged to use pencil and paper techniques to solve the problems in the text. Numerous examples are worked out in full and an explanation of the procedure is included. The student is then encouraged to apply the calculator to the solution of these problems and compare the two sets of answers. Students are also encouraged to estimate the answer before attempting a solution. This is helpful in avoiding gross errors. Some of the numbers used in the problems have more than 10 places and therefore will not fit on most calculators. I have challenged the students to devise some method of applying the calculator to the solution of these problems.

The calculator described in this edition is a scientific calculator. I believe that the functions on the calculators that students use will be the same as or similar to the functions on the calculator described in the text. Many people who use the text suggested that problem solving with the calculator be introduced only after the material has first been learned, and practice provided, using pencil and paper methods for solving the problems. We have adopted this suggestion. Apart from addition, subtraction, multiplication, and division, calculator functions are explained as each function is introduced. If the material presented involves a function not previously used, e.g., y^x, the calculator section is placed after the problem set in which the function first appears. If the functions needed are the same as functions previously used, then there are no calculator exercises, since calculator skills are cumulative and the student should be able to apply the calculator to the solution of those problems.

All chapters from the seventh edition have been retained as valuable tools in the study of vocational and technical mathematics. The round-off rule now used is the round-off rule most users apply.

We would appreciate further comments from users of the book, including comments concerning possible new material for future editions, new or different methods of solving problems, material that you may feel is not necessary to the text, and any errors you might find.

We thank the many people who offered suggestions and criticisms as the work progressed.

John G. Boyce

Contents

1

Whole Numbers

1-1 Introduction

The set of *whole numbers* consists of the set of counting numbers (i.e., 1, 2, 3, 4, 5, . . .) and the number zero (0). Many of our everyday mathematical activities use whole numbers exclusively. When someone asks you how old you are or how much you weigh, you usually round the answer off to a whole number. If you buy an item that is priced at two for 25 cents, then the cost of one is usually 13 cents, a whole number. The post office deals in whole numbers when you mail a letter or package. If your package weighs $10\frac{1}{4}$ ounces, you pay for 11 ounces. Can you think of other examples where whole numbers are used?

1-2 Addition of Whole Numbers

We use the *decimal system* with the Arabic numerals 0, 1, 2, 3, 4, 5, 6, 7, 8, and 9 in our everyday work. These ten numerals can be arranged to represent numbers of any size, and in every number the position of each numeral determines its value. For example, in the number 265, the rightmost numeral (the 5) means one times five. The number immediately to the left of the rightmost numeral (the 6) means ten times six, and the numeral immediately to its left (the 2) means one hundred times two. The number 265 then reads two hundred sixty-five. This positioning is called *place value* and is summarized in the following chart (Fig. 1-1) for numbers up to, but not including, ten million.

Place Value Chart						
Millions	Hundred thousands	Ten thousands	Thousands	Hundreds	Tens	Ones
5	4	7	4	8	2	1

Fig. 1-1

Because of the position of the digits in this number, we read the number as five million, four hundred seventy-four thousand, eight hundred twenty-one.

Example
     Add 123, 254, 52, and 964.

SOLUTION   Arrange the addends in vertical format as follows:

$$\left. \begin{array}{r} 123 \\ 254 \\ 52 \\ + \ 964 \end{array} \right\} \text{addends}$$

Add the ones column (the rightmost): $3 + 4 + 2 + 4 = 13$. Write the 3 under the ones column and carry the tens digit (the numeral 1) over to the tens column. Add the tens column, including the carried 1: $1 + 2 + 5 + 5 + 6 = 19$. Write the 9 under the tens column and carry the hundreds digit, 1, over to the hundreds column (the 19 really means 19 tens or 190). Add the hundreds column, including the carried 1: $1 + 1 + 2 + 9 = 13$. Write the 3 under the hundreds column and carry the thousands digit, 1, over to the thousands column. Since there are no thousands in the problem, write the 1 to the left of the 3 in the answer (sum). Your work will look like this:

$$\begin{array}{r} 1\ 1\ 1 \quad \leftarrow \text{carries} \\ 1\ 2\ 3 \\ 2\ 5\ 4 \\ 5\ 2 \\ + \quad 9\ 6\ 4 \\ \hline 1\ 3\ 9\ 3 \quad \leftarrow \text{sum} \end{array} \Big\} \text{addends}$$

Addition problems are usually added in vertical format as just shown. All of the addends are aligned to the right so that the units, or ones, (the right-most digits) line up one under the other. Carefully aligning the columns is an important part of neat and accurate work! Check your addition by adding the numbers again. If you add downward, then add upward for the check. Your answer should be the same both times.

Problems 1-2

Add the following numbers as indicated.

1.	278	2.	473	3.	772	4.	225
	805		380		82		298
	399		222		993		531
	258		714		588		759
	320		26		438		539
	+ 927		+ 404		+ 406		+ 617

5.	131	6.	24	7.	674	8.	752
	887		910		273		791
	511		802		281		32
	705		19		898		429
	760		568		750		565
	+ 954		+ 625		+ 23		+ 492

9.	570	10.	54	11.	848	12.	223
	982		592		657		621
	594		654		346		60
	172		336		580		206
	639		293		691		963
	+ 746		+ 469		+ 386		+ 822

13.	167	14.	847	15.	326	16.	196
	791		169		430		588
	427		864		864		914
	299		71		457		738
	222		122		362		299
	+ 543		+ 408		+ 103		+ 7

17.	537	18.	362	19.	436	20.	152
	626		121		643		476
	709		64		682		686
	715		737		588		30
	572		393		911		813
	+ 946		+ 396		+ 392		+ 182

21.	847	22.	727	23.	303	24.	995
	907		160		710		552
	562		935		322		768
	531		458		763		949
	50		369		36		915
	+ 413		+ 374		+ 579		+ 718

25.	308	26.	5,741	27.	6,912	28.	4,911
	664		3,996		2,498		2,801
	29		4,304		5,310		9,601
	844		1,580		3,103		9,835
	812		5,285		516		4,373
	+ 764		1,187		8,001		8,711
			9,572		6,135		1,522
			3,754		5,086		4,634
			4,855		5,131		4,389
			5,470		579		2,281
			3,451		3,023		3,660
			2,964		7,787		3,689
			4,470		5,265		7,074
			2,365		7,710		7,822
			+ 3,748		+ 5,988		+ 8,455

29.	9,002	30.	3,228	31.	1,672	32.	2,996
	9,531		3,713		6,052		699
	8,897		5,068		9,925		8,450
	5,448		9,692		6,483		4,942
	4,249		5,938		7,179		6,238
	2,448		2,732		4,470		849
	34		9,165		6,964		1,942
	3,622		5,850		3,861		6,124
	5,240		9,573		5,028		3,370
	6,191		4,007		2,150		3,934
	5,946		6,703		4,489		1,626
	1,248		4,369		9,637		7,742
	7,896		7,990		3,292		6,203
	4,397		3,575		6,036		5,957
	+ 8,252		+ 342		+ 6,155		+ 5,033

33.	34.	35.
8,566	2,130	2,982
7,945	1,238	1,976
5,450	8,273	2,100
9,447	3,544	4,568
3,569	5,565	96
4,293	8,111	828
7,694	8,109	8,373
6,751	2,208	2,647
6,893	4,798	6,381
2,029	3,450	9,479
6,435	1,673	7,474
5,774	3,900	1,639
2,320	9,868	5,194
5,112	4,278	6,681
+ 3,106	+ 2,575	+ 7,170

36.	37.	38.
7,940,149,875	68,067,239,787	771,852,260,717
27,570,978,261	22,140,698,112	5,841,725,401
292,754,944,000	81,182,737,000	87,765,160,384
52,692,198,749	190,109,375	335,702,375
328,223,422,792	143,795,466,919	68,870,582,299
582,810,602,977	51,936,866,891	769,330,693,747
83,797,669,376	5,322,708,936	426,957,777
423,395,573,229	85,358,358,827	251,956,962,496
40,318,322,589	381,551,383,277	120,553,784
551,368	889,999,523,659	26,357,170,176
83,970,131,939	678,687,959,872	62,005,093,912
39,200,051,773	96,071,912	276,276,047,697
299,955,703,384	88,002,363,392	2,851,206,632
121,287,375	650,812,752,296	165,650,055,488
+ 177,504,328	+ 154,076,860,881	+ 22,882,115,719

39.	40.
250,881,416,443	257,380,823,881
2,347,334,289	53,327,207,744
3,334,661,784	1,481,544
208,422,380,089	472,729,139
331,373,888,000	54,224,761,625
369,219,064,024	149,721,291,000
15,870,019,697	382,499,074,979
257,623,674,147	935,441,352,000
668,774,093,625	269,711,350,181
340,799,721,625	33,138,024,128
623,711,453,184	186,365,104,128
75,098,035,459	150,993,703,125
5,969,141,144	182,863,555,776
212,883,113,611	411,830,784
+ 384,240,583,000	+ 35,611,289

Rearrange the numbers so that they are in vertical format and add.

41. 6 + 87 + 92 + 34 + 12 + 4 + 197
42. 145 + 465 + 437 + 32 + 96 + 389 + 1,008 + 345
43. 1,987 + 2,543 + 3,678 + 156 + 12 + 5,432 + 98,135 + 6,543
44. 198 + 1,923 + 19,245 + 19,876 + 123 + 9,854 + 2,654 + 198,999
45. 1,009 + 1,000 + 18,763 + 1,965 + 18,639 + 1,859 + 16,846 + 198,871
46. 834 + 8,345 + 37,445 + 24,654 + 2,765 + 24,655 + 29,864 + 468
47. 123 lb + 145 lb + 98 lb + 133 lb + 165 lb + 170 lb
48. 1,678 cars + 1,986 cars + 1,776 cars + 1,492 cars + 2,001 cars
49. 19,876 tons + 18,872 tons + 17,655 tons + 28,775 tons
50. 155 miles + 345 miles + 132 miles + 188 miles + 200 miles
51. If a contractor poured 12 cubic yards of concrete on Monday, 15 cubic yards of concrete on Tuesday, 10 cubic yards of concrete on Wednesday, 15 cubic yards of concrete on Thursday, and spent the rest of the week removing the forms and cleaning up, how many cubic yards were poured for the week?
52. If a sheetrock mechanic has three jobs that require 120 4-by-8 sheets, 115 4-by-8 sheets, and 130 4-by-8 sheets of sheetrock respectively, how many 4-by-8 sheets of sheetrock are needed to complete the three jobs?
53. If an electrician needs to install 5 receptacles in the living room, 7 receptacles in the study, 2 receptacles in the bathroom, 6 receptacles in the master bedroom, 4 receptacles in the smaller bedroom, and 10 receptacles in the kitchen, how many receptacles must be installed?
54. If the monthly production of cars was as follows: January, 4,356; February, 4,353; March, 4,400; April, 4,290; May, 4,425; June, 4,287; July, 4,456; August, 4,223; September, 4,265; October, 4,365; November, 4,109; and December, 4,001; how many cars were produced for the year?
55. The number of people who immigrated to the United States during the 1800s is as follows: 1800–1820, 8,385; 1821–1830, 143,439; 1831–1840, 599,125; 1841–1850, 1,713,251; 1851–1860, 2,598,214; 1861–1870, 2,314,824; 1871–1880, 2,812,191; 1881–1890, 5,246,613; 1891–1900, 3,687,564. How many people immigrated to the United States in the 1800s? (Source: U.S. Immigration and Naturalization Service.)

 1-3 Calculator Addition

The *scientific calculator* is a great help to the vocational–technical person. The danger in using a calculator to solve problems is that the user often becomes dependent upon the answers given by the calculator, and if the wrong sequence of keys is pressed on the calculator, the answer can be incorrect. Care must be taken to enter the correct values, and care must be taken to press the keys in the correct sequence. Each entry must be analyzed by the operator, and the results of the operation must be considered carefully. Always remember that a calculator is a machine and therefore follows your directions blindly. It can respond only to the keys you press and the sequence in which you press them. If you press the wrong key, you'll probably get the wrong answer.

Example
    Using a calculator, do Problem 1 from Problem Set 1-2 again.

    1.   278
           805
           399
           258
           320
      + 927

SOLUTION   Use the calculator in the following manner.

	The display shows
Turn on the calculator	0.
Enter 278	278.
Press +	278.
Enter 805	805.
Press +	1083.
Enter 399	399.
Press +	1482.
Enter 258	258.
Press +	1740.
Enter 320	320.
Press +	2060.
Enter 927	927.
Press =	2987. (answer)

Problems 1-3

Do all the rest of Problem Set 1-2 in the same manner. Check your calculator answers against the answers you got with pencil and paper. You will notice that you cannot do Problems 36–40 with your calculator. Most calculators have eight-digit or ten-digit capability, and these problems have more than ten digits in the addends. You might figure out some way of doing these problems with the calculator by doing each addend in two steps. Good luck.

1-4   Subtraction of Whole Numbers

Subtraction can be thought of as the opposite of addition. It is the process of determining the difference between two numbers. The number that is to have another number subtracted from it is called the ***minuend***, while the number that is to be subtracted is called the ***subtrahend***. The answer to the subtraction process is called the ***difference***.

Example
Subtract 126 from 543

SOLUTION   The problem indicates that 543 is the minuend and 126 is the subtrahend. Write the numbers in vertical format.

$$543 \leftarrow \text{minuend}$$
$$-\ 126 \leftarrow \text{subtrahend}$$

Start subtracting in the units column, that is, subtract the 6 from the 3. Since 6 cannot be subtracted from 3, you must borrow a ten from the tens column and subtract 6 from 13, which equals 7. Place the 7 in the units column in the answer. The 4 in the tens column is now 3, since you borrowed a ten from it. Now subtract the 2 from the 3, which gives a remainder of 1. Place this 1 in the tens column in the answer. Lastly, in the hundreds column, subtract the 1 from the 5 which equals 4. Place this difference in the hundreds column of the answer. The work will look like this:

$$\begin{array}{r} \overset{3}{5}\overset{1}{4}3 \leftarrow \text{minuend} \\ -\ 126 \leftarrow \text{subtrahend} \\ \hline 417 \leftarrow \text{difference} \end{array}$$

To check your answer, add the difference and the subtrahend, and that sum should be the minuend.

Problems 1-4

1. Subtract 381 from 1,895.
2. Subtract 146 from   850.

3. Subtract 852 from 1,682.
4. Subtract 132 from   957.

5. Subtract 665 from 1,084.
6. Subtract 670 from 2,064.
7. Subtract 443 from   694.

8. Subtract 739 from 1,591.
9. Subtract 154 from 1,284.
10. Subtract 754 from   772.

11.   10,327
    − 7,477

12.   8,851
    − 8,453

13.   7,561
    − 5,339

14.   9,443
    − 4,045

15.   7,609
    − 6,957

16.   8,905
    − 1,846

17.   14,652
    − 9,195

18.   8,858
    −   182

19.   19,338
    − 7,353

20.   19,919
    − 8,881

21.   13,488
    − 3,497

22.   8,888
    − 1,015

23.   8,322
    − 4,090

24.   1,224
    −   595

25.   17,431
    − 4,548

26.   13,380
    − 3,121

27.   5,715
    − 4,183

28.   18,446
    − 5,461

29.   13,896
    − 3,849

30.   4,430
    − 1,250

31. In the 1986 Congressional election, Arlen Spector received 1,902,126 votes, while his opponent, Bob Edgar, received 1,438,393 votes. How many more votes did Spector receive than Edgar? (Source: News Election Service.)

32. In May of 1985 there were 36,686,000 beneficiaries in the Social Security program, while in May of 1986 there were 37,344,000 beneficiaries. What was the increase in beneficiaries in that one year? (Source: Social Security Administration.)

33. The net receipts for individual income taxes for 1986 were $416,568,384,000; in 1985 they were $396,659,558,000. What was the increase in net receipts from 1985 to 1986? (Source: U.S. Treasury Department.)

34. In 1985 South Africa mined 48,216,620 troy ounces of gold, while the United States mined 2,475,436 troy ounces of gold. How many more troy ounces of gold were mined by South Africa than by the United States? (Source: IMF, International Financial Statistics.)

35. In 1986 General Motors had total sales of $102,813,700,000, while Ford Motors had total sales of $62,715,800,000. Compared to Ford, how much greater were the total sales of General Motors? (Source: *Fortune Magazine*.)

36. The U.S. farm debt in 1984 was $111,637,035,999, and in 1985 the farm debt was $105,410,951,000. What was the dollar decrease in the debt between 1984 and 1985? (Source: U.S. Department of Agriculture.)

37. The U.S. Bureau of the Census estimates that the population of the United States is now 241,078,000, and that 117,360,000 of these are male. How many females are there, according to this estimate? How many more females are there than males?

38. In 1945 there were 8,266,373 U.S. Army personnel on active duty. In 1986 there were 781,023 U.S. Army personnel on active duty. Compared to 1986, how many more people on active duty did the Army have in 1945? (Source: Department of the Army.)

39. In 1986 *Reader's Digest* had a circulation of 16,609,847, and *TV Guide* had a circulation of 16,800,441. How many fewer issues of *Reader's Digest* were circulated than those of *TV Guide?* (Source: Audit Bureau of Circulations; FAS-FAX Report.)

40. Mount Everest, on the border of Nepal and Tibet, is 29,028 feet high, while Mount McKinley in Alaska is 20,320 feet high. How much higher is Mount Everest than Mount McKinley? (Source: National Geographic Society, Washington, D.C.)

1-5   Calculator Subtraction

Subtraction on the calculator depends on accurate keying in of the subtrahend and minuend. Make it a practice to read the number in the display after you enter it on the keyboard.

Example

Using a calculator, do Problem 1 from Problem Set 1-4 again. Subtract 381 from 1,895.

SOLUTION   Use the calculator in the following manner.

The display shows

Turn on the calculator	$0.$
Enter 1895	$1895.$
Press −	$1895.$
Enter 381	$381.$
Press =	$1514.$ (answer)

Problems 1-5

Do the rest of Problem Set 1-4 in the same manner. Check your calculator answers against the answers you got with pencil and paper. If your answers do not agree, try to analyze why the answers are different so that you do not make the same mistake in the future.

1-6 Multiplication of Whole Numbers

Multiplication can be thought of as addition repeated a given number of times. For example, 3 times 5 can be solved by 5 + 5 + 5. The 3 means that the 5 is to be used a total of three times. The same problem can also be thought of as 5 times 3, or 3 + 3 + 3 + 3 + 3. Written this way, the 3 is used a total of five times. In either case the solution is 15. When two numbers are placed in vertical format for multiplication, the number in the upper position is called the *multiplicand* and the number in the lower position is called the *multiplier*. The answer to the multiplication problem is called the *product*. Partial products, shown in the following example, are intermediate results of the multiplication process.

Example

Find the product of 234 and 46.

SOLUTION   Place the numbers in vertical format, such that 234 is the multiplicand and 46 is the multiplier. Starting from the units column, multiply the 6 by the multiplicand's 4. The product of 6 times 4 is 24, which can be thought of as 2 tens and 4 ones. Write the 4 in the answer's unit column and carry over the 2 to the tens column. Now multiply the 6 by the multiplicand's 3, which equals 18. Add the carried-over 2 to the 18 for a sum of 20. Write the 0 in the answer's tens column and carry over the 2 to the hundreds column. Now multiply the 6 by the multiplicand's 2, which equals 12. Add the carried-over 2 to the 12 for a sum of 14. Write 14 into the answer, with the 4 in the hundreds column and the 1 in the thousands column. What now appears is 1,404, the product of 6 times 234, which is the first partial product of this problem.

$$
\begin{array}{r}
2\,2 \leftarrow \text{carries} \\
2\,3\,4 \leftarrow \text{multiplicand} \\
\times \quad 4\,6 \leftarrow \text{multiplier} \\
\hline
1\,4\,0\,4 \leftarrow \text{first partial product}
\end{array}
$$

The next step is to multiply the multiplicand, 234, by the multiplier's tens column, 4. Multiply this 4 by the multiplicand's 4, which equals 16. Write the 6 in the answer's ten column (remember, the 16 is really 16 tens) and carry over the 1 to the tens column. Now multiply the 4 by the multiplicand's 3, which equals 12. Add the carried-over 1 to the 12 for a sum of 13. Write the 3 in the answer's hundreds column and carry over the 1 to the hundreds column. Now multiply the 4 by the multiplicand's 2, which equals 8. Add the carried-over 1 for a sum of 9. Write the 9 in the answer's thousands column. Imagining the zero that belongs in the units column, you can see that the second partial product is 9,360.

$$
\begin{array}{r}
1\,1 \leftarrow \text{carries} \\
2\,3\,4 \leftarrow \text{multiplicand} \\
\times \quad 4\,6 \leftarrow \text{multiplier} \\
\hline
1\,4\,0\,4 \leftarrow \text{first partial product} \\
9\,3\,6 \leftarrow \text{second partial product}
\end{array}
$$

The final step to this multiplication process is to add the partial products to get the final product.

Therefore, the product of 46 and 234 is 10,764.

$$
\begin{array}{r}
234 \\
\times \quad 46 \\
\hline
1404 \leftarrow \text{first partial product} \\
+ \quad 939 \leftarrow \text{second partial product} \\
\hline
10764 \leftarrow \text{product}
\end{array}
$$

Problems 1-6

1. Find the product of   852 and 10,000.
2. Find the product of   132 and   100.
3. Find the product of   665 and   100.
4. Find the product of   670 and 10,000.
5. Find the product of   694 and   10.
6. Find the product of   739 and   1,000.
7. Find the product of   154 and   1,000.

8. Find the product of 754 and 100.
9. Find the product of 393 and 100.
10. Find the product of 948 and 1,000.
11. Find the product of 446 and 1,000.
12. Find the product of 791 and 100.
13. Find the product of 323 and 1,000.
14. Find the product of 160 and 10.
15. Find the product of 1,063 and 10,000.
16. Find the product of 267 and 10,000.
17. Find the product of 527 and 100.
18. Find the product of 322 and 1,000.
19. Find the product of 947 and 10.
20. Find the product of 964 and 10.
21. Find the product of 493 and 496.
22. Find the product of 536 and 743.
23. Find the product of 782 and 688.
24. Find the product of 1,011 and 492.
25. Find the product of 252 and 576.
26. Find the product of 786 and 130.
27. Find the product of 913 and 282.
28. Find the product of 947 and 1,007.
29. Find the product of 662 and 631.
30. Find the product of 150 and 513.
31. Find the product of 827 and 260.
32. Find the product of 1,035 and 558.
33. Find the product of 469 and 474.
34. Find the product of 403 and 810.
35. Find the product of 422 and 863.
36. Find the product of 136 and 679.
37. Find the product of 1,095 and 652.
38. Find the product of 868 and 1,049.
39. Find the product of 1,015 and 818.
40. Find the product of 408 and 764.
41. If a mechanic makes $400 a week, what is the mechanic's yearly salary? (52 weeks = one year.)
42. If a bricklayer can lay 165 bricks in one hour, how many bricks can this bricklayer lay in an eight-hour day?
43. If 1 cubic yard of concrete costs $45, how much would 13 cubic yards cost?
44. If your car's average speed is 50 miles per hour, how far can you drive in nine hours?
45. If a worker can make 357 bolts in one hour, how many bolts can this worker make in eight hours?
46. If oak flooring costs two dollars a square foot, then how much would 1,350 square feet of oak flooring cost?

47. If light travels 186,000 miles per second, how far does light travel in one year? (60 seconds = one minute; 60 minutes = one hour; 24 hours = one day; 365 days = one year.)

48. If the satellite Voyager II, launched September 20, 1977, travels at 17,000 miles per hour, how far does it travel in one year? In 10 years? How far has it traveled by today's date?

49. Sound travels approximately 1,100 feet per second in air. How far will sound travel in two minutes?

50. One cubic foot contains 1,728 cubic inches. How many cubic inches are contained in 25 cubic feet?

 ## 1-7 Calculator Multiplication

Multiplication on the calculator depends upon accurate keying in of the multiplier and the multiplicand. The operator must be careful not to exceed the display capability of the calculator in use. Many calculators can hold 10 place numbers, that is, numbers up to but not including 10 billion. If the calculator answer to a problem is greater than the display capability, then the calculator may display an error message, or it might display the answer in scientific notation. (Scientific notation will be discussed later in this book.)

Example
Using a calculator, do Problem 1 from Problem Set 1-6 again.
Find the product of 852 and 10,000.

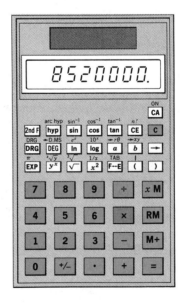

SOLUTION   Use the calculator in the following manner.

                          The display shows

    Turn on the calculator   0.
    Enter 852   852.
    Press ×   852.
    Enter 10000   10000.
    Press =   8520000. (answer)

Problems 1-7

Do the rest of Problem Set 1-6 in the same manner. Check your calculator answers against the answers you got with pencil and paper. If your answers do not agree, analyze where the error might be so that you will not make the same mistake again.

1-8   Division of Whole Numbers

Division can be thought of as repeated subtraction. For example, 12 divided by 4 can be solved by subtracting the 4 from the 12 for a remainder of 8, subtracting the 4 from the remaining 8 for a new remainder of 4, and subtract the 4 from the remaining 4 for a final remainder of zero. It took three steps of subtracting 4 from 12 to reach zero, so 12 divided by 4 is 3. The 12 in this problem is called the **dividend**, the 4 is called the **divisor**, and the 3 is called the **quotient**. Signs that indicate division are $\div$, $\overline{)}$, and /. Therefore, dividing 12 by 4 can be indicated by $12 \div 4$ or $4\overline{)12}$ or 12/4 or $\frac{12}{4}$.

Division can be done by repeated subtraction as just demonstrated, but this process can be unwieldy. An algorithm (steps to follow for a solution) is used to speed up the division process.

Example
Divide 1,924 by 37.

SOLUTION   Using the division symbol $\overline{)}$, enter 1,924 as the dividend and 37 as the divisor.

$$\text{divisor} \rightarrow 37\overline{)1924} \leftarrow \text{dividend}$$

First determine how many times 37 may be divided into the first few digits (from the left) of the dividend. The 37 cannot be divided into (or subtracted from) the 1 or the 19 of the dividend, but it can be divided into the 192 of the dividend. The question that must now be answered is: How many times can 37 be subtracted from 192 before the remainder is less than 37? If the 37 is thought of as 40 and the 192 is thought of as 200, then you can mentally think of the number of subtractions that can be made. In this case you should figure that 40 can be subtracted from 200 five times. (The 40 and 200 were chosen because they are nearly equal to the 37 and the 192 but are much easier to use mentally.) So, try 5 as the number of times that 37 can be subtracted from 192.

$$37)\overline{1\,9\,2\,4}\quad 5$$

Because the 192 means 192 tens, the 5 means 5 tens, so write the 5 in the quotient's tens place. To check if this is correct, multiply the 5 by the divisor 37. If 5 times 37 is greater than 192, then 5 is too great a number for the quotient's tens place. In this case, 5 is not too great a number because the product of 5 and 37 is 185. This 185 means 185 tens, so align it properly under the 192, as shown.

$$37)\overline{1\,9\,2\,4}\quad 5$$
$$-1\,8\,5$$
$$7$$

Now subtract the 185 from the 192. If the remainder is greater than the divisor, 37, the 5 is too small a number for the quotient's tens place. In this case, 5 is not too small a number, because the remainder of 185 from 192 is 7. This 7 means 7 tens, so write it in the tens column under the subtraction bar.

To determine the rest of the quotient, bring down the dividend's units amount, 4, to be beside the remainder 7. Now you will be dividing the 37 into the 74. Again, you can make a reasonable guess by thinking of 37 and 74 as 40 and 80. You should be able to figure that 40 can be subtracted from 80 two times.

$$37)\overline{1\,9\,2\,4}\quad 5$$
$$-1\,8\,5$$
$$7\,4$$

Write the 2 in the quotient's units place. To check if this is correct, multiply the 2 by the divisor, 37. As previously explained, the product must be greater than 74. In this case, 2 times 37 is exactly 74, and when 74 is subtracted from 74, the remainder is zero.

$$
\begin{array}{r}
52 \leftarrow \text{quotient} \\
\text{divisor} \rightarrow 37)\overline{1924} \leftarrow \text{dividend} \\
-185 \\
\hline
74 \\
74 \\
\hline
0 \leftarrow \text{remainder}
\end{array}
$$

Therefore, the quotient of 1,924 divided by 37 is 52.

1. Divide 4,330 by 10.	21. Divide 64,897 by 73.
2. Divide 52,400 by 100.	22. Divide 58,158 by 81.
3. Divide 17,500 by 100.	23. Divide 7,704 by 18.
4. Divide 7,260,000 by 10,000.	24. Divide 31,248 by 36.
5. Divide 8,790 by 10.	25. Divide 13,621 by 53.
6. Divide 4,450,000 by 10,000.	26. Divide 12,731 by 29.
7. Divide 5,370,000 by 10,000.	27. Divide 26,345 by 55.
8. Divide 18,100 by 100.	28. Divide 11,842 by 62.
9. Divide 7,000 by 100.	29. Divide 15,120 by 16.
10. Divide 9,170 by 10.	30. Divide 44,732 by 53.
11. Divide 6,320 by 10.	31. Divide 11,496 by 958.
12. Divide 29,800 by 100.	32. Divide 18,468 by 486.
13. Divide 6,520 by 10.	33. Divide 19,030 by 346.
14. Divide 55,000 by 100.	34. Divide 13,410 by 447.
15. Divide 6,310 by 10.	35. Divide 9,000 by 375.
16. Divide 150,000 by 1,000.	36. Divide 17,500 by 250.
17. Divide 692,000 by 1,000.	37. Divide 16,794 by 311.
18. Divide 252,000 by 1,000.	38. Divide 4,806 by 801.
19. Divide 875,000 by 1,000.	39. Divide 31,558 by 509.
20. Divide 85,500 by 100.	40. Divide 3,016 by 58.

41. A sheetrock contractor agrees to install and spackle 150 sheets of $\frac{1}{2}$-inch 4-by-8 sheetrock for $1,200. What is the installation cost per sheet?

42. A carpenter is to install a staircase that will have 8-inch risers (the height between steps), and the total height of the staircase is to be 8 feet. How many steps will the staircase have? (12 inches = 1 foot.)

43. How many 15-inch long pieces of wire can be cut from a 100-foot roll of wire?

44. A contractor agrees to build a 3,500 square foot house for $77,000. What is the cost per square foot?

45. A motorist drives 376 miles in 8 hours. What is the motorist's average speed per hour?

46. An airplane travels 3,630 miles in 6 hours. What is the airplane's average speed per hour?

47. A gallon of water has a volume of 231 cubic inches. How many gallons are in a full tank that holds 5,775 cubic inches?

48. A car company made 31,025 cars in 1986. What was its average number of cars made per day for the entire year? (365 days = 1 year.)

49. Last year Alabama received $281,060 in motor fuel taxes. If the drivers in Alabama bought 2,162,000 gallons of fuel, what was Alabama's tax per gallon in cents?

50. If the area of the United States is 3,618,770 square miles and there are 241,078,000 people in the United States, about how many people are there per square mile?

1-9   Calculator Division

Division on the calculator not only involves careful entry of the dividend and the divisor, it also requires that you enter the dividend before you enter the divisor. Read the problem thoroughly and determine which number is the dividend and which number is the divisor. If the answer seems improbable, you may have entered the numbers in the wrong sequence.

Example

Using a calculator, do Problem 1 from Problem Set 1-8 again.
Divide 4,330 by 10.

SOLUTION   Use the calculator in the following manner.

The display shows

Turn on the calculator	$0.$
Enter the dividend, 4330	$4330.$
Press $\div$	$4330.$
Enter the divisor, 10	$10.$
Press $=$	$433.$ (answer)

Problems 1-9

Do the rest of Problem Set 1-8 in the same manner. Be watchful of the dividend and the divisor in the word problems. Check your calculator answers against the answers you got with pencil and paper. If they do not agree, analyze where you might have made a mistake so that you will not make that mistake again.

Do the following problems. Check your answers with the answers in the back of the book.

1. Add: 231 + 345 + 1,000 + 10,000
2. Add: 3,459 + 8,973 + 3,456
3. Subtract: 3,478 − 2999
4. Subtract: 3,450,435 − 9,839
5. Multiply: 549 × 10,000
6. Multiply: 5,469 × 3,981
7. Divide: 11,349 ÷ 873
8. Divide: 10,000 ÷ 400
9. The United States Treasury determined that it has in circulation $3,571,913,726 in one-dollar bills; $707,773,472 in two-dollar bills; $4,939,073,340 in five-dollar bills; $11,363,371,940 in ten-dollar bills; $51,586,205,780 in twenty-dollar bills; $21,715,575,800 in fifty-dollar bills; $76,516,974,400 in hundred-dollar bills; $154,472,000 in one-thousand-dollar bills; and $3,470,000 in ten-thousand-dollar bills. How much paper money is in circulation in the United States?
10. Based on the information in Problem 9, how many individual bills are in circulation in the United States?

Chapter Test

Add:

1.		2.	
	152,818		153,045
	157,699		254,515
	58,830		540,678
	513,162		311,581
	438,256		531,488
	125,424		77,740
	28,512		270,882
+	210,912	+	72,416

3.		4.	
	549,400		148,509
	98,208		135,408
	416,816		565,972
	11,533		189,720
	524,095		342,160
	657,020		381,034
	360,079		113,040
+	245,971	+	57,420

5. 119,883
    62,548
    494,262
    190,512
    8,804
    420,920
    669,536
    + 635,024

6. 111,408
    120,560
    651,900
    237,216
    243,089
    383,959
    128,478
    + 62,037

7. 230,868
    403,323
    104,300
    182,535
    41,904
    9,190
    221,628
    + 408,285

8. 298,073
    6,435
    57,304
    163,784
    104,974
    865,488
    150,378
    + 275,010

Subtract:

9.   98,362
     − 6,052

10. 168,997
     − 93,755

11. 161,211
     − 130,441

12. 155,465
     − 121,907

13. 112,846
     − 52,241

14. 143,463
     − 4,794

15. 137,938
     − 129,863

16. 126,633
     − 124,763

Multiply:

17. 4,060
     × 827

18. 1,437
     × 913

19. 1,127
     ×   49

20. 2,841
     × 984

21. 3,640
     × 936

22. 2,088
     × 358

23. 5,166
     × 512

24. 2,805
     × 198

Divide:

25. $490\overline{)34,790}$ 26. $60\overline{)5,820}$

27. $343\overline{)11,662}$ 28. $273\overline{)18,291}$

29. $182\overline{)9,464}$ 30. $96\overline{)5,952}$

31. $644\overline{)16,100}$ 32. $72\overline{)4,968}$

Answer the following questions:

33. In 1985 the U.S. Mint in Philadelphia minted no silver dollars; $9,353,481 in half dollars; $193,954,740 in quarters; $70,520,096 in dimes; $32,355,748 in nickels; and $49,519,048 in pennies. How much money in coins was minted in Philadelphia in 1985? How many coins did the Philadelphia Mint make? (Source: U.S. Treasury Department.)

34. In 1985 8,002,259 passenger cars and 3,356,905 trucks were sold in the United States. How many more passenger cars were sold than trucks? (Source: Motor Vehicle Manufacturer Association.)

35. The French airliner Concorde can fly 3,060 miles in 3 hours. How far can the Concorde fly in 1 minute?

2

Common Fractions

2-1 Introduction

The changeover to the metric system, when it occurs, will tend to lessen the need for common fractions. Until the changeover is complete, fractions will be used where it is still convenient to use them. Goods produced using English measure will need a set of tools different from the tools made to work on goods produced using metric measure. The mechanic is going to need two sets of tools during the changeover period. To lessen the cost, some manufacturers are making inserts that convert sockets from English to metric. There are also metric sockets with a $\frac{1}{2}''$ drive so ratchets and torque bars will work with the metric sockets.

Common fractions will be needed for some time and everyone who works with numbers should know how to use common fractions in solving everyday problems.

2-2 Definitions

If an inch is divided into eight equal parts, each of the parts is one-eighth of an inch, written $\frac{1}{8}$ of an inch. Two such parts would be two-eighths of an

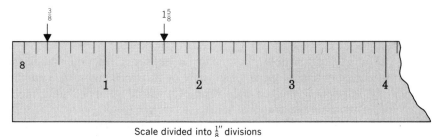

Scale divided into $\frac{1''}{8}$ divisions

Fig. 2-1

inch, written $\frac{2}{8}$ of an inch; three such parts would be three-eighths of an inch, written $\frac{3}{8}$ of an inch; and so on. In each instance the numeral 8, written below the short horizontal line, shows into how many parts the inch has been divided; the numeral written above the line shows how many of the parts have been taken. Thus in the expression $\frac{3}{8}$ of an inch, the numeral 8 below the line shows that the inch has been divided into eight equal parts; the numeral 3 above the line shows that three of these equal parts are taken. The inches on the ruler in Fig. 2-1 have been divided into eighths.

To show 15 divided by 3, you may write $15 \div 3 = 5$. Another way to indicate the division of 15 by 3 is to write $\frac{15}{3} = 5$.

Such expressions as $\frac{1}{8}$, $\frac{2}{8}$, $\frac{3}{8}$, and $\frac{15}{3}$ are called *fractions*. The numeral below the fraction line is called the ***denominator***. The numeral above the fraction line is called the ***numerator***. The numerator and denominator are called the ***terms*** of the fraction.

$$\text{terms} \underbrace{\begin{array}{l} \nearrow 3 \leftarrow \text{numerator} \\ \searrow 8 \leftarrow \text{denominator} \end{array}}$$

A fraction like $\frac{5}{8}$, whose value is less than one, is called a ***proper fraction***. In a proper fraction the numerator is less than the denominator. A fraction like $\frac{9}{8}$, whose value is greater than one, is called an ***improper fraction***. An improper fraction is a fraction whose numerator is equal to or greater than its denominator. A fraction like $\frac{8}{8}$, whose value is equal to one, is also called an improper fraction.

2-3   Changing Whole or Mixed Numbers to Improper Fractions

The ***counting numbers*** are those numbers used to count objects, such as the number of pencils in a box. There is no greatest counting number, but there is a least counting number, namely the number one. This collection of numbers is called the set of counting numbers. The word ***set*** implies the collection of all the numbers described.

Whole numbers include the counting numbers and the number zero, (0), which is the least number of the set of whole numbers. The whole numbers, then, is the set of counting numbers and zero.

$$0, 1, 2, 3, 4, 5, \ldots$$

The changing of a whole number to an improper fraction makes these numbers easier to use in operations involving fractions and mixed numbers. A ***mixed number*** is a number consisting of a whole number and a fraction, such as $5\frac{1}{2}$, read as five and one-half.

Any whole number can be written as an improper fraction by using the whole number as the numerator and the number 1 as the denominator. For instance, $4 = \frac{4}{1}$ and $199 = \frac{199}{1}$.

It is possible to write whole numbers as improper fractions with denominators other than 1. The improper fractions $\frac{8}{1}$ and $\frac{16}{2}$ are equivalent fractions. The bar indicates that the numerator is to be divided by the denominator. In this case $8 \div 1$ is the same as $16 \div 2$; they are therefore called equivalent fractions. They are also both improper fractions.

As previously stated, changing whole numbers to improper fractions makes these numbers easier to use when working with fractions and mixed numbers. A whole number can be written as an equivalent improper fraction with any counting number as its denominator by writing the whole number as a numerator and 1 as its denominator and then multiplying the numerator and denominator by the desired denominator.

Example
Change 3 to fifths.

SOLUTION $\frac{3}{1} \times \frac{5}{5} = \frac{15}{5}$

EXPLANATION   Write the whole number as an improper fraction using 1 as a denominator, that is, $\frac{3}{1}$. Multiply the numerator and denominator each by 5, obtaining $\frac{15}{5}$.

Example 1
Change $5\frac{3}{8}$ to eighths.

SOLUTION $5\frac{3}{8} = \frac{5}{1} + \frac{3}{8} = \frac{40}{8} + \frac{3}{8} = \frac{43}{8}$

EXPLANATION   In 1 there are 8 eighths, in 5 there are 5 times 8 eighths, or 40 eighths. In $5\frac{3}{8}$ there are 40 eighths plus 3 eighths or a total of 43 eighths, written $\frac{43}{8}$.

Multiply the whole number by the denominator of the fraction, add the numerator, and write the result over the denominator.

$$5\frac{3}{8} = (8 \times 5 + 3) \text{ eighths} = \frac{43}{8}$$

Example 2
Change $6\frac{7}{8}$ to an improper fraction.

SOLUTION $6\frac{7}{8} = (8 \times 6 + 7)$ eighths or $\frac{55}{8}$

EXPLANATION Multiplying the 6 by the denominator 8 is 48 (the number of eighths in 6). Adding the 7 eighths is a total of 55 eighths; this is written as $\frac{55}{8}$.

Problems 2-3

Answer the following questions involving fractions.

1. Is $\frac{5}{8}$ a proper fraction or an improper fraction?
2. Is $\frac{8}{3}$ a proper fraction or an improper fraction?
3. Is $\frac{6}{8}$ a proper fraction or an improper fraction?
4. In the fraction $\frac{3}{4}$ the numeral 3 is called the _____ .
5. In the fraction $\frac{7}{8}$ the numeral 8 is called the _____ .
6. In the fraction $\frac{1}{2}$ the numerals 1 and 2 are called the _____ of the fraction.
7. How many halves in 3? in 4? in 5? in 12?
8. How many quarters in $2\frac{1}{2}$? in $3\frac{1}{2}$? in $7\frac{1}{2}$?
9. How many eighths in 4? in $5\frac{1}{8}$? in $6\frac{7}{8}$?
10. Change the following mixed numbers to improper fractions:

(a) $1\frac{1}{8}$	(b) $3\frac{3}{4}$	(c) $5\frac{1}{2}$	(d) $6\frac{7}{8}$
(e) $3\frac{7}{16}$	(f) $1\frac{1}{32}$	(g) $5\frac{15}{16}$	(h) $7\frac{3}{32}$
(i) $5\frac{2}{3}$	(j) $8\frac{7}{10}$	(k) $12\frac{5}{8}$	(l) $15\frac{17}{64}$
(m) $6\frac{3}{10}$	(n) $15\frac{9}{10}$	(o) $1\frac{1}{10}$	(p) $4\frac{31}{100}$

2-4 Changing Improper Fractions to Whole or Mixed Numbers

Improper fractions are used in multiplying and dividing mixed numbers. The answers to these problems are then changed back to whole or mixed numbers. This changing back process is also called *reducing.*

Example 1
Change $\frac{24}{8}$ to a whole number.

SOLUTION $\frac{24}{8} = 24 \div 8 = 3$

EXPLANATION $\frac{24}{8}$ means 24 divided by 8. To change this fraction, carry out the indicated division. The answer is the whole number 3.

Example 2
Change $\frac{13}{4}$ to a mixed number.

SOLUTION $\frac{13}{4} = 13 \div 4 = 3\frac{1}{4}$

EXPLANATION $\frac{13}{4}$ means that 13 is to be divided by 4. When 13 is divided by 4, there is a remainder of 1. This 1 indicates that the remainder is 1 part of 4 and must be added to the quotient 3. Therefore $\frac{13}{4} = 3\frac{1}{4}$.

Problems 2-4

Change the following improper fractions to whole or mixed numbers.

1. $\frac{4}{2}$	2. $\frac{8}{4}$	3. $\frac{16}{8}$	4. $\frac{8}{2}$
5. $\frac{4}{3}$	6. $\frac{8}{3}$	7. $\frac{16}{3}$	8. $\frac{8}{8}$
9. $\frac{24}{4}$	10. $\frac{5}{2}$	11. $\frac{9}{8}$	12. $\frac{10}{3}$
13. $\frac{55}{8}$	14. $\frac{125}{100}$	15. $\frac{100}{64}$	16. $\frac{27}{8}$
17. $\frac{31}{4}$	18. $\frac{18}{10}$	19. $\frac{15}{3}$	20. $\frac{27}{12}$
21. $\frac{10}{4}$	22. $\frac{35}{16}$	23. $\frac{17}{8}$	24. $\frac{33}{32}$
25. $\frac{7}{2}$	26. $\frac{33}{12}$	27. $\frac{25}{10}$	28. $\frac{150}{100}$
29. $\frac{18}{16}$	30. $\frac{84}{64}$	31. $\frac{71}{32}$	32. $\frac{31}{8}$
33. $\frac{62}{32}$	34. $\frac{33}{8}$	35. $\frac{200}{100}$	36. $\frac{37}{12}$
37. $\frac{130}{60}$	38. $\frac{37}{24}$	39. $\frac{32}{16}$	40. $\frac{69}{64}$
41. $\frac{27}{10}$	42. $\frac{121}{100}$	43. $\frac{400}{100}$	44. $\frac{6750}{1000}$
45. $\frac{163}{100}$	46. $\frac{63}{10}$	47. $\frac{6380}{1000}$	48. $\frac{47}{10}$
49. $\frac{142}{100}$	50. $\frac{880}{100}$		

2-5 Changing a Fraction to Lowest Terms

A fraction is in its *lowest terms* when the numerator and denominator are *prime* to each other, that is, they share no common factor other than 1. For example, 5 and 8 are prime to each other because 1 is the only number that will divide *both* of them without a remainder. The numbers 10 and 12 are *not* prime to each other because both can be divided by 2 without a remainder.

The student will find it helpful, when working with fractions, to memorize the first seven or eight prime numbers. A *prime number* is a number that can be divided only by itself and one; the number 1, however, is not a prime number. The first ten prime numbers are:

2 (the only even prime), 3, 5, 7, 11, 13, 17, 19, 23, 29

RULE

To change a fraction to lower terms, divide the numerator and denominator by a factor common to both. A fraction is in its lowest terms when the numerator and denominator have no common factor other than 1.

Example 1

Change $\frac{6}{8}$ to lowest terms.

SOLUTION $\dfrac{6}{8} = \dfrac{6 \div 2}{8 \div 2} = \dfrac{3}{4}$

EXPLANATION   Dividing both terms of the given fraction by 2 gives the fraction $\frac{3}{4}$. Since 3 and 4 contain no common divisor, the fraction $\frac{3}{4}$ is in its lowest terms.

This problem could be also solved by factoring the numerator and denominator into their prime factors and cancelling as shown.

$$\text{SOLUTION} \quad \frac{6}{8} = \frac{\overset{1}{\cancel{2}} \times 3}{\underset{1}{\cancel{2}} \times 2 \times 2} = \frac{3}{4}$$

EXPLANATION   The prime factors of 6 are 2 and 3, while the prime factors of 8 are 2, 2, and 2. The common 2 in the numerator and denominator can be cancelled and the remaining factors multiplied to give $\frac{3}{4}$.

Example 2
Change $\frac{70}{105}$ to lowest terms.

SOLUTION BY DIVISION METHOD

$$\frac{70 \div 5}{105 \div 5} = \frac{14}{21} \qquad \frac{14}{21} = \frac{14 \div 7}{21 \div 7} = \frac{2}{3} \quad \text{or} \quad \frac{70 \div 35}{105 \div 35} = \frac{2}{3}$$

EXPLANATION   Dividing both terms of the given fraction by 5 gives the fraction $\frac{14}{21}$; dividing both terms of $\frac{14}{21}$ by 7 gives $\frac{2}{3}$. Or divide both terms of $\frac{70}{105}$ by 35, giving at once $\frac{2}{3}$. Since 2 and 3 contain no common divisor, the fraction $\frac{2}{3}$ is in its lowest terms.

SOLUTION BY PRIME FACTOR AND CANCEL METHOD   The problem could also be solved by finding all the prime factors of 70 and 105 as shown and cancelling the common 5 and 7.

$$\frac{70}{105} = \frac{2 \times \overset{1}{\cancel{5}} \times \overset{1}{\cancel{7}}}{3 \times \underset{1}{\cancel{5}} \times \underset{1}{\cancel{7}}} = \frac{2}{3}$$

Changing a fraction to its lowest terms is also called *simplification* of the fraction. You can see that it is easier to read $\frac{2}{3}$ than $\frac{70}{105}$ even though $\frac{2}{3}$ and $\frac{70}{105}$ are equal to each other and represent the same number.

Problems 2-5

Simplify the following fractions. (All answers should be given with fractions in their lowest terms and improper fractions changed to whole or mixed numbers.)

1. $\frac{3}{6}$ 2. $\frac{2}{4}$ 3. $\frac{8}{12}$ 4. $\frac{12}{8}$
5. $\frac{18}{24}$ 6. $\frac{12}{32}$ 7. $\frac{24}{64}$ 8. $\frac{4}{16}$
9. $\frac{8}{32}$ 10. $\frac{30}{32}$ 11. $\frac{64}{128}$ 12. $\frac{125}{1000}$
13. $\frac{100}{10}$ 14. $\frac{16}{64}$ 15. $\frac{96}{144}$ 16. $\frac{375}{1000}$
17. $\frac{157}{314}$ 18. $\frac{900}{360}$ 19. $\frac{216}{1728}$ 20. $\frac{448}{128}$

21. What part of a foot is 6 inches? (12 inches = 1 foot)
22. What part of a foot is 9 inches?
23. Thirty inches is what part of a yard? (36 inches = 1 yard)
24. Six inches is what part of a yard?
25. What part of a pound is 8 ounces? (16 ounces = 1 pound)
26. Four ounces is what part of a pound?
27. What part of a pound is 10 ounces?
28. What part of a mile is 440 yards? (1,760 yards = 1 mile)
29. What part of a yard is 18 inches?
30. What part of a yard is 21 inches?
31. What part of 12 is 3?
32. Fifteen is what part of 100?
33. Three is what part of 8?
34. Nine is what part of 16?
35. What part of 64 is 32?

2-6 Changing a Fraction with a Given Denominator to One with a Higher Denominator

When adding, subtracting, or comparing fractions it is often necessary to change the denominator of a fraction to some other denominator, usually a higher or greater denominator. This is done by multiplying both terms of the fraction by the same number; that is, multiplying the fraction by 1. Multiplying any number by 1 does not change the value of the number.

RULE

Multiplying both terms of a fraction by the same number gives a fraction equal in value to the original fraction.

Example 1
Change $\frac{3}{4}$ to a fraction having 20 for its denominator.

SOLUTION $\quad \frac{3}{4} = \frac{3 \times 5}{4 \times 5} = \frac{15}{20}$

EXPLANATION Since the required denominator of the new fraction is 20, multiply both numerator and denominator by a number that will make the new denominator 20 (5 is the required number). The original fraction and the fraction obtained by this process are called ***equivalent fractions***.

Example 2
Change $\frac{5}{16}$ to a fraction having 64 for its denominator.

SOLUTION $\dfrac{5}{16} = \dfrac{5 \times 4}{16 \times 4} = \dfrac{20}{64}$

EXPLANATION Since the required denominator of the new fraction is 64, multiply both the numerator and denominator by 4. (The denominator is the key! You must ask yourself, "What number multiplies 16 to give an answer of 64?")

Problems 2-6

Change the given fractions to equivalent fractions having the required higher denominators.

1. $\frac{1}{2}$ to 4ths
2. $\frac{1}{2}$ to 10ths
3. $\frac{3}{8}$ to 32nds
4. $\frac{5}{8}$ to 64ths
5. $\frac{2}{3}$ to 12ths
6. $\frac{3}{5}$ to 10ths
7. $\frac{3}{5}$ to 15ths
8. $\frac{5}{7}$ to 35ths
9. $\frac{11}{12}$ to 180ths
10. $\frac{5}{13}$ to 52nds
11. $\frac{23}{24}$ to 144ths
12. $\frac{23}{24}$ to 360ths
13. $\frac{11}{18}$ to 360ths
14. $\frac{11}{18}$ to 144ths
15. $\frac{9}{16}$ to 64ths
16. $\frac{9}{16}$ to 32nds
17. $\frac{9}{16}$ to 144ths
18. $\frac{5}{6}$ to 60ths
19. $\frac{3}{4}$ to 60ths
20. $\frac{2}{5}$ to 60ths
21. $\frac{5}{4}$ to 16ths
22. $\frac{7}{10}$ to 100ths
23. $\frac{13}{8}$ to 32nds
24. $\frac{7}{5}$ to 60ths

2-7 Changing Two or More Fractions to Equivalent Fractions Having a Common Denominator

When fractions have a ***common denominator,*** it is easy to compare them and determine which one is the greater by comparing their numerators—the greater numerator is the greater fraction. To add or subtract fractions with common denominators, add or subtract their numerators and place the result over their common denominator.

The lowest common denominator ***(l.c.d.)*** of the desired fractions will be the least common multiple ***(l.c.m.)*** of the given denominators. Very often the least common multiple can be obtained at sight. For example, the l.c.m. of 2, 4, 8, and 16 is 16; that is, 16 is the lowest number that 2, 4, 8 and 16 will divide evenly.

Example 1
Change $\frac{1}{2}$, $\frac{3}{4}$, $\frac{5}{8}$, and $\frac{5}{16}$ to equivalent fractions having the lowest common denominator.

SOLUTION
$$\frac{1}{2} = \frac{1}{2} \times \frac{8}{8} = \frac{8}{16}$$
$$\frac{3}{4} = \frac{3}{4} \times \frac{4}{4} = \frac{12}{16}$$
$$\frac{5}{8} = \frac{5}{8} \times \frac{2}{2} = \frac{10}{16}$$
$$\frac{5}{16} = \frac{5}{16} \times \frac{1}{1} = \frac{5}{16}$$

EXPLANATION Beside each of the given fractions write a fraction line with the l.c.d. 16 below it. To obtain the new numerator, divide the l.c.d. by the given denominator and multiply the result by the given numerator. Thus,

16 divided by 2 is 8; 8 times 1 is 8, the new numerator.
16 divided by 4 is 4; 4 times 3 is 12, the new numerator.
16 divided by 8 is 2; 2 times 5 is 10, the new numerator.
16 divided by 16 is 1; 1 times 5 is 5, the new numerator.

When the least common multiple of two or more denominators cannot be obtained on sight, the following method may be used: Find all the prime factors of each denominator. (A prime factor is a factor that is divisible only by itself and 1.) Make a list of the prime factors according to the number of times the prime factor appears in any one denominator, that is, if the prime factor appears twice in a denominator, then list it twice, and so on. Every different prime factor must appear at least once on the list. The lowest common denominator (l.c.d.) is the product of all the prime factors on the list.

Example 2

Reduce $\frac{5}{8}$, $\frac{2}{3}$, and $\frac{1}{10}$ to equivalent fractions with the lowest common denominator.

SOLUTION Factor the denominators 8, 3, and 10 into prime factors.

$$8 = 2 \times 2 \times 2$$
$$3 = 3$$
$$10 = 2 \times 5$$

Every different prime factor must be used, and if a prime factor appears more than once in any number, then it is used as many times as it appears. The 8 in this problem had the prime number 2 appear three times, therefore this factor must be used three times.

The prime factors 3 and 5 appear once. The lowest common denominator is therefore $2 \times 2 \times 2 \times 3 \times 5$, or 120.

$$\frac{5}{8} \times \frac{15}{15} = \frac{75}{120}$$
$$\frac{2}{3} \times \frac{40}{40} = \frac{80}{120}$$
$$\frac{1}{10} \times \frac{12}{12} = \frac{12}{120}$$

Problems 2-7

Change the following given fractions to equivalent fractions having the lowest common denominator:

1. $\frac{1}{2}, \frac{3}{4}$ 2. $\frac{1}{2}, \frac{1}{3}$ 3. $\frac{1}{2}, \frac{2}{3}$ 4. $\frac{1}{2}, \frac{3}{8}$

5. $\frac{1}{3}, \frac{1}{4}$ 6. $\frac{1}{3}, \frac{1}{5}$ 7. $\frac{1}{3}, \frac{5}{6}$ 8. $\frac{1}{2}, \frac{1}{3}, \frac{1}{4}$

9. $\frac{1}{2}, \frac{2}{3}, \frac{3}{4}$ 10. $\frac{1}{2}, \frac{1}{3}, \frac{1}{6}$ 11. $\frac{2}{3}, \frac{3}{4}, \frac{5}{6}$ 12. $\frac{1}{2}, \frac{2}{3}, \frac{5}{6}, \frac{7}{8}$

13. $\frac{3}{8}, \frac{5}{16}$ 14. $\frac{5}{8}, \frac{1}{4}, \frac{1}{16}$ 15. $\frac{3}{4}, \frac{7}{8}, \frac{5}{16}$ 16. $\frac{3}{10}, \frac{1}{2}, \frac{3}{4}$

17. $\frac{5}{12}, \frac{5}{24}$ 18. $\frac{15}{16}, \frac{7}{8}, \frac{3}{4}, \frac{1}{2}$ 19. $\frac{7}{8}, \frac{1}{10}, \frac{7}{12}$ 20. $\frac{5}{32}, \frac{3}{64}, \frac{1}{2}$

21. $\frac{1}{13}, \frac{1}{11}$ 22. $\frac{1}{5}, \frac{1}{7}, \frac{3}{17}$ 23. $\frac{1}{7}, \frac{3}{35}, \frac{1}{2}$ 24. $\frac{5}{64}, \frac{1}{100}, \frac{3}{10}$

25. Write the following fractions in ascending (smallest first) order: $\frac{2}{3}, \frac{5}{7}, \frac{5}{8}$

26. Write the following fractions in descending order: $\frac{5}{16}, \frac{5}{32}, \frac{11}{64}, \frac{1}{8}$

27. Which is greater: $\frac{3}{5}$ or $\frac{2}{3}$?

28. Which is greater: $\frac{7}{10}$ or $\frac{3}{4}$?

29. Is $\frac{13}{32}$ larger than $\frac{25}{64}$?

30. Is $\frac{2}{3}$ less than $\frac{43}{64}$?

2-8   Addition of Fractions

To add fractions having the same denominator, add their numerators and place the sum over the common denominator. Thus $\frac{3}{8} + \frac{5}{8} + \frac{7}{8}$ means 3 eighths + 5 eighths + 7 eighths = 15 eighths = $\frac{15}{8} = 1\frac{7}{8}$. To add fractions with different denominators, such as $\frac{3}{4} + \frac{7}{8} + \frac{5}{6}$, first change the given fractions to equivalent fractions having a common denominator. Then add the fractions by adding the numerators and placing the sum over the common denominator.

Example

Find the sum of $\frac{3}{4}, \frac{7}{8}, \frac{5}{6}$.

SOLUTION $\frac{3}{4} = \frac{18}{24}$

$\frac{7}{8} = \frac{21}{24}$

$\frac{5}{6} = \frac{20}{24}$

$\frac{59}{24} = 2\frac{11}{24}$

EXPLANATION   The sum of 18 twenty-fourths and 21 twenty-fourths and 20 twenty-fourths is 59 twenty-fourths, written $\frac{59}{24}$, and reduced to the mixed number $2\frac{11}{24}$.

Problems 2-8

Find the sums of the following fractions. Your answers should be in lowest terms.

1. $\frac{1}{2}, \frac{3}{4}$ 2. $\frac{1}{3}, \frac{1}{2}$ 3. $\frac{2}{3}, \frac{3}{4}$ 4. $\frac{1}{5}, \frac{3}{10}$

5. $\frac{2}{3}, \frac{5}{8}$ 6. $\frac{5}{6}, \frac{3}{4}$ 7. $\frac{7}{8}, \frac{7}{16}$ 8. $\frac{5}{16}, \frac{5}{32}$

9. $\frac{3}{4}, \frac{7}{8}$ 10. $\frac{3}{32}, \frac{15}{16}$ 11. $\frac{15}{16}, \frac{31}{32}$ 12. $\frac{29}{32}, \frac{63}{64}$

13. $\frac{1}{4} + \frac{3}{8} + \frac{5}{8} + \frac{7}{8}$ 14. $\frac{29}{21} + \frac{13}{21} + \frac{5}{21} + \frac{38}{21}$

15. $\frac{3}{16} + \frac{7}{16} + \frac{5}{16} + \frac{9}{16} + \frac{11}{16}$ 16. $\frac{1}{2} + \frac{3}{4} + \frac{5}{8} + \frac{7}{16}$

17. $\frac{7}{12} + \frac{3}{10} + \frac{8}{15}$ 18. $\frac{3}{7} + \frac{5}{14} + \frac{7}{18} + \frac{1}{9}$

19. $\frac{7}{8} + \frac{2}{3} + \frac{1}{6} + \frac{1}{2}$ 20. $\frac{5}{6} + \frac{3}{14} + \frac{3}{4} + \frac{11}{12}$

21. $\frac{15}{16} + \frac{8}{3} + \frac{2}{3}$ 22. $\frac{9}{16} + \frac{7}{8} + \frac{15}{32}$

23. $\frac{1}{8} + \frac{5}{12} + \frac{11}{18} + \frac{23}{24}$ 24. $\frac{2}{3} + \frac{3}{4} + \frac{5}{7} + \frac{7}{12}$

25. $\frac{2}{5} + \frac{5}{9} + \frac{1}{8} + \frac{7}{20}$ 26. $\frac{3}{5} + \frac{3}{7} + \frac{3}{8} + \frac{3}{10}$

27. $\frac{9}{28} + \frac{31}{49} + \frac{16}{21} + \frac{32}{35}$ 28. $\frac{1}{10} + \frac{1}{100} + \frac{1}{1000}$

29. $\frac{1}{10} + \frac{3}{100} + \frac{17}{1000}$ 30. $\frac{7}{10} + \frac{21}{100} + \frac{7}{1000}$

2-9 Addition of Mixed Numbers

When adding mixed numbers it is often easier to add the mixed numbers if their fractional parts have a common denominator. Remember that the mixed number represents a fraction added to a whole number, that is, $1\frac{1}{2}$ means $\frac{1}{2}$ added to 1.

Example

Find the sum of $5\frac{3}{8}$, $6\frac{1}{4}$, $2\frac{1}{3}$, $7\frac{1}{2}$.

SOLUTION The sum of the mixed numbers is equal to the sum of the whole numbers added to the sum of the fractions. The work may conveniently be arranged as shown.

$$
\begin{aligned}
5\tfrac{3}{8} &= 5\tfrac{9}{24}\\
6\tfrac{1}{4} &= 6\tfrac{6}{24}\\
2\tfrac{1}{3} &= 2\tfrac{8}{24}\\
7\tfrac{1}{2} &= 7\tfrac{12}{24}\\
\hline
20\tfrac{35}{24} &= 21\tfrac{11}{24}
\end{aligned}
$$

EXPLANATION The sum of the fractions is $\frac{35}{24}$, but $\frac{35}{24} = 1\frac{11}{24}$. The 1 is added to the 20 which is the sum of the whole number part. The complete answer is $21\frac{11}{24}$.

Problems 2-9

Find the sums of the following mixed numbers. Your answers should be in lowest terms.

1. $2\frac{1}{2} + 4\frac{3}{4}$ 2. $5\frac{1}{3} + 2\frac{1}{2}$ 3. $1\frac{3}{4} + 3\frac{3}{8}$

4. $3\frac{1}{2} + 5\frac{2}{3}$ 5. $6\frac{2}{3} + 2\frac{1}{6}$ 6. $10\frac{1}{8} + 9\frac{15}{16}$

7. $3\frac{5}{8} + 1\frac{5}{16}$ 8. $5\frac{7}{8} + 12\frac{7}{16}$ 9. $7\frac{1}{10} + 5\frac{4}{5}$

10. $2\frac{5}{16} + 7\frac{3}{4}$ 11. $11\frac{1}{3} + 15\frac{1}{8}$ 12. $17\frac{1}{2} + 10\frac{9}{10}$

13. $2\frac{1}{2} + 5\frac{1}{2} + 3\frac{1}{2} + 7\frac{1}{2}$ 14. $3\frac{7}{16} + 4\frac{5}{16} + 2\frac{11}{16} + 8\frac{9}{16}$

15. $5\frac{3}{4} + 2\frac{1}{8} + 3\frac{1}{16} + 3\frac{7}{32}$ 16. $2\frac{1}{5} + 8\frac{2}{3} + 10\frac{3}{4} + 4\frac{2}{7}$

17. $6\frac{7}{16} + \frac{7}{8} + \frac{2}{3} + 3\frac{3}{4}$ 18. $4 + 11\frac{3}{8} + \frac{7}{8} + 3\frac{1}{2}$

19. $14\frac{3}{10} + 8\frac{7}{15} + 6\frac{9}{20} + 2\frac{17}{30}$ 20. $\frac{19}{32} + \frac{11}{16} + 18 + 12$

21. $1\frac{3}{8} + 5\frac{7}{8} + 3\frac{5}{8} + 6\frac{7}{8}$ 22. $9\frac{3}{4} + 6\frac{1}{2} + 4\frac{3}{8}$

23. $5\frac{3}{8} + 4\frac{1}{4} + 8\frac{11}{12}$ 24. $9\frac{5}{6} + 12\frac{3}{4} + 10\frac{15}{16} + 4\frac{11}{12}$

25. $8\frac{1}{2} + 6\frac{1}{3} + 5\frac{1}{4} + 7\frac{1}{8} + 12\frac{1}{18}$ 26. $8\frac{1}{2} + 4\frac{5}{8} + 3\frac{2}{3} + 2\frac{1}{8}$

27. $1\frac{11}{16} + 15 + 12 + 3\frac{7}{8}$ 28. $5\frac{2}{3} + \frac{49}{240} + 4\frac{7}{120} + 8\frac{21}{80}$

29. Add the following dimensions: $1\frac{7}{8}''$, $2\frac{1}{4}''$, $\frac{9}{64}''$, $\frac{11}{32}''$, $3\frac{1}{2}''$.

30. Add the following weights: $1\frac{1}{4}$ lb, $\frac{1}{2}$ lb, $\frac{7}{16}$ lb, $1\frac{5}{16}$ lb, $\frac{7}{8}$ lb.

31. Add the following dimensions: $6\frac{1}{4}''$, $1\frac{3}{64}''$, $3\frac{7}{8}''$, $2\frac{11}{32}''$, $\frac{1}{2}''$, $1\frac{9}{16}''$, $5\frac{5}{8}''$.

32. Add the following dimensions: $\frac{15}{64}''$, $\frac{13}{32}''$, $\frac{5}{8}''$, $\frac{53}{64}''$, $1\frac{7}{16}''$, $\frac{35}{64}''$, $1\frac{9}{16}''$.

33. Cheryl worked $5\frac{1}{2}$ hours on Monday, $6\frac{1}{4}$ hours on Tuesday, 8 hours on Wednesday, $6\frac{1}{2}$ hours on Thursday, and $7\frac{1}{4}$ hours on Friday. What is the total number of hours she worked?

34. John worked $4\frac{1}{2}$ hours on Monday and $8\frac{1}{2}$ hours on Tuesday. How many hours did he work?

35. Betty worked $6\frac{1}{2}$ hours on Monday, $6\frac{1}{2}$ hours on Tuesday, $6\frac{1}{2}$ hours on Wednesday, $6\frac{1}{2}$ on Thursday, and 8 hours on Friday. How many hours did she work this week?

2-10 Subtraction of Fractions

To subtract two fractions you must have a common denominator. The procedure is much like adding fractions, except that the numerators are subtracted and the difference is placed over the common denominator.

Example 1

Take $\frac{3}{9}$ from $\frac{7}{9}$.

SOLUTION $\frac{7}{9}$ (minuend)

     $- \frac{3}{9}$ (subtrahend)

     $\frac{4}{9}$ (difference)

EXPLANATION   Since the two fractions have a common denominator, subtract the subtrahend's numerator, 3, from the minuend's numerator, 7, and write the result over the common denominator. As shown, 3 ninths subtracted from 7 ninths is 4 ninths.

Example 2

Take $\frac{2}{3}$ from $\frac{7}{8}$.

SOLUTION $\frac{7}{8} = \frac{21}{24}$

     $- \frac{2}{3} = \frac{16}{24}$

           $\frac{5}{24}$

EXPLANATION   Since the fractions have different denominators, change them to fractions having a common denominator. The lowest common denominator of the two fractions is the least common multiple of 8 and 3, namely, 24. The given fractions become $\frac{21}{24}$ and $\frac{16}{24}$. The problem is now similar to Example 1.

Do the indicated subtractions. Simplify your answers.

1. Take $\frac{5}{16}$ from $\frac{9}{16}$.   2. Take $\frac{3}{8}$ from $\frac{7}{8}$.
3. From $\frac{3}{4}$ take $\frac{1}{2}$.   4. From $\frac{29}{32}$ take $\frac{5}{8}$.
5. $\frac{1}{2} - \frac{1}{3}$ 6. $\frac{4}{5} - \frac{3}{4}$ 7. $\frac{7}{8} - \frac{5}{16}$ 8. $\frac{13}{15} - \frac{9}{20}$
9. $\frac{14}{15} - \frac{7}{12}$ 10. $\frac{5}{12} - \frac{5}{18}$ 11. $\frac{7}{8} - \frac{3}{10}$ 12. $\frac{1}{12} - \frac{1}{144}$
13. How much more is $\frac{5}{8}$ than $\frac{1}{2}$?
14. How much more is $\frac{3}{4}$ than $\frac{2}{3}$?
15. How much less is $\frac{1}{4}$ than $\frac{1}{3}$?
16. How much less is $\frac{3}{8}$ than $\frac{5}{7}$?
17. A planer takes a $\frac{3}{32}''$ cut from a piece of steel $\frac{3}{8}''$ thick. What is the remaining thickness?
18. By how much must the diameter of a $\frac{3}{4}''$ shaft be reduced to bring it down to a diameter of $\frac{41}{64}''$?
19. What is the difference in thickness between a $\frac{9}{16}''$-thick steel plate and a $\frac{7}{8}''$-thick steel plate?
20. A steel casting $\frac{7}{8}''$ thick is finished by taking a cut $\frac{3}{64}''$ deep. What is the final thickness?

2-11 Subtraction of Mixed Numbers

To subtract two mixed numbers, common denominators must be used. The fractional parts are subtracted first and then the whole number parts are subtracted. It may be necessary to borrow a 1 from the greater whole number and add it to the fractional part so the fractions can be subtracted. Usually, the **minuend** is the larger number in a subtraction problem and the **subtrahend** is the smaller number. The answer is called the **difference**.

Example 1
Take $7\frac{3}{4}$ from $15\frac{5}{6}$.

SOLUTION $15\frac{5}{6} = \quad 15\frac{10}{12}$ (minuend)
$\underline{- \quad 7\frac{3}{4} = - \quad 7\frac{9}{12}}$ (subtrahend)
$8\frac{1}{12}$ (difference)

EXPLANATION This case is similar to that of adding mixed numbers. First find the difference between the fractions and then find the difference between the whole numbers. The complete answer consists of the two results.

Example 2
Take $5\frac{2}{3}$ from $8\frac{3}{8}$.

SOLUTION $8\frac{3}{8} = \quad 8\frac{9}{24} = (7\frac{24}{24} + \frac{9}{24}) = 7\frac{33}{24}$
$\underline{- \quad 5\frac{2}{3} = - \quad 5\frac{16}{24} = \qquad\qquad\qquad -5\frac{16}{24}}$
$2\frac{17}{24}$

EXPLANATION It is not possible to take $\frac{2}{3}$ from $\frac{3}{8}$, because $\frac{2}{3}$ is larger than $\frac{3}{8}$. When written with a common denominator, $\frac{3}{8} = \frac{9}{24}$, and $\frac{2}{3} = \frac{16}{24}$. Borrow 1 from the whole number 8 and change this borrowed 1 to twenty-four twenty-fourths, $\frac{24}{24}$. Add $\frac{24}{24}$ to $\frac{9}{24}$, giving $(24 + 9)/24 = \frac{33}{24}$. Instead of writing $8\frac{9}{24}$ in the minuend, write its equivalent number $7\frac{33}{24}$. To take $5\frac{16}{24}$ from $7\frac{33}{24}$, take $\frac{16}{24}$ from $\frac{33}{24}$, giving $\frac{17}{24}$; and 5 from 7, giving 2. The answer is $2\frac{17}{24}$.

Example 3

A steel bar 6' long has four pieces each $4\frac{5}{8}''$ long cut from it. The saw kerf (width of the saw cut) is $\frac{1}{8}''$. How long is the remainder of the bar?

SOLUTION $4\frac{5}{8}'' + \frac{1}{8}'' = 4\frac{3}{4}''$ (length of one piece plus the saw cut)
$4\frac{3}{4}'' + 4\frac{3}{4}'' + 4\frac{3}{4}'' + 4\frac{3}{4}'' = 19''$ or $1'7''$ (length of four pieces plus four saw cuts)

$$6' - 1'7'' = 5'12'' - 1'7''$$
$$= 4'5'' \text{ (length of remaining bar)}$$

EXPLANATION The kerf (width of saw cut) must be added to the length of each cut. The sum of the lengths of all the pieces plus the width of the saw cuts must be subtracted from the length of the original bar to find the length of the bar that would be left in stock. Note that $12''$ is borrowed from the 6' to enable the subtraction of $1'7''$ from $6'0''$.

Problems 2-11

Do the indicated subtractions. Simplify your answers.

1. Take 5 from $9\frac{7}{8}$.
2. Take 2 from $6\frac{3}{4}$.
3. Take $2\frac{13}{32}$ from $5\frac{19}{32}$.
4. From $18\frac{15}{16}$ take $4\frac{9}{16}$.
5. From $12\frac{5}{8}$ take $8\frac{1}{4}$.
6. Take $\frac{1}{3}$ from 2.

7. $4 - \frac{3}{8}$
8. $9 - \frac{7}{16}$
9. $6\frac{1}{2} - 2\frac{1}{4}$
10. $3\frac{2}{5} - 1\frac{1}{3}$
11. $10\frac{1}{3} - 7\frac{3}{4}$
12. $5\frac{2}{5} - 3\frac{7}{8}$
13. $14\frac{7}{18} - 8\frac{19}{24}$
14. $32\frac{5}{16} - 21\frac{11}{12}$
15. $5\frac{3}{8} - \frac{7}{8}$
16. $6\frac{3}{4} - \frac{15}{16}$
17. $8\frac{3}{7} - 6\frac{7}{9}$
18. $16\frac{25}{64} - 10\frac{7}{8}$
19. $4\frac{1}{4} - 2\frac{3}{5}$
20. $8\frac{1}{3} - 6\frac{5}{7}$
21. $15\frac{3}{4} - 12\frac{7}{8}$
22. $19\frac{1}{16} - 10\frac{17}{32}$
23. $12\frac{3}{4} - 7\frac{15}{16}$
24. $75\frac{1}{2} - 37\frac{63}{64}$
25. $16\frac{1}{8} - 7\frac{3}{16}$
26. $12\frac{2}{3} - 8\frac{1}{4}$
27. $9\frac{3}{4} - 2\frac{1}{4}$
28. $3\frac{5}{8} - 1\frac{5}{16}$
29. $7\frac{13}{32} - 3\frac{13}{32}$
30. $4\frac{17}{64} - 2\frac{5}{8}$
31. $2\frac{7}{8} - 2\frac{3}{4}$
32. $8\frac{1}{2} - 6$
33. $17\frac{15}{16} - 11\frac{7}{8}$
34. $23\frac{1}{16} - 15\frac{15}{16}$
35. $14\frac{7}{16} - 11\frac{23}{32}$
36. $29\frac{3}{4} - 21\frac{7}{8}$
37. $67\frac{1}{8} - 37\frac{1}{4}$
38. $43\frac{5}{16} - 23\frac{9}{16}$
39. $22\frac{13}{64} - 12\frac{1}{2}$
40. $104\frac{1}{2} - 75\frac{3}{4}$

Do the indicated subtractions and additions in the order that they appear in the problem.

41. Simplify $16\frac{3}{4} - 2\frac{1}{2} - 4\frac{3}{16} + 2$.
42. Find the value of $5\frac{3}{7} + 8\frac{7}{10} - 3\frac{1}{3} - 1\frac{1}{5}$.

43. Find the value of $8\frac{3}{4} - 1\frac{7}{16} + 4 - 2\frac{1}{3}$.
44. From a bar of brass $16\frac{1}{2}''$ long the following three pieces are cut: $1\frac{1}{8}''$, $3\frac{1}{2}''$, and $3\frac{3}{32}''$. What is the final length of the bar, allowing $\frac{1}{16}''$ for each cut?
45. Allowing $\frac{1}{8}''$ for each cut, what must be the length of a bar of brass to provide the following four pieces: $1\frac{7}{8}''$, $3\frac{3}{16}''$, $\frac{11}{32}''$, and $2\frac{1}{4}''$?
46. Add the following dimensions: $2\frac{7}{16}''$, $3\frac{1}{8}''$, $5\frac{9}{64}''$, $4\frac{1}{32}''$, $\frac{1}{2}''$.
47. A shaft required to be $\frac{11}{16}''$ in diameter measures $\frac{49}{64}''$. How much oversized is it?
48. A planer takes a $\frac{3}{32}''$ cut on a plate that is $1\frac{3}{4}''$ thick. What is the final thickness of the plate?
49. A piece of tapered stock has a diameter of $1\frac{1}{16}''$ at one end and $\frac{57}{64}''$ at the other end. Find the difference between the two diameters.
50. Sam could have worked 40 hours this week. He worked $7\frac{1}{2}$ hours on Monday, $5\frac{1}{4}$ hours on Tuesday, and $8\frac{1}{2}$ hours on Wednesday. He did not work the rest of the week. How many hours was he short of 40 hours?

2-12 Multiplication of Fractions

To multiply two or more fractions, you multiply the numerators and use that answer as the new numerator, then multiply the denominators and use that answer as the new denominator. It is not necessary to find common denominators when multiplying fractions. The product should be expressed in lowest terms.

RULE
The product of two or more fractions is the product of the numerators over the product of the denominators.

Example 1
Multiply $\frac{3}{4}$ by 5.

SOLUTION $\quad \frac{3}{4} \times \frac{5}{1} = \frac{3 \times 5}{4 \times 1} = \frac{15}{4} = 3\frac{3}{4}$

or, more briefly, $\frac{3}{4} \times \frac{5}{1} = \frac{15}{4} = 3\frac{3}{4}$.

EXPLANATION The multiplier is 5, a whole number. To obtain the product, multiply the numerator of the given fraction by the whole number and place the result over the denominator. Thus $3 \times 5 = 15$, the numerator of the product that is written over the denominator 4, giving $\frac{15}{4}$ or $3\frac{3}{4}$.

Example 2
Multiply $\frac{2}{3}$ by $\frac{5}{7}$.

SOLUTION $\quad \frac{2}{3} \times \frac{5}{7} = \frac{2 \times 5}{3 \times 7} = \frac{10}{21}$

or, in one step, $\frac{2}{3} \times \frac{5}{7} = \frac{10}{21}$.

EXPLANATION   To multiply one fraction by another, multiply the numerators together and place the result over the product of the denominators. In the example, $2 \times 5 = 10$, the numerator of the product, and $3 \times 7 = 21$, the denominator of the product.

Example 3

Multiply $\frac{3}{4}$ by $\frac{8}{9}$.

This example, being similar to Example 2, may be done in the same way.

SOLUTION $\dfrac{3}{4} \times \dfrac{8}{9} = \dfrac{3 \times 8}{4 \times 9} = \dfrac{24}{36}$

EXPLANATION   The answer can be reduced to lowest terms by dividing the numerator and denominator by 12; thus $\dfrac{24}{36} = \dfrac{24 \div 12}{36 \div 12} = \dfrac{2}{3}$. It is not necessary to do the multiplication first and then the division to reduce a fraction to its lowest terms. The division can be done before the multiplication, and this often makes the work much easier. Thus,

$$\frac{\overset{1}{\cancel{3}}}{\underset{1}{\cancel{4}}} \times \frac{\overset{2}{\cancel{8}}}{\underset{3}{\cancel{9}}} = \frac{2}{3}$$

The numerator and the denominator each have the common factors 3 and 4. Divide both the numerator and the denominator by 3. Then divide both the numerator and denominator by 4. The answer is simply the product of 1 and 2, that is, 2 divided by the product of 1 and 3, that is, 3.

The numerator and the denominator are divided by factors common to both. Before doing the actual multiplication, always look for common factors in the numerator and the denominator and divide the numerator and the denominator by them. If all the common factors are found, the answer will be in the lowest terms.

Example 4

Multiply $\frac{8}{11} \times \frac{5}{12}$.

SOLUTION $\dfrac{\overset{2}{\cancel{8}}}{11} \times \dfrac{5}{\underset{3}{\cancel{12}}} = \dfrac{2 \times 5}{11 \times 3} = \dfrac{10}{33}$

EXPLANATION   This case is similar to Example 3. The 8 in the numerator and the 12 in the denominator are each divided by their common factor 4. The answer is obtained by multiplying all the remaining numerators for the new numerator and multiplying all the remaining denominators for the new denominator.

Example 5
 Multiply $2\frac{1}{2} \times 3\frac{1}{4}$.

SOLUTION $2\frac{1}{2} \times 3\frac{1}{4} = \frac{5}{2} \times \frac{13}{4} = \frac{65}{8} = 8\frac{1}{8}$

EXPLANATION   When the quantities to be multiplied are mixed numbers, change them to improper fractions and then proceed as in previous examples.

Example 6
 Find $\frac{2}{3}$ of $\frac{4}{5}$.

SOLUTION $\dfrac{2}{3} \times \dfrac{4}{5} = \dfrac{2 \times 4}{3 \times 5} = \dfrac{8}{15}$

EXPLANATION   The word "of" has the same meaning as the multiplication sign ($\times$), therefore $\frac{2}{3}$ of $\frac{4}{5}$ means $\frac{2}{3} \times \frac{4}{5} = \frac{8}{15}$.

Example 7
 Multiply $\frac{3}{4} \times \frac{5}{8} \times \frac{2}{7}$.

SOLUTION $\dfrac{3}{\underset{2}{4}} \times \dfrac{5}{8} \times \dfrac{\overset{1}{2}}{7} = \dfrac{3 \times 5 \times 1}{2 \times 8 \times 7} = \dfrac{15}{112}$

EXPLANATION   The 2 in the numerator and the 4 in the denominator are each divided by their common factor 2. The product of the remaining numerators, $3 \times 5 \times 1 = 15$, is the numerator of the answer; the product of the remaining denominators, $2 \times 8 \times 7 = 112$, is the denominator of the answer. The answer is $\frac{15}{112}$.

Problems 2-12

Do the indicated multiplications. Make sure your answers are reduced to lowest terms.

1. $\frac{1}{3} \times 5$	2. $\frac{1}{2} \times 6$	3. $\frac{3}{4} \times 8$
4. $\frac{5}{8} \times 16$	5. $\frac{1}{2} \times \frac{2}{3}$	6. $\frac{2}{5} \times \frac{10}{11}$
7. $\frac{3}{4} \times \frac{5}{16}$	8. $\frac{3}{8} \times \frac{16}{27}$	9. $1\frac{1}{2} \times 3$
10. $1\frac{2}{3} \times 1\frac{1}{2}$	11. $2\frac{1}{2} \times 1\frac{1}{4}$	12. $\frac{1}{2} \times 3\frac{1}{4} \times 4$
13. $\frac{16}{21} \times \frac{3}{4}$	14. $\frac{13}{51} \times \frac{17}{39}$	15. $\frac{1}{3} \times \frac{1}{3}$
16. $\frac{7}{12} \times \frac{36}{49}$	17. $1\frac{1}{3} \times 8\frac{3}{5}$	18. $\frac{2}{7} \times 28$
19. $2\frac{1}{16} \times 1\frac{1}{16}$	20. $12\frac{1}{2} \times 7\frac{1}{2} \times \frac{2}{3}$	21. $\frac{1}{2} \times \frac{1}{2} \times \frac{1}{2}$
22. $\frac{3}{4} \times \frac{1}{3}$	23. $\frac{5}{6} \times \frac{6}{15}$	24. $\frac{9}{16} \times \frac{2}{3} \times 10\frac{1}{2} \times 12\frac{7}{8}$

25. What is the total length of 8 pieces of steel each $5\frac{1}{2}''$ long?
26. What is the total length of 25 pieces of drill rod each $5\frac{1}{16}''$ long?
27. The volume of a rectangular block of wood is found by multiplying the length by the width by the height. What is the volume of a rectangular block of wood $15\frac{1}{2}''$ long, $4\frac{1}{4}''$ wide, and $3''$ high? (Your answer will be in cubic inches.)
28. To find the approximate circumference of a circle, we multiply the diameter by $3\frac{1}{7}$. Find the approximate circumference of a circle whose diameter is $2\frac{7}{8}''$.
29. What is the volume of a block of steel $8\frac{1}{2}''$ long, $2\frac{1}{8}''$ wide, and $1\frac{3}{4}''$ thick? (Your answer will be in cubic inches.)
30. What is $\frac{1}{2}$ the volume of the block of steel in Problem 29? (Your answer will be in cubic inches.)

2-13   Division of Fractions

To divide one fraction by another, multiply the dividend (the number to be divided) by the **reciprocal** of the divisor. The reciprocal of a fraction is formed by using the numerator of the original fraction as the denominator and by using the denominator of the original fraction as the numerator, (inverting the fraction). For instance, the reciprocal of $\frac{2}{3}$ is $\frac{3}{2}$.

Example 1
   Divide $\frac{5}{7}$ by $\frac{3}{4}$. (The reciprocal of $\frac{3}{4}$ is $\frac{4}{3}$.)

SOLUTION $\frac{5}{7} \div \frac{3}{4} = \frac{5}{7} \times \frac{4}{3} = \frac{20}{21}$

EXPLANATION   The division problem is changed to a multiplication problem by changing the divisor to its reciprocal and proceeding as in multiplication. The reciprocal of a whole number is 1 written over the whole number, such as $\frac{1}{6}$, which is the reciprocal of 6. This procedure is commonly called **inverting.** Thus for division involving fractions, invert the divisor and multiply.

RULE _____
   To divide fractions, invert the divisor (second fraction) and proceed as in multiplication.

Example 2
   Find $\frac{2}{3} \div 5$.

SOLUTION   Invert the 5 and multiply.

$$\frac{2}{3} \div \frac{5}{1} = \frac{2}{3} \times \frac{1}{5} = \frac{2}{15}$$

Example 3
   Divide $4\frac{5}{8}$ by $2\frac{1}{3}$.

SOLUTION   Change each of the mixed numbers to an improper fraction. $4\frac{5}{8} = \frac{37}{8}$
   and $2\frac{1}{3} = \frac{7}{3}$. Invert the divisor and multiply.

$$4\frac{5}{8} \div 2\frac{1}{3} = \frac{37}{8} \div \frac{7}{3} = \frac{37}{8} \times \frac{3}{7} = \frac{111}{56} = 1\frac{55}{56}$$

Example 4
   Divide 7 by $\frac{2}{5}$.

SOLUTION   Invert the divisor, $\frac{2}{5}$, and multiply.

$$7 \div \frac{2}{5} = \frac{7}{1} \times \frac{5}{2} = \frac{35}{2} = 17\frac{1}{2}$$

Problems 2-13

Do the indicated divisions. Simplify your answers.

1. $\frac{18}{25} \div 3$	2. $\frac{12}{9} \div 4$	3. $\frac{7}{8} \div 5$
4. $\frac{3}{4} \div 7$	5. $\frac{9}{16} \div \frac{3}{8}$	6. $\frac{11}{12} \div \frac{1}{2}$
7. $7 \div \frac{1}{2}$	8. $4\frac{1}{2} \div \frac{1}{4}$	9. $3\frac{3}{4} \div \frac{1}{5}$
10. $8 \div \frac{2}{3}$	11. $15 \div \frac{3}{8}$	12. $18\frac{2}{3} \div 6$
13. $12\frac{3}{4} \div 8$	14. $10\frac{7}{16} \div \frac{3}{4}$	15. $4\frac{1}{3} \div \frac{2}{5}$
16. $5 \div 4\frac{2}{3}$	17. $7 \div 3\frac{1}{8}$	18. $6 \div 2\frac{1}{4}$
19. $2\frac{1}{3} \div 3\frac{1}{2}$	20. $4\frac{1}{16} \div 8\frac{1}{8}$	21. $12\frac{5}{6} \div 4\frac{5}{18}$
22. $3\frac{1}{10} \div 18\frac{1}{3}$	23. $8\frac{8}{9} \div 6\frac{2}{3}$	24. $20\frac{1}{2} \div 16\frac{2}{3}$
25. $6\frac{1}{5} \div 5\frac{1}{4}$	26. $3\frac{1}{16} \div 4\frac{7}{8}$	27. $7\frac{3}{10} \div 12\frac{3}{4}$
28. $6\frac{7}{8} \div 2\frac{1}{2}$	29. $6\frac{1}{2} \div 2\frac{1}{4}$	30. $15\frac{7}{10} \div 5\frac{1}{10}$
31. $11\frac{2}{3} \div 4\frac{5}{6}$	32. $7\frac{1}{8} \div \frac{1}{16}$	33. $10\frac{1}{6} \div 3\frac{1}{4}$
34. $9\frac{3}{16} \div 5\frac{1}{8}$	35. $28\frac{1}{10} \div 14\frac{1}{20}$	36. $40\frac{7}{8} \div 20\frac{7}{16}$

37. The approximate diameter of a circle is found by dividing the circumference by $3\frac{1}{7}$. What is the diameter of a circle whose circumference is $26\frac{3}{4}''$?

38. A space $50\frac{3}{4}''$ long is to be divided into 14 equal parts. What is the length of each part?

39. A cubic foot contains approximately $7\frac{1}{2}$ gallons. How many cubic feet are there in a 100-gallon tank?

40. The diagonal of a square is approximately the length of a side divided by $\frac{7}{10}$. How long is the diagonal of a square whose side is $1\frac{3}{4}''$?

2-14   Problems Involving Multiplication and Division

Sometimes a problem involves more than two numbers and more than one operation. If the operations are addition and subtraction and there are no parentheses, then the addition and subtraction can be done from left to right as the operations and the numbers are met. If there are parentheses in the problem, then the operations inside the parentheses are done first. If the operations are a combination of multiplication and division, it is necessary to know when each operation is to be done. Parentheses show the order of the operations. Operations inside the parentheses are to be done before operations outside the parentheses.

Example

Find the value of $(\frac{2}{3} \times \frac{7}{8}) \div 3\frac{1}{2}$.

SOLUTION $(\overset{1}{\underset{}{\frac{2}{3}}} \times \overset{}{\underset{4}{\frac{7}{8}}}) \div 3\frac{1}{2} = \frac{7}{12} \div 3\frac{1}{2} = \frac{7}{12} \div \frac{7}{2} = \overset{1}{\underset{6}{\frac{7}{12}}} \times \overset{1}{\underset{1}{\frac{2}{7}}} = \frac{1}{6}$

EXPLANATION   Do the operation inside the parentheses first, $(\frac{2}{3} \times \frac{7}{8}) = \frac{7}{12}$; then divide the $\frac{7}{12}$ by $3\frac{1}{2}$.

Problems 2-14

Do the following problems involving multiplication and division. Make sure you perform the operations inside the parentheses before any operations outside the parentheses. Reduce your answers to lowest terms.

1. $(\frac{2}{5} \times \frac{1}{3}) \div \frac{3}{4}$

2. $(\frac{7}{8} \times \frac{1}{2}) \div \frac{1}{7}$

3. $(2\frac{1}{3} \times \frac{5}{8}) \div 1\frac{1}{4}$

4. $(3\frac{1}{5} \times 1\frac{1}{2}) \times (2\frac{1}{2} \div 4\frac{2}{3})$

5. $(\frac{2}{3} \times \frac{15}{16}) \times (\frac{7}{8} \div \frac{1}{2})$

6. $(\frac{2}{7} \times \frac{5}{9}) \times (\frac{3}{10} \div 4)$

7. $(2 \times \frac{5}{8}) \times (6 \div \frac{7}{10})$

8. $(3\frac{1}{7} \times 2) \div 4$

9. $(4\frac{1}{2} \times 4\frac{1}{2} \times 16) \div 144$

10. $(18\frac{2}{3} \times 16\frac{1}{2}) \div 144$

11. $(5\frac{1}{3} \times 5\frac{1}{4} \times 4\frac{1}{2}) \div 1,728$

12. $(3\frac{3}{4} \times 3\frac{3}{4} \times 3\frac{1}{5}) \div 144$

13. $(\frac{7}{8} \times \frac{3}{8}) \div \frac{21}{64}$

14. $(\frac{3}{4} \times \frac{7}{8}) \div \frac{7}{8}$

15. $(1\frac{1}{2} \times \frac{5}{16}) \div 15$

16. $(2\frac{3}{4} \times 1\frac{2}{3}) \div \frac{5}{12}$

17. $(1\frac{5}{8} \times \frac{5}{8}) \div \frac{7}{8}$

18. $(2\frac{7}{16} \times \frac{7}{16}) \div \frac{1}{144}$

19. $(3\frac{1}{8} \times 1\frac{1}{2}) \div \frac{3}{16}$

20. $(1\frac{1}{2} \times 2\frac{3}{4}) \div \frac{3}{64}$

21. $(\frac{5}{8} \div \frac{1}{7}) \times 14$

22. $(\frac{7}{16} \div \frac{3}{4}) \times \frac{1}{2}$

23. $(\frac{1}{64} \div \frac{7}{8}) \times \frac{3}{4}$

24. $(2\frac{1}{64} \div \frac{1}{32}) \times (2\frac{1}{64} \div 1\frac{1}{2})$

2-15 Complex Fractions

Sometimes an arithmetic operation involves a fraction where a fraction or a mixed number is the numerator and a fraction or a mixed number is the denominator. These fractions are called *complex fractions.*

    A complex fraction is one in which the numerator, the denominator, or both, are fractions or mixed numbers. It was pointed out that a fraction may be regarded as an indicated division in which the numerator is the dividend and the denominator is the divisor. To evaluate the given expression, perform the indicated division.

Example 1

Find the value of the expression $\dfrac{\frac{2}{3}}{\frac{5}{7}}$.

SOLUTION $\dfrac{\frac{2}{3}}{\frac{5}{7}} = \frac{2}{3} \div \frac{5}{7} = \frac{2}{3} \times \frac{7}{5} = \frac{14}{15}$

EXPLANATION   This complex fraction indicates that $\frac{2}{3}$ is to be divided by $\frac{5}{7}$. Proceed as in division of fractions by inverting the divisor ($\frac{5}{7}$) and then multiplying the two fractions.

Example 2

Find the value of $\dfrac{2\frac{3}{4}}{16}$.

SOLUTION $2\frac{3}{4} \div \frac{16}{1} = \frac{11}{4} \times \frac{1}{16} = \frac{11}{64}$

EXPLANATION   This complex fraction indicates that $2\frac{3}{4}$ is to be divided by 16. Change the $2\frac{3}{4}$ to an improper fraction and proceed as in the division of fractions by inverting the divisor to $\frac{1}{16}$ and then multiplying the two fractions.

Example 3

Find the value of $\dfrac{3\frac{1}{4} - 2\frac{1}{8}}{4\frac{1}{8} - 1\frac{1}{2}}$.

SOLUTION   Subtraction is involved in both the numerator and the denominator. To solve this problem, simplify the numerator and denominator separately before simplifying the resulting fraction. $3\frac{1}{4} - 2\frac{1}{8} = 1\frac{1}{8}$ and $4\frac{1}{8} - 1\frac{1}{2} = 2\frac{5}{8}$.

$$\frac{3\frac{1}{4} - 2\frac{1}{8}}{4\frac{1}{8} - 1\frac{1}{2}} = \frac{1\frac{1}{8}}{2\frac{5}{8}} = \frac{\overset{3}{\cancel{9}}}{\underset{1}{\cancel{8}}} \times \frac{1}{\underset{7}{\cancel{21}}} = \frac{3}{7}$$

EXPLANATION   First simplify the numerator by performing the indicated subtraction, giving the mixed number $1\frac{1}{8}$. Then simplify the denominator in the same way, obtaining $2\frac{5}{8}$. Finally, simplify the resulting fraction $\dfrac{1\frac{1}{8}}{2\frac{5}{8}}$ by carrying out the indicated division of $1\frac{1}{8}$ divided by $2\frac{5}{8}$.

Example 4

Find the value of $\dfrac{\frac{3}{4} \text{ of } 2\frac{3}{8}}{3\frac{1}{4} \div 14\frac{1}{3}}$.

SOLUTION $\frac{3}{4}$ of $2\frac{3}{8} = \frac{3}{4} \times \frac{19}{8} = \frac{57}{32}$

$3\frac{1}{4} \div 14\frac{1}{3} = \frac{13}{4} \times \frac{3}{43} = \frac{39}{172}$

$$\dfrac{\frac{3}{4} \text{ of } 2\frac{3}{8}}{3\frac{1}{4} \div 14\frac{1}{3}} = \frac{57}{32} \div \frac{39}{172}$$

$$\frac{57}{32} \div \frac{39}{172} = \frac{\overset{19}{\cancel{57}}}{\underset{8}{\cancel{32}}} \times \frac{\overset{43}{\cancel{172}}}{\underset{13}{\cancel{39}}} = \frac{817}{104} = 7\frac{89}{104}$$

Problems 2-15

Simplify the following complex fractions. Reduce answers to lowest terms.

1. $\dfrac{\frac{1}{2}}{\frac{1}{3}}$ 2. $\dfrac{\frac{3}{5}}{4\frac{1}{8}}$ 3. $\dfrac{6}{3\frac{2}{3}}$ 4. $\dfrac{12\frac{1}{2}}{16\frac{2}{3}}$

5. $\dfrac{2\frac{5}{8}}{4}$ 6. $\dfrac{4\frac{9}{16} - 3\frac{1}{8}}{2\frac{1}{4} + 1\frac{3}{16}}$ 7. $\dfrac{5\frac{1}{2} - 4\frac{3}{4}}{3\frac{7}{8} - 2\frac{15}{16}}$ 8. $\dfrac{3\frac{1}{2} + 5\frac{1}{4}}{4\frac{1}{4} + 2\frac{1}{2}}$

9. $\dfrac{\frac{2}{3} \text{ of } \frac{3}{4}}{\frac{1}{2} \text{ of } \frac{7}{8}}$ 10. $\dfrac{2\frac{1}{3} \times 3\frac{1}{3}}{7 \times 6\frac{2}{3}}$ 11. $\dfrac{8\frac{4}{5} \div 3\frac{1}{7}}{10\frac{1}{2} \times 6\frac{2}{3}}$ 12. $\dfrac{\frac{7}{16} \text{ of } 4\frac{2}{5}}{\frac{2}{5} \text{ of } \frac{9}{14}}$

13. $\dfrac{4\frac{1}{4} - 2\frac{1}{2}}{3\frac{1}{2} + 2\frac{1}{2}}$ 14. $\dfrac{3\frac{1}{8} + 4\frac{3}{8}}{5 - 1\frac{1}{2}}$ 15. $\dfrac{\frac{7}{8} - \frac{1}{2}}{\frac{7}{8} - \frac{1}{2}}$ 16. $\dfrac{\frac{5}{8} \times \frac{8}{9}}{\frac{3}{4} \div \frac{3}{8}}$

17. $\dfrac{\frac{5}{16} \div \frac{3}{16}}{4 \times \frac{1}{2}}$ 18. $\dfrac{3\frac{1}{2} \times 4\frac{1}{2}}{\frac{5}{8} \times \frac{3}{8}}$ 19. $\dfrac{\frac{2}{3} \div 1\frac{2}{3}}{\frac{2}{5} \div 3\frac{1}{5}}$ 20. $\dfrac{4 + 1\frac{2}{3}}{5 + 3\frac{1}{3}}$

21. $\dfrac{\frac{4}{5} + \frac{5}{16}}{\frac{3}{20} + 1\frac{3}{4}}$ 22. $\dfrac{1\frac{4}{5} + 3\frac{1}{4}}{5\frac{3}{8} - 1\frac{1}{10}}$ 23. $\dfrac{\frac{3}{8} \times 3\frac{1}{16}}{\frac{5}{16} \div \frac{3}{4}}$ 24. $\dfrac{3\frac{1}{2} \div 4\frac{1}{2}}{5\frac{1}{2} \times 6\frac{1}{2}}$

Self-Test

Do the following ten problems. Check your answers with the answers in the back of the book.

1. $\frac{1}{8} + \frac{3}{4}$ 2. $2\frac{1}{2} + 4\frac{3}{4}$ 3. $\frac{3}{8} - \frac{1}{16}$ 4. $3\frac{1}{4} - 1\frac{3}{8}$

5. $\frac{2}{3} \times \frac{4}{5}$ 6. $1\frac{3}{4} \times 1\frac{1}{2}$ 7. $\frac{1}{3} \div \frac{1}{2}$ 8. $2\frac{1}{2} \div 1\frac{1}{2}$

9. $\dfrac{1\frac{1}{2} + \frac{1}{4}}{3\frac{1}{2} \times \frac{1}{2}}$ 10. $\dfrac{1\frac{1}{3} - \frac{1}{2}}{4\frac{1}{2} \div 2}$

Chapter Test

Rewrite the following examples. Do the indicated operations. Reduce answers to lowest terms.

Add:

1. $\frac{1}{2} + \frac{3}{4}$ 2. $1\frac{2}{3} + \frac{2}{5}$ 3. $\frac{3}{8} + 2\frac{3}{4}$ 4. $1\frac{5}{8} + 2\frac{3}{4}$

5. $2\frac{1}{16} + \frac{3}{32}$ 6. $5\frac{15}{16} + \frac{7}{8}$ 7. $3\frac{7}{8} + \frac{31}{32}$ 8. $2\frac{3}{4} + 5\frac{63}{64}$

Subtract:

9. $\frac{3}{4} - \frac{2}{3}$ 10. $1\frac{3}{4} - \frac{7}{8}$ 11. $4 - \frac{15}{16}$ 12. $2\frac{1}{2} - 1\frac{7}{8}$

13. $12\frac{1}{2} - 3\frac{7}{16}$ 14. $3\frac{1}{4} - 2\frac{15}{16}$ 15. $1\frac{1}{2} - \frac{31}{32}$ 16. $3\frac{1}{8} - 1\frac{21}{64}$

Multiply:

17. $\frac{1}{2} \times \frac{3}{4}$ 18. $\frac{7}{8} \times 1\frac{1}{2}$ 19. $2\frac{1}{2} \times \frac{3}{4}$ 20. $2\frac{3}{8} \times 5$

21. $2\frac{1}{8} \times 3\frac{3}{4}$ 22. $4\frac{1}{2} \times 3\frac{1}{8}$ 23. $5\frac{1}{2} \times 6\frac{2}{3}$ 24. $\frac{33}{77} \times \frac{7}{16}$

Divide:

25. $1\frac{1}{2} \div 4$ 26. $2\frac{3}{4} \div \frac{7}{8}$ 27. $3\frac{1}{2} \div \frac{1}{2}$ 28. $5\frac{3}{8} \div \frac{1}{3}$

29. $2\frac{1}{2} \div 3\frac{1}{4}$ 30. $\frac{1}{2} \div 2\frac{1}{2}$ 31. $4\frac{1}{4} \div 2\frac{3}{8}$ 32. $1\frac{1}{2} \div 1\frac{15}{16}$

Simplify:

33. $\dfrac{3\frac{1}{2} + \frac{7}{8}}{4\frac{1}{2} + 3\frac{3}{4}}$
      34. $\dfrac{2\frac{1}{2} - 1\frac{1}{4}}{3\frac{1}{8} - \frac{7}{8}}$
      35. $\dfrac{2\frac{1}{2} \times 3\frac{1}{4}}{2\frac{1}{2} \times 4\frac{1}{8}}$
      36. $\dfrac{1\frac{3}{4} \div 4}{3\frac{1}{16} \div 2\frac{3}{16}}$

Miscellaneous Problems in Common Fractions

1. Find the total length of the shaft in Fig. 2-2.

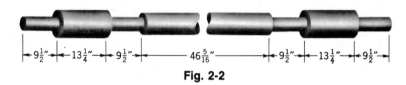

Fig. 2-2

2. Find the total width of the boiler plate in Fig. 2-3.

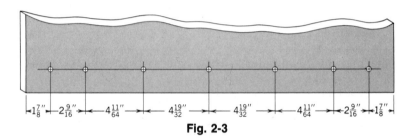

Fig. 2-3

3. What is the outside diameter of the 1 in. I.D. tubing in Fig. 2-4?

Fig. 2-4

4. Find the distance between the end holes of the splice plate in Fig. 2-5.

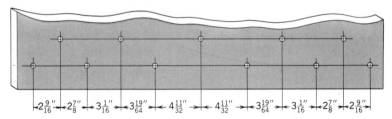

Fig. 2-5

5. Find the total length of the bolt in Fig. 2-6.

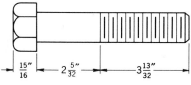

Fig. 2-6

6. Find the total length of the spindle shown in Fig. 2-7.

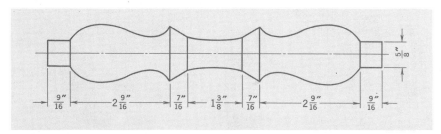

Fig. 2-7

7. What is the total length of the piece needed to make the handle for the gavel shown in Fig. 2-8?

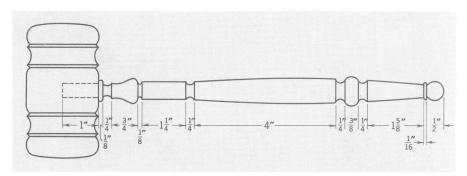

Fig. 2-8

8. What is the length of stock needed to make the bracket shown in Fig. 2-9? (Add $\frac{1}{2}$ of radius, R, for each bend.)

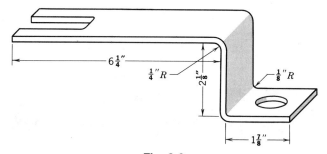

Fig. 2-9

9. Compute the overall length of the bevel gear blank in Fig. 2-10.
10. Compute the overall length of the gland in Fig. 2-11.

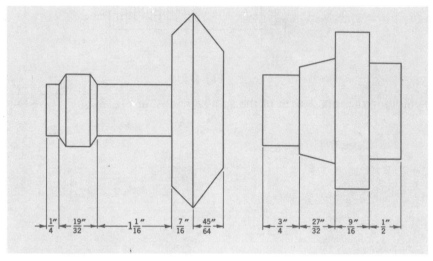

        Fig. 2-10   **Fig. 2-11**

11. Compute the overall length of the wheel hub in Fig. 2-12.

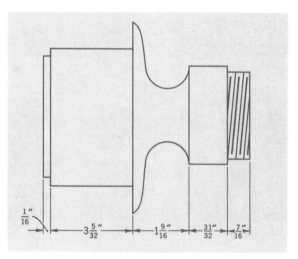

Fig. 2-12

12. Find the difference between the diameters at the ends of the tapered piece in Fig. 2-13.

Fig. 2-13

13. Supply the missing dimension in Fig. 2-14 if the overall length is $1\frac{5}{8}''$.

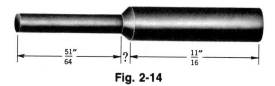

$\frac{51''}{64}$? $\frac{11''}{16}$

Fig. 2-14

14. Find the center-to-center distance of holes in the crankcase cover in Fig. 2-15 if the holes are spaced evenly.

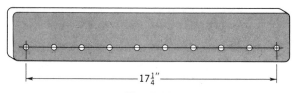

$17\frac{1}{4}''$

Fig. 2-15

15. Find the distance between end holes in the girder connection in Fig. 2-16.

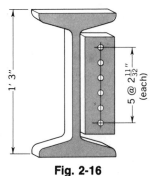

1' 3"

5 @ $2\frac{11}{32}''$ (each)

Fig. 2-16

16. How many binding posts $1\frac{7}{8}''$ long can be cut from a piece of brass rod $27\frac{1}{2}''$ long? Allow $\frac{1}{8}''$ waste for each cut.

17. How long a piece of drill rod is required to make 15 drills $3\frac{1}{16}''$ long? Allow $\frac{1}{8}''$ waste for each cut.

18. Four standard distance blocks are used to measure a certain space. The blocks are $\frac{5}{16}''$, $\frac{7}{64}''$, $\frac{3}{8}''$, and $\frac{1}{32}''$. How wide is the space?

19. How many holes spaced $1\frac{7}{16}''$ center to center can be drilled in an angle iron $22\frac{5}{8}''$ long, allowing $1\frac{1}{4}''$ distance on each end?

20. Find the length of the steel plate required to make the drill jig shown in Fig. 2-17.

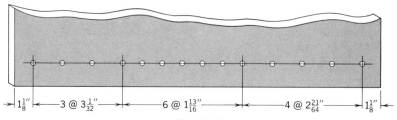

Fig. 2-17

21. Seven pins, each $6\frac{7}{8}''$ long, were cut from a piece of drill rod 72″ long. Allowing $\frac{3}{32}''$ waste for each cut, what was the length of the piece left?
22. How many washers, each $\frac{3}{32}''$ thick, can be made from a piece of stock $25\frac{1}{2}''$ long, allowing $\frac{1}{16}''$ waste for each cut?
23. Allowing $1\frac{1}{4}$ minutes for putting away the finished pin and placing the stock in the lathe, how long will it take to machine 25 pins if each pin requires $10\frac{1}{2}$ minutes machining time?
24. Find the length of the piece of stock required for eight taper keys each $6\frac{1}{2}''$ long, allowing $\frac{1}{8}''$ waste for each cut.
25. A machinist spends $2\frac{1}{4}$ hours at the lathe, $4\frac{1}{2}$ hours at the planer, and the rest of the day at the shaper. If the working day is 8 hours, how much time is spent at the shaper? If the pay is $9.75 per hour, how much is paid for the time spent at each machine?
26. Find the two missing dimensions in Fig. 2-18.

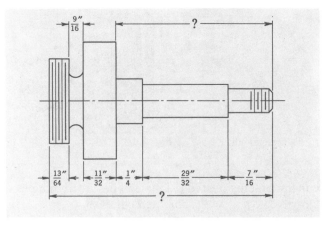

Fig. 2-18

27. If the run of each step in the flight of stairs shown in Fig. 2-19 is $9\frac{3}{8}''$, what is the total run of the flight of stairs?

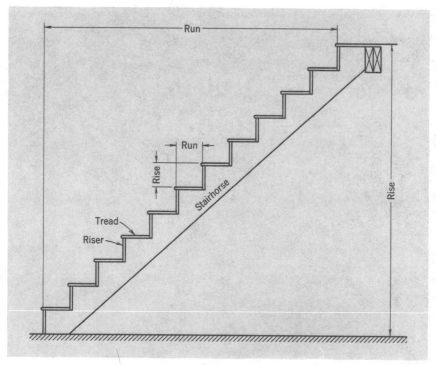

Fig. 2-19

28. Find the inside diameter of the pipe shown in cross section in Fig. 2-20.
29. Find the missing dimension in Fig. 2-21.

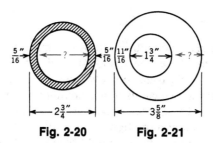

Fig. 2-20   **Fig. 2-21**

30. Compute the overall dimension in Fig. 2-22.

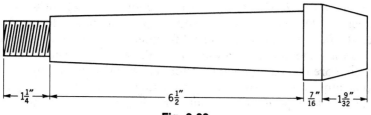

Fig. 2-22

3

Decimal Fractions

3-1 Introduction

Decimal fractions are frequently used by the machinist in daily work. Look at any drawing or machine design and you will see that decimal fractions, or *decimals,* are commonly used to show the various dimensions.

There are still some occasions to use common fractions (halves, quarters, eighths, sixteenths, etc.), since dimensions involving drill sizes and so on are generally given in common fractions. It is easy for the mechanic to visualize these dimensions. The mechanic should be able to work with both kinds of fractions and know the decimal equivalents of halves, quarters, eighths, sixteenths, and so forth.

A *decimal fraction* is a fraction whose denominator is 10 or some power of 10, such as 100, 1,000, or 10,000. Thus $\frac{3}{10}$, $\frac{25}{100}$, and $\frac{625}{1000}$ are decimal fractions. Since the denominator of a decimal fraction is always 10 or a power of 10 (that is the denominator is always a 1 with zeros), we can write a decimal fraction more compactly, and therefore more conveniently, by omitting the denominator entirely. Thus $\frac{3}{10}$ can be written 0.3; $\frac{25}{100}$, 0.25; and $\frac{625}{1000}$, 0.625.

The use of the period (.) in a number indicates that the number following the period is a fraction with a power of ten as its denominator. This period is called the *decimal point.* Therefore any number with a decimal point in front of it is a fraction whose numerator is the number after the point and whose denominator is a 1 with as many zeros after it as there are figures in the number to the right of this point. Thus 0.3 means $\frac{3}{10}$; 0.35 means $\frac{35}{100}$; 0.03 means $\frac{3}{100}$; 0.003 means $\frac{3}{1000}$, etc.

The relative values of the figures in any number are shown in the following table.

VALUES OF FIGURES	
100	= One hundred
10	= Ten
1	= One
0.1	= One tenth
0.01	= One hundredth
0.001	= One thousandth
0.0001	= One ten-thousandth
0.00001	= One hundred-thousandth

When the mechanic reads a decimal fraction such as 4.01, he or she should read the whole number part and then count decimal places to read the decimal part. Two places indicate hundredths; therefore, 4.01 is read as "four and one one-hundredth." The word "and" indicates that one one-hundredth is to be *added* to the whole number 4. As another example, 0.132 is read as "one hundred thirty-two thousandths."

Problems 3-1

Read the following numerals.

1. 0.1   2. 0.01   3. 0.001   4. 0.011
5. 0.15   6. 0.83   7. 0.083   8. 2.016

Write the following numerals.

9. Six tenths.
10. Two and five tenths.
11. Three hundredths.
12. Twenty-five hundredths.
13. One and thirty-three hundredths.
14. Three thousandths.

15. Ten and one one-hundredth.
16. One and one one-thousandth.
17. Three hundred two thousandths.
18. Two hundred five and sixty-four thousandths.
19. One thousand one ten-thousandths.
20. Four and one hundred twenty-five ten-thousandths.
21. Ten and one tenth.
22. One hundred one and one hundred one ten-thousandths.
23. Two and six hundred twenty-five ten-thousandths.
24. One and three thousand one hundred twenty-five hundred-thousandths.
25. Five hundred twelve and seventy-five ten-thousandths.
26. Sixty-five ten-thousandths.
27. Five and forty-two millionths.
28. One hundred twenty-five thousand and one hundred twenty-five thousandths.

3-2   Changing a Decimal Fraction to a Common Fraction

It is sometimes convenient to change a decimal fraction to a common fraction. The procedure is quite easy to follow. The numerator of the common fraction is all the numerals that are present in the decimal fraction and the denominator of the common fraction is a 1 followed by as many zeros as there are places to the right of the decimal point. The fraction is then reduced to lowest terms following the procedures in Chapter 2.

Example 1
Change 0.4 to a common fraction.

SOLUTION $0.4 = \dfrac{4}{10} = \dfrac{4 \div 2}{10 \div 2} = \dfrac{2}{5}$

EXPLANATION   The number after the decimal point is the numerator of the fraction; the denominator is a 1 with one zero because there is one figure to the right of the decimal point. Thus $0.4 = \frac{4}{10}$. This is changed to the lowest terms by dividing both numerator and denominator by 2.

Example 2
Change 0.75 to a common fraction.

SOLUTION $0.75 = \dfrac{75}{100} = \dfrac{75 \div 5}{100 \div 5} = \dfrac{15}{20} = \dfrac{15 \div 5}{20 \div 5} = \dfrac{3}{4}$

or $\dfrac{75}{100} = \dfrac{\overset{1}{\cancel{5}} \times \overset{1}{\cancel{5}} \times 3}{\underset{1}{\cancel{5}} \times \underset{1}{\cancel{5}} \times 4} = \dfrac{3}{4}$

EXPLANATION   The numerator of the common fraction is 75 and the denominator is 1 with two zeros because there are two figures after the decimal point. Thus $0.75 = \frac{75}{100}$. To reduce $\frac{75}{100}$ to its lowest terms, divide both terms by 5 and get $\frac{15}{20}$; dividing each term of $\frac{15}{20}$ by 5 gives $\frac{3}{4}$. The desired result could be obtained in one step by dividing both terms of the fraction $\frac{75}{100}$ by 25. Thus $\dfrac{75 \div 25}{100 \div 25} = \dfrac{3}{4}$.

Example 3
Change 0.024 to a common fraction.

SOLUTION $0.024 = \dfrac{24}{1,000} = \dfrac{24 \div 2}{1,000 \div 2} = \dfrac{12}{500}$

$\dfrac{12 \div 2}{500 \div 2} = \dfrac{6}{250} \quad \dfrac{6 \div 2}{250 \div 2} = \dfrac{3}{125}$

These steps can be condensed as follows:

$$\dfrac{\begin{matrix}3\\ \cancel{6}\\ \cancel{12}\\ \cancel{24}\end{matrix}}{\begin{matrix}\cancel{1,000}\\ \cancel{500}\\ \cancel{250}\\ 125\end{matrix}} = \dfrac{3}{125}$$

EXPLANATION   To reduce $\frac{24}{1000}$ to lowest terms, proceed as shown in the previous chapter by finding all the common factors of 24 and 1,000 and cancelling them. If the student found 8 to be the highest common factor of both these numbers, the common fraction could be found by dividing 8 into both of these terms in one step. Thus $\dfrac{24 \div 8}{1,000 \div 8} = \dfrac{3}{125}$.

Problems 3-2

Change the following decimal fractions to common fractions and write each answer in its lowest terms.

1. 0.5	2. 0.15	3. 0.125	4. 0.25
5. 0.75	6. 0.375	7. 0.875	8. 0.16
9. 0.8	10. 0.06	11. 0.0625	12. 0.187
13. 0.042	14. 0.625	15. 0.03125	16. 0.015625
17. 0.0125	18. 0.0375	19. 0.9625	20. 0.005
21. 0.0075	22. 0.002	23. 0.0875	24. 0.982
25. 0.96875	26. 0.984375	27. 0.953125	28. 0.844
29. 0.9375	30. 0.6875	31. 0.8125	32. 0.7875

3-3   Changing a Common Fraction to a Decimal Fraction

Round-off Rule

All common fractions can be written as repeating decimal fractions. A decimal approximation of the common fraction however is often sufficiently accurate for solving a shop problem. It will be necessary to adopt a rule to use when expressing a decimal equivalent of common fractions. The mechanic will determine how many decimal places are needed to solve the problem. A carpenter certainly would seldom use an accuracy greater than one-hundredth, while a machinist would sometimes want to work to the nearest ten-thousandth.

In this book the following rules will be used to round off decimal fractions.

1. Determine the number of places that are necessary for the particular problem.
2. Now look at the remaining places. If the remaining places start with a 5 or a number greater than 5, add 1 to the last place of the decimal that is to be kept. For example, if **7.36482** is to be rounded to **three** decimal places, then observe that the numeral after the third place is an 8, which is greater than 5. Therefore, 1 is added to the third place (4), and the number 7.36482, to three decimal places, is 7.365.

   If **344.06250** is to be rounded to **three** decimal places, then observe that the numeral after the third place is a 5. Therefore, 1 is added to the third place (2), and the number 344.06250, to three decimal places, is 344.063.
3. Suppose the mechanic wanted to round **6.28236** to **three** decimal places. Observe that the numeral directly after the third place is a 3, which is less than 5. Therefore, all the numerals after the third place are ignored, and the numeral 6.28236, rounded to three decimal places, is 6.282.

To change a common fraction, such as halves, quarters, eighths, sixteenths, and so on, to a decimal fraction, divide the denominator into the numerator as shown in the following examples.

Example 1

Change $\frac{7}{8}$ to a decimal fraction.

SOLUTION

$$
\begin{array}{r}
0.875 \\
8\overline{)7.000} \\
\underline{6\ 4} \\
60 \\
\underline{56} \\
40 \\
\underline{40}
\end{array}
$$

EXPLANATION   Since a fraction is an indicated division, divide 7 by 8. Place a decimal point after the 7 and annex zeros. The division is carried out as shown $\frac{7}{8} = 0.875$.

Example 2

Change $\frac{2}{3}$ to a decimal fraction accurate to the hundredth place.

SOLUTION

$$
\begin{array}{r}
0.66\ldots \\
3\overline{)2.000} \\
\underline{1\,8} \\
20 \\
\underline{18} \\
2
\end{array}
$$

EXPLANATION   Proceeding as in Example 1, it is found that no matter how many zeros are added after the decimal point, the quotient will not come out evenly. In such a case, it must be decided how far the answer should be carried out. The answer to the nearest hundredth, which means two places to the right of the decimal point, would be $\frac{2}{3} = 0.67$.

Example 3

Change $\frac{3}{16}$ to a decimal of three places.

SOLUTION

$$
\begin{array}{r}
0.1875 \\
16\overline{)3.0000} \\
\underline{1\,6} \\
1\,40 \\
\underline{1\,28} \\
120 \\
\underline{112} \\
80 \\
\underline{80} \\
0
\end{array}
$$

EXPLANATION   Carrying out the indicated division, the digit after the third place is exactly 5. Add 1 to the third place, since the round-off rule says to add 1 if the remaining places start with a 5 or a numeral greater than 5. $\frac{3}{16} = 0.188$ to three decimal places.

Problems 3-3

Change the following common fractions to decimal fractions of three places.

NOTE.   If the digit in the fourth place is a numeral equal to or greater than 5, add 1 to the third place; thus 8.7236 would be 8.724 to three decimal places, and 4.3425 would be 4.343 to three decimal places. If the numeral in the fourth place is less than 5, leave the third numeral unchanged; thus 6.1482, to three decimal places, would be 6.148.

1. $\frac{1}{4}$ 2. $\frac{3}{8}$ 3. $\frac{7}{16}$ 4. $\frac{1}{3}$ 5. $\frac{2}{3}$

6. $\frac{3}{5}$ 7. $\frac{1}{80}$ 8. $\frac{3}{20}$ 9. $\frac{11}{64}$ 10. $\frac{3}{32}$

11. $\frac{7}{8}$ 12. $\frac{13}{16}$ 13. $\frac{5}{6}$ 14. $\frac{1}{11}$ 15. $\frac{1}{12}$

16. $\frac{1}{25}$ 17. $\frac{1}{250}$ 18. $\frac{1}{7}$ 19. $\frac{1}{9}$ 20. $\frac{9}{10}$

21. $\frac{19}{144}$ 22. $\frac{450}{1728}$ 23. $\frac{490}{1728}$ 24. $\frac{11}{17}$ 25. $\frac{631}{5280}$

26. $\frac{87}{144}$ 27. $\frac{625}{1728}$ 28. $\frac{26}{27}$ 29. $\frac{440}{5280}$ 30. $\frac{158}{314}$

31. $\frac{158}{3141}$ 32. $\frac{158}{31416}$ 33. $\frac{1}{144}$ 34. $\frac{235}{350}$ 35. $\frac{187}{250}$

36. What decimal part of a dollar is 42 cents?

37. What decimal fraction of a pound is 13 ounces?

38. What decimal fraction of a foot is 8"?

39. Change 100 sq in. into a decimal fraction of a square foot.

40. Change 500 cu in. into a decimal fraction of a cubic foot.

3-4   Using a Table of Decimal Equivalents

The answer to a problem may often be a fraction such as $\frac{5}{7}$ in. or $\frac{3}{13}$ in. To lay off such a dimension with the ordinary scale, change the answer to the nearest eighth, sixteenth, thirty-second, or sixty-fourth. This is done most conveniently by means of a *table of decimal equivalents*. See Table 4 at the back of this book.

Example 1

On a scale divided into sixty-fourths of an inch, what is the dimension nearest to $\frac{13}{37}$ in.?

SOLUTION $\frac{13}{37}$ in. = 0.35135 in. $\approx \frac{22}{64}$ in. (the symbol "$\approx$" means *is approximately equal to.*)

EXPLANATION   First change the common fraction $\frac{13}{37}$ to the decimal 0.35135. Looking down the table of decimal equivalents, observe that 0.35135 lies between 0.34375 and 0.359375. By subtracting, find that 0.35135 − 0.34375 = 0.0076, whereas 0.359375 − 0.35135 = 0.008025. Therefore, 0.35135 is nearer to 0.34375, and its nearest equivalent in sixty-fourths of an inch is $\frac{22}{64}$ in.

To find the equivalent of $\frac{13}{37}$ in sixteenths: $\frac{5}{16}$ = 0.3125; $\frac{6}{16}$ or $\frac{3}{8}$ = 0.375. Since 0.35135 is nearer to 0.375, the nearest equivalent of $\frac{13}{37}$ in. is $\frac{6}{16}$ in.

Where great accuracy is not required, it is sufficient to convert the given fraction into a decimal of three places and use only the first three figures of the table.

If a table of decimal equivalents is not available, any fraction can be converted into any other fraction with the desired denominator as shown in the following example.

Example 2

Change $\frac{13}{37}$ to sixty-fourths.

SOLUTION $\frac{13}{37} \times 64 = \frac{832}{37} = 22\frac{18}{37} \approx 22$. Therefore $\frac{13}{37} \approx \frac{22}{64}$ to the nearest sixty-fourth.

EXPLANATION Since in one unit there are 64 sixty-fourths, in $\frac{13}{37}$ of a unit there will be $\frac{13}{37}$ of 64 or $22\frac{18}{37}$ sixty-fourths. Since $\frac{18}{37}$ is less than 0.5, round off to $\frac{22}{64}$.

Example 3

Change $\frac{2}{7}$ to sixteenths.

SOLUTION $\frac{2}{7} \times 16 = \frac{32}{7} = 4\frac{4}{7} \approx 5$

$\frac{2}{7} \approx \frac{5}{16}$ to the nearest sixteenth.

EXPLANATION Since in one unit there are 16 sixteenths, in $\frac{2}{7}$ of a unit there will be $\frac{2}{7}$ of 16 or $4\frac{4}{7}$ sixteenths. Rounded off, that is $\frac{5}{16}$.

Problems 3-4

Use Table 4 at the back of the book to find the decimal equivalent of each of the following fractions to the nearest sixteenth of an inch.

1. $\frac{1}{2}$	2. $\frac{3}{4}$	3. $\frac{5}{8}$	4. $\frac{29}{32}$
5. $\frac{8}{17}$	6. $\frac{5}{21}$	7. $\frac{2}{3}$	8. $\frac{2}{9}$
9. $1\frac{9}{21}$	10. $3\frac{5}{9}$	11. $4\frac{7}{11}$	12. $5\frac{5}{12}$
13. $2\frac{1}{3}$	14. $6\frac{3}{11}$	15. $8\frac{6}{19}$	16. $7\frac{5}{14}$

Using Table 4, find the decimal equivalent of each of the following fractions to the nearest thirty-second of an inch.

17. $\frac{2}{3}$	18. $\frac{3}{5}$	19. $\frac{3}{10}$	20. $\frac{15}{31}$
21. $2\frac{8}{17}$	22. $5\frac{1}{8}$	23. $3\frac{6}{19}$	24. $2\frac{2}{7}$
25. $6\frac{5}{8}$	26. $1\frac{11}{13}$	27. $8\frac{19}{144}$	28. $2\frac{4}{5}$
29. $4\frac{325}{1728}$	30. $12\frac{880}{5280}$	31. $10\frac{1}{144}$	32. $3\frac{1320}{1728}$

Challenge   Using Table 4, find the decimal equivalent of each of the following fractions to the nearest sixty-fourth of an inch.

33. $\frac{2}{3}$ 34. $\frac{3}{4}$ 35. $\frac{3}{5}$ 36. $\frac{7}{10}$

37. $2\frac{9}{17}$ 38. $5\frac{5}{8}$ 39. $3\frac{16}{19}$ 40. $2\frac{6}{7}$

41. $6\frac{5}{8}$ 42. $1\frac{11}{13}$ 43. $8\frac{143}{144}$ 44. $2\frac{4}{5}$

45. $4\frac{1000}{1728}$ 46. $12\frac{880}{1760}$ 47. $10\frac{440}{5280}$ 48. $3\frac{121}{144}$

3-5   Conversion of Dimensions

It is sometimes necessary in computation to convert inches and fractions of an inch into feet and decimal fractions of a foot. The following examples will illustrate the procedure in problems of this type. (The examples use the table of decimal equivalents, Table 4, in the back of the book.)

RULE
Dividing inches by 12 will change inches to feet.

Example 1
Convert $3\frac{3}{4}$ inches into a decimal fraction of a foot. State the answer to three decimal places.

SOLUTION $3\frac{3}{4}$ in. = 3.75 in. $\frac{3.75}{12}$ = 0.3125 ft

$3\frac{3}{4}$ in. = 0.313 ft (to three decimal places)

Example 2
Convert $46\frac{1}{8}$ inches into feet and a decimal fraction of a foot to three decimal places.

SOLUTION $46\frac{1}{8}$ in. = 46.125 in. $\frac{46.125}{12}$ = 3.84375 ft

$46\frac{1}{8}$ in. = 3.844 ft (to three decimal places)

Example 3
Convert 43.172 inches into feet and inches and the nearest sixty-fourth of an inch. (Use the table of decimal equivalents.)

SOLUTION   43.172 in. = 43 in. + 0.172 in.

$\frac{43\text{ in.}}{12}$ = $3\frac{7}{12}$ ft =   3 ft 7 in.

0.172 in. = + $\frac{11}{64}$ in.

3ft $7\frac{11}{64}$ in.

EXPLANATION For easier solving, break 43.172 into its two parts—a whole number (43) and a decimal fraction (0.172). Divide 43 by 12 to find the number of feet in 43 inches. Because there are 12 inches in one foot, $3\frac{7}{12}$ ft = 3 ft 7 in. To convert 0.172 to 64ths of an inch, refer to the table, which shows that the decimal nearest in value is 0.171875, which is equivalent to $\frac{11}{64}$. Finally, add 3 ft 7 in. to $\frac{11}{64}$ in. for an answer of 3 ft $7\frac{11}{64}$ in.

RULE——————————————————————————————————————
   Multiplying feet by 12 will change feet to inches.

Example 4
   Convert 0.319 foot into inches and a fraction of an inch. (Use the table of decimal equivalents.)

SOLUTION 0.319 ft $\times$ 12 = 3.828 in. The figure in the table of decimal equivalents nearest to 0.828 is 0.828125, which is the decimal equivalent of $\frac{53}{64}$. 0.319 ft = $3\frac{53}{64}$ in.

Problems 3-5

Convert the following dimensions from inches into feet. State the answers in whole numbers and/or decimals to three decimal places.

1. $7\frac{1}{2}''$	2. $9\frac{3}{4}''$	3. $8\frac{7}{16}''$	4. $9\frac{5}{8}''$
5. $\frac{7}{8}''$	6. $6\frac{9}{16}''$	7. $71\frac{3}{4}''$	8. $5\frac{11}{16}''$
9. $2\frac{1}{4}''$	10. $53\frac{1}{2}''$	11. $75\frac{5}{8}''$	12. $96\frac{1}{8}''$
13. $27\frac{5}{8}''$	14. $33\frac{3}{4}''$	15. $25\frac{7}{16}''$	16. $16\frac{7}{8}''$
17. $21\frac{1}{4}''$	18. $42\frac{1}{2}''$	19. $37\frac{9}{16}''$	20. $120\frac{3}{8}''$

Convert the following dimensions into feet, inches, and the nearest sixty-fourth of an inch. (Use Table 4.)

21. 219.31″	22. 28.19″	23. 39.11″	24. 165.3″
25. 73.26″	26. 326.83″	27. 172.8″	28. 91.3″
29. 53.17″	30. 220.94″	31. 14.116″	32. 19.1″
33. 217.6″	34. 32.31″	35. 56.26″	36. 181.33″
37. 234.54″	38. 121.116″	39. 200.017″	40. 147.32″

Convert the following dimensions into inches and the nearest sixty-fourth of an inch. (Use Table 4.)

41. 0.91′	42. 0.81′	43. 0.864′	44. 0.395′
45. 0.217′	46. 0.09′	47. 0.186′	48. 0.78′
49. 0.146′	50. 0.391′	51. 0.82′	52. 0.211′
53. 1.91′	54. 2.864′	55. 1.23′	56. 1.065′
57. 3.021′	58. 2.11′	59. 2.36′	60. 3.421′

3-6 Addition of Decimals

Addition of decimal fractions is much like addition of whole numbers. The difference is that when adding decimal fractions, the decimal points must be placed one under the other. The decimal point appears in the answer in the same column as it is in the problem.

Example 1

Add 3.25, 72.004, 864.0725, 647, and 0.875.

SOLUTION

$$
\begin{array}{r}
3.25 \\
72.004 \\
864.0725 \\
647. \\
+\quad 0.875 \\
\hline
1587.2015
\end{array}
$$

$$
\begin{array}{r}
3.2500 \\
72.0040 \\
864.0725 \\
647.0000 \\
+\quad 0.8750 \\
\hline
1587.2015
\end{array}
$$

EXPLANATION To add the given numbers, write them so that the decimal points are in the same column. Then proceed as in the addition of ordinary whole numbers and place the decimal point in the answer in the same column as the other decimal points.

To facilitate the work and avoid errors, it is advisable to annex zeros to the numbers with the fewer decimal places, so that all the numbers will have an equal number of places after the decimal point. The work then appears as shown.

Example 2

Find the sum of $\frac{1}{2}$ + 0.662 + $\frac{7}{8}$.

SOLUTION $\frac{1}{2}$ = 0.5, $\frac{7}{8}$ = 0.875

$$
\begin{array}{r}
0.500 \\
0.662 \\
+0.875 \\
\hline
2.037
\end{array}
$$

EXPLANATION First change $\frac{1}{2}$ and $\frac{7}{8}$ to decimals, obtaining 0.5 and 0.875, and then add the numbers as in Example 1.

Find the sums of the following numbers. Make sure that your answers are in decimal fraction form.

1. $3 + 5.6$
2. $46.2 + 37.5$
3. $3.1 + \frac{1}{2}$
4. $4.2 + \frac{1}{2} + 1.3$
5. $4.6 + .6 + \frac{7}{8}$
6. $5.31 + 4.21 + \frac{1}{4}$
7. $27.62 + .62 + \frac{3}{4}$
8. $\frac{3}{4} + \frac{1}{2} + 4.9$
9. $\frac{1}{2} + \frac{1}{4} + \frac{1}{8}$
10. $\frac{1}{2} + \frac{1}{4} + \frac{1}{8} + \frac{1}{16} + \frac{1}{32}$
11. $235 + 49.2$
12. $321.42 + 69.73$
13. $1,624.08 + 12.236$
14. $1.728 + 0.0084 + 6.52$
15. $932.04 + 93.204 + 9.3204 + 0.93204$
16. $0.732 + 4.896 + 0.153 + 18.654 + 2.404$
17. $1.335 + \frac{5}{8} + 1\frac{1}{2}$
18. $2\frac{1}{3} + \frac{3}{4} + 0.625$
19. $3.482 + \frac{1}{6} + 0.025$
20. $4.603 + 2.136 + \frac{2}{3}$
21. $3.201 + \frac{1}{8} + \frac{1}{32}$
22. $5.206 + \frac{1}{64}$
23. $8.0123 + \frac{15}{16}$
24. $\frac{1}{2} + \frac{1}{4} + \frac{1}{8} + \frac{1}{16} + \frac{1}{32} + \frac{1}{64}$
25. $.563 + 1.72 + 3.6131 + \frac{1}{8}$
26. $1.362 + 1.11 + 3.315 + \frac{7}{16}$
27. $\frac{3}{16} + 1.365 + \frac{7}{8} + 2.812 + 3$
28. $6.21 + 532.8 + 11.1 + 87.631$
29. $3.214 + 3.1416 + \frac{7}{16} + 3.1182$
30. $6.315 + \frac{15}{16} + 5.312 + 4.117 + 8$
31. $37.2 + 463.118 + 17.163 + 81.223$

32. 264.	33. 1.508	34. 0.8435	35. 0.5957
0.065	4.391	2.6264	0.4056
30.832	0.484	5.1706	2.9540
716.45	0.745	0.2035	3.3470
+ 9.7	+0.656	+8.6302	+0.4941

36. $\$4.35 + \$6.75 + \$3.00 + \$5.75$
37. $\$6.76 + \$9.70 + \$8.65 + \$5.40$
38. $\$2.05 + \$18.99 + \$36.18 + \$45.17$
39. $\$629.03 + \$875.92 + \$492.85 + \$867.62$
40. $\$103.00 + \$4,621.01 + \$2,317.05 + \$6,417.07$

3-7 Subtraction of Decimals

Subtraction of decimal fractions follows the same rules as subtraction of whole numbers. The decimal points in the problem must be in the same column. The decimal point in the answer appears in the same column as it is in the problem.

Example 1

Take 18.275 from 42.63.

SOLUTION
$$
\begin{array}{ll}
42.630 & \text{(minuend)} \\
-18.275 & \text{(subtrahend)} \\
\hline
24.355 & \text{(difference)}
\end{array}
$$

EXPLANATION   Write the numbers so that the decimal points are under each other. If the minuend contains fewer figures after the decimal point than the subtrahend, annex zeros and proceed as in subtraction of whole numbers. Place the decimal point in the difference under the other decimal points.

Example 2
From $2\frac{1}{3}$ take 0.675.

SOLUTION $2\frac{1}{3} = 2.333$ (minuend)
$\underline{\quad -0.675\quad}$ (subtrahend)
1.658 (difference)

EXPLANATION   Changing $\frac{1}{3}$ to a decimal fraction, the number $2\frac{1}{3}$ becomes 2.333 (to three decimal places, to match the subtrahend). Perform the subtraction as in Example 1.

Problems 3-7

Do the following indicated subtractions. All answers are to be in decimal fraction form.

1. Take 234 from 625.75.
2. Take 148.35 from 436.62.
3. Take 82.875 from 125.
4. Take 139.25 from 218.08.
5. Take 0.4025 from 0.4929.
6. Take 0.6535 from 1.
7. Take 0.40875 from 2.856.
8. Take 0.545 from 5.9205.
9. Take 0.6346 from 0.8964.
10. Take 0.4058 from 0.4948.
11. From $4\frac{5}{8}$ take 2.326.
12. From 5.604 take $1\frac{2}{3}$.
13. From 3.862 take $2\frac{1}{6}$.
14. From $5\frac{3}{8}$ take 3.078.
15. From $6\frac{3}{8}$ take 4.286.

3-8   Multiplication of Decimals

Multiplication of decimal fractions follows the same rules as multiplication of whole numbers. The product (answer) will have as many decimal places as the total number of decimal places in both the multiplicand and multiplier. Study the example problem below. Notice that unlike in addition or subtraction, it is not necessary to align decimal points in multiplication. In fact, when multiplying decimal fractions, align the rightmost numerals, regardless of the position of the decimal points.

Example 1
Multiply 43.286 by 6.04.

SOLUTION   43.286   (multiplicand)
            × 6.04   (multiplier)
            ‾‾‾‾‾‾‾‾
            173144
            2597160
            ‾‾‾‾‾‾‾‾
            261.44744   (product)

EXPLANATION   Multiply as in whole numbers; then, beginning at the right, point off in the product as many decimal places as there are decimal places in both the multiplier and the multiplicand. Thus in the given example, the multiplicand has three places after the decimal point and the multiplier has two, a total of five places. Hence the product must have five decimal places.

The decimal part of this product denotes hundred-thousandths. But the machinist, the cabinet maker, and the electrician are not interested in such small fractions of a unit. Therefore, rewrite the result so that it will be of practical value to the person in the shop or factory. The machinist may want measurements given to the nearest thousandth of an inch; therefore, the answer to a machinist's problem will be stated with three decimal places. The answer would then read 261.447. The figures after the third decimal place are dropped because their value is less than one-half of a thousandth.

Likewise the answer to two decimal places would be 261.45; to one decimal place, it would read 261.4.

It is therefore important to decide in advance the degree of accuracy required in the answer to any given problem. After performing the indicated operations and obtaining the exact answer, rewrite the answer with as many decimal places as are required to bring it to the degree of accuracy previously determined for the given problem. The degree of accuracy, however, cannot be greater than the least accurate measure in the problem.

Example 2
Multiply 0.36 by 0.24.

SOLUTION   0.36
            ×0.24
            ‾‾‾‾‾
              144
               72
            ‾‾‾‾‾
            0.0864

EXPLANATION   Multiplying as in whole numbers, there are only three figures in the product, although, according to the rule given previously, the product must have four places after the decimal point. The fourth place is supplied by a zero written in front of the three figures, thus 0.0864.

Example 3

Multiply 0.85 by $1\frac{3}{4}$.

SOLUTION
$$
\begin{array}{r}
1.75 \\
\underline{0.85} \\
875 \\
\underline{1400} \\
1.4875
\end{array}
$$

EXPLANATION   Change the common fraction $\frac{3}{4}$ to a decimal fraction: $1\frac{3}{4} = 1.75$. Then multiply 0.85 by 1.75 as in Example 2.

RULE _____

To multiply a number by 10; 100; 1,000; 10,000; and so on, move the decimal point in the multiplicand as many places to the right as there are zeros in the multiplier; if necessary, annex zeros.

Thus $10 \times 3.25 = 32.5$; $100 \times 3.25 = 325$; $10,000 \times 3.25 = 32,500$; and so forth.

Example 4

Multiply 1.0356 by 1,000.

SOLUTION $1.0356 \times 1,000 = 1035.6$

EXPLANATION   Note that there are three zeros in 1,000. To multiply 1.0356 by 1,000, move the decimal point in 1.0356 three places to the right. The answer is 1,035.6

Problems 3-8

Find the following products. All answers are to be in decimal fraction form.

1. 0.25×10	2. 0.375×100	3. 0.5×100
4. $0.625 \times 1,000$	5. 0.3125×10	6. $0.4375 \times 10,000$
7. 1.875×10	8. $2.75 \times 1,000$	9. 15.25×100
10. $1.75 \times 1,000$	11. $6.375 \times 1,000$	12. 4.9375×100
13. $0.25 \times \frac{1}{4}$	14. $0.375 \times \frac{3}{4}$	15. 0.5×0.5
16. $0.5 \times \frac{1}{2}$	17. 0.5×1.5	18. $\frac{1}{2} \times \frac{1}{2}$
19. 0.8×0.9	20. 0.32×0.4	21. 0.57×0.61

22. 0.256 × 0.4   23. 0.25 × 0.375   24. 0.8125 × 4
25. 0.831 × 0.3   26. 4.26 × 3.7   27. 5.79 × 0.625
28. 0.0625 × 5   29. 0.031 × .062   30. 3.1416 × 5
31. 8 × 0.5346   32. 8.75 × 3.5   33. 8.94808 × 0.6
34. 86.4 × 0.0458   35. 1.125 × 3.375   36. 0.7854 × 3.25 × 3.25
37. 5,280 × 0.876   38. 16 × 0.1875   39. 0.125 × 0.36
40. 0.008 × 0.072   41. 5.84 × 0.0059   42. 2.25 × 0.0125
43. 1,728 × 0.621   44. 231 × 3.125   45. 0.2465 × 12
46. 2.457 × 2.568   47. 0.457 × 12.36   48. 0.008 × 24.86
49. 3.1416 × 2.5 × 2.5   50. 144 × 2.368
51. Restate the answers to Problems 21–25 with three decimal places.
52. Restate the answers to Problems 26–30 with two decimal places.
53. Restate the answers to Problems 31–35 with four decimal places.
54. Restate the answers to Problems 36–40 with two decimal places.
55. 0.062 × $2\frac{5}{8}$ 56. 2.246 × $4\frac{1}{3}$ 57. $5\frac{2}{3}$ × 0.425
58. $4\frac{2}{9}$ × 3.065   59. $3\frac{5}{6}$ × 2.083   60. 72 × 10
61. 18 × 100   62. 92 × 1,000   63. 6.86 × 10
64. 7.2 × 100   65. 1.37 × 1,000   66. 0.721 × 10
67. 0.063 × 100   68. 0.375 × 1,000

3-9 Division of Decimals

Division of decimal fractions follows the same rules as division of whole numbers except for the placement of the decimal point in the quotient (answer). The divisor is converted to a whole number and the dividend is changed correspondingly. Study the following example problems.

Example 1
Divide 76.125 by 24. State the answer to three decimal places.

SOLUTION (quotient) $3.171\frac{21}{24}$ or 3.172
        (divisor)   24)76.125 (dividend)
                        72
                        4 1
                        2 4
                        1 72
                        1 68
                          45
                          24
                          21

EXPLANATION Divide as in whole numbers and place the decimal point in the quotient above the point in the dividend. Since the remainder after the third place is $\frac{21}{24}$, or more than $\frac{1}{2}$, increase the third place by 1, giving the quotient 3.172.

Example 2
Divide 432 by 0.625.

SOLUTION

$$
\begin{array}{r}
691.2 \\
625.\overline{)432{,}000.0} \\
375\ 0 \\
\hline
57\ 00 \\
56\ 25 \\
\hline
750 \\
625 \\
\hline
1250 \\
1250 \\
\hline
\end{array}
$$

EXPLANATION   When the dividend is a whole number and the divisor is a decimal fraction, convert the divisor into a whole number by moving the decimal point to the right of it; in this instance, three places to the right. Then move the decimal point in the dividend also three places to the right. To do this, annex three zeros after 432, giving the number 432,000, with the decimal point after the third zero. Proceed to divide as in ordinary division. Place the decimal point in the quotient directly above the decimal point in the dividend, annex zeros after the decimal point in the dividend, and continue the division to as many decimal places as is required. In this instance there is no remainder after the first decimal place.

Example 3
Divide 0.78 by 0.964.

SOLUTION

$$
\begin{array}{r}
0.809 \\
964.\overline{)780.000} \\
771\ 2 \\
\hline
8\ 800 \\
8\ 676 \\
\hline
124 \\
\end{array}
$$

EXPLANATION   First convert the divisor into a whole number by moving the decimal point to the right three places. Then move the decimal point in the dividend an equal number of places to the right. If, as in this instance, the dividend has fewer figures than the divisor, annex zeros. Then proceed to divide as explained in the preceding examples.

RULE
   To divide a number by 10; 100; 1,000; 10,000; and so on, move the decimal point in the dividend as many places to the left as there are zeros in the divisor; if necessary, prefix zeros.

   Thus $56.4 \div 10 = 5.64$; $56.4 \div 100 = 0.564$; and $56.4 \div 1{,}000 = 0.0564$; and so forth.

Example 4
Divide 924.3 by 100.

SOLUTION   924.3 ÷ 100 = 9.243

EXPLANATION   Note that there are two zeros in 100. To divide 924.3 by 100, move the decimal point in 924.3 two places to the left. The answer is 9.243.

Problems 3-9

Give the answers to the following division problems to three places after the decimal point.

1. 25 ÷ 10	2. 3.75 ÷ 10	3. 4.375 ÷ 100
4. 7.125 ÷ 10	5. 625 ÷ 100	6. 3125 ÷ 1,000
7. 93.75 ÷ 100	8. 312.5 ÷ 10	9. 25.625 ÷ 100
10. 35 ÷ 1,000	11. 6.25 ÷ 10	12. 0.0625 ÷ 1,000
13. 1 ÷ 16	14. 1 ÷ 32	15. 10 ÷ 16
16. 10 ÷ 32	17. 100 ÷ 16	18. 100 ÷ 32
19. 1,000 ÷ 32	20. 1 ÷ 3.14	21. 3.216 ÷ 0.9
22. 4.36 ÷ .07	23. 37 ÷ 0.063	24. 15.07 ÷ $\frac{2}{3}$
25. 6.21 ÷ 1$\frac{1}{4}$	26. 3.21 ÷ 1.10	27. 5.623 ÷ 1.21
28. 1.23 ÷ 0.625	29. 124.625 ÷ 48	30. 0.2086 ÷ 32
31. 81.63 ÷ 0.09	32. 0.875 ÷ 0.634	33. 0.454 ÷ 6.05
34. 72 ÷ 0.625	35. 80 ÷ 4.05	36. 2.36 ÷ 5.48
37. 6.455 ÷ 0.008	38. 0.565 ÷ 8.673	39. 0.11 ÷ 1,848.43
40. 42 ÷ 0.0075	41. 0.762 ÷ 888	42. 1 ÷ 1,728
43. 7 ÷ 144	44. 1 ÷ 3.1416	45. 1 ÷ 0.7854
46. 450 ÷ 1,728	47. 233 ÷ 5,280	48. 0.0406 ÷ 40.08
49. $\frac{4}{7}$ ÷ 3.45	50. $\frac{5}{9}$ ÷ 2.083	51. $\frac{5}{12}$ ÷ 4.96
52. 6.38 ÷ $\frac{2}{3}$	53. 5.09 ÷ $\frac{3}{7}$	54. 766 ÷ 10
55. 13 ÷ 100	56. 7,216 ÷ 1,000	57. 1.28 ÷ 10
58. 42.6 ÷ 100	59. 11.9 ÷ 1,000	60. 0.153 ÷ 100

 ## 3-10   Using the Calculator to Solve Decimal Problems

The calculator is a great help in solving decimal problems quickly and accurately. In order not to become too dependent upon the calculator, the operator should not only understand the problem to be solved, but should also analyze the result to judge whether the answer is reasonable for the problem at hand. A good habit is to check the answer after the problem is solved. The operator must make sure that the keys are pressed in the correct sequence. In using calculators and computers, the operator must remember *GIGO— Garbage In = Garbage Out.*

The problems demonstrated here for use with the calculator were selected from the problems in the previous sections of this chapter. Compare how you solved the problems with pencil and paper with how you solve the problems using the calculator.

Example 1

Change $\frac{7}{8}$ to a decimal fraction.

SOLUTION

	The display shows
Turn on the calculator	$0.$
Enter 7	$7.$
Press ÷	$7.$
Enter 8	$8.$
Press =	0.875 (answer)

This is the decimal equivalent to $\frac{7}{8}$.

Example 2

Change $\frac{2}{3}$ to a decimal fraction accurate to the hundredths place.

SOLUTION

	The display shows
Turn on the calculator	$0.$
Enter 2	$2.$
Press ÷	$2.$
Enter 3	$3.$
Press =	0.666666666

Use the round-off rule and write 0.67 as your answer.

Example 3

Change $\frac{3}{16}$ to a decimal of three places.

SOLUTION

The display shows

Turn on the calculator	0.
Enter 3	3.
Press ÷	3.
Enter 16	16.
Press =	0.1875

Use the round-off rule and write 0.188 as your answer.

Example 4

On a scale divided into sixty-fourths of an inch, what is the dimension nearest $\frac{13}{37}$ in.?

SOLUTION

The display shows

Turn on the calculator	0.
Enter 13	13.
Press ÷	13.
Enter 37	37.
Press ×	0.351351351
Enter 64	64.
Press =	22.48648649

Round off to the nearest whole number (22), and use this whole number as the numerator and the 64 as the denominator. This fraction reduces to $\frac{11}{32}$. Therefore $\frac{13}{37}$ in. is approximately equal to $\frac{11}{32}$ in.

Example 5

Change $\frac{2}{7}$ to sixteenths.

SOLUTION

The display shows

Turn on the calculator	0.
Enter 2	2.
Press ÷	2.
Enter 7	7.
Press ×	0.2871142BS
Enter 16	16.
Press =	4.571142B571

Round off to the nearest whole number (5), and use this whole number as the numerator and the 16 as the denominator. This fraction is $\frac{5}{16}$. Therefore $\frac{2}{7}$ is approximately equal to $\frac{5}{16}$.

Example 6

Convert 3¾ in. into a decimal fraction of a foot. State the answer to three decimal places.

SOLUTION

The display shows

Turn on the calculator	0.
Enter numerator 3	3.
Press ÷	3.
Enter 4	4.

The fraction part of the mixed number is usually done first on calculators.

| Press + | 0.75 |
| Enter 3 | 3. |

The whole number part (3) is now added.

Press =	3.75
Press ÷	3.75
Enter 12	12.
Press =	0.3125

To convert inches to feet, divide by 12.

Using the round-off rule makes the answer 0.313 feet.

Example 7

Convert 46⅛ in. into feet and a decimal fraction of a foot to three decimal places.

SOLUTION

The display shows

Turn on the calculator	0.
Enter 1	1.
Press ÷	1.
Enter 8	8.
Press +	0.125
Enter 46	46.
Press =	46.125
Press ÷	46.125
Enter 12	12.
Press =	3.84375 (answer)

Round this display answer off to three decimal places for an answer of 3.844 ft.

Example 8

Convert 43.172 in. into feet and inches and the nearest sixty-fourth of an inch.

SOLUTION

The display shows

Turn on the calculator	0.
Enter 43.172	43.172
Press ÷	43.172
Enter 12	12.
Press =	3.597666667 (feet part)

The whole number part shows us that we have 3 feet in 43.172 inches. Now write down the 3 feet and subtract 3 from our display to determine how many inches we have in the remainder of 0.597666667 feet.

Press −	3.597666667
Enter 3	3.
Press =	0.597666667
Press ×	0.597666667
Enter 12	12.
Press =	7.172 (inches part)

The whole number part shows us that we have 7 inches in 0.597666667 feet. Write down 7 inches and subtract 7 from our display to determine how many sixty-fourths are in the remainder 0.172 inches.

Press −	7.172
Enter 7	7.
Press =	0.172
Press ×	0.172
Enter 64	64.
Press =	11.008 (64ths part)

The whole number part shows us that we have approximately 11 sixty-fourths in 0.172 inches. Now we combine all the subtotals that we have and write 43.172 in. = 3 ft $7\frac{11}{64}$ in.

Example 9

Convert 0.319 ft into inches and a fraction of an inch (to the nearest sixty-fourth inch).

SOLUTION

The display shows

Turn on the calculator	0.
Enter 0.319	0.319
Press ×	0.319
Enter 12	12.
Press =	3.828 (inches part)

This number tells us that we have 3 inches and a fraction of an inch remaining. Write down the 3 and subtract 3 from our display to work on the fractional part.

Press −	3.828
Enter 3	3.
Press =	0.828
Press ×	0.828
Enter 64	64.
Press =	52.992 (64ths part)

This number rounds off to 53, the approximate number of sixty-fourths that are remaining. The answer is $3\frac{53}{64}$ in.

Example 10
Add 3.25, 72.004, 864.0725, 647, and 0.875.

SOLUTION

The display shows

Turn on the calculator	0.
Enter 3.25	3.25
Press +	3.25
Enter 72.004	72.004
Press +	75.254
Enter 864.0725	864.0725
Press +	939.3265
Enter 647	647.
Press +	1586.3265
Enter 0.875	0.875
Press =	1587.2015 (answer)

Example 11
Find the sum of $\frac{1}{2}$ + 0.662 + $\frac{7}{8}$.

SOLUTION

The display shows

Turn on the calculator	0.
Enter 1	1.
Press ÷	1.
Enter 2	2.
Press =	0.5
Press +	0.5
Enter .662	0.662
Press +	1.162
Enter 7	7.
Press ÷	7.
Enter 8	8.
Press =	2.037 (answer)

Example 12
Take 18.275 from 42.63.

SOLUTION

	The display shows
Turn on the calculator	0.
Enter 42.63	42.63
Press −	42.63
Enter 18.275	18.275
Press =	24.355 (answer)

Example 13
From $2\frac{1}{3}$ take 0.675. Round off the answer to three decimal places.

SOLUTION

	The display shows
Turn on the calculator	0.
Enter 2	2.
Press +	2.
Enter 1	1.
Press ÷	1.
Enter 3	3.
Press −	2.333333333
Enter .675	0.675
Press =	1.658333333 (answer)

Round off your answer to 1.658.

Example 14
Multiply 43.286 by 6.04.

SOLUTION

	The display shows
Turn on the calculator	0.
Enter 43.286	43.286
Press ×	43.286
Enter 6.04	6.04
Press =	261.44744 (answer)

Example 15
Multiply 0.85 by $1\frac{3}{4}$. (It is usually easier to enter the mixed number first. Special care is needed when entering common fractions into a calculator.)

SOLUTION

The display shows

Turn on the calculator	0.
Enter 1	1.
Press +	1.
Enter 3	3.
Press ÷	3.
Enter 4	4.
Press =	1.75
Press ×	1.75
Enter .85	0.85
Press =	1.4875 (answer)

Example 16
Divide 78.125 by 24. State the answer to three decimal places.

SOLUTION

The display shows

Turn on the calculator	0.
Enter 76.125	76.125
Press ÷	76.125
Enter 24	24.
Press =	3.171875 (answer)

This answer is 3.172 when rounded to three decimal places.

Problems 3-10

Fill out the following equivalent chart using the calculator. Write the decimal equivalent of the fraction in inches to five decimal places and the millimeter equivalent to four places. One inch is exactly equal to 25.4 millimeters.

Common Fraction (inches)	Decimal Equiv-alent	mm Equiv-alent	Common Fraction (inches)	Decimal Equiv-alent	mm Equiv-alent
1. $\frac{1}{32}$	___	___	11. $\frac{11}{32}$	___	___
2. $\frac{1}{16}$	___	___	12. $\frac{3}{8}$	___	___
3. $\frac{3}{32}$	___	___	13. $\frac{13}{32}$	___	___
4. $\frac{1}{8}$	___	___	14. $\frac{7}{16}$	___	___
5. $\frac{5}{32}$	___	___	15. $\frac{15}{32}$	___	___
6. $\frac{3}{16}$	___	___	16. $\frac{1}{2}$	___	___
7. $\frac{7}{32}$	___	___	17. $\frac{17}{32}$	___	___
8. $\frac{1}{4}$	___	___	18. $\frac{9}{16}$	___	___
9. $\frac{9}{32}$	___	___	19. $\frac{19}{32}$	___	___
10. $\frac{5}{16}$	___	___	20. $\frac{5}{8}$	___	___

Common Fraction (inches)	Decimal Equiv- alent	mm Equiv- alent		Common Fraction (inches)	Decimal Equiv- alent	mm Equiv- alent
21. $\frac{21}{32}$	————	————	27.	$\frac{27}{32}$	————	————
22. $\frac{11}{16}$	————	————	28.	$\frac{7}{8}$	————	————
23. $\frac{23}{32}$	————	————	29.	$\frac{29}{32}$	————	————
24. $\frac{3}{4}$	————	————	30.	$\frac{15}{16}$	————	————
25. $\frac{25}{32}$	————	————	31.	$\frac{31}{32}$	————	————
26. $\frac{13}{16}$	————	————	32.	1	1.00000	25.4000

Use the calculator to change each of the following decimal fractions to the nearest 16th, 32nd, and 64th. Round each answer to the nearest whole number.

	Decimal Fraction	Nearest 16th	Nearest 32nd	Nearest 64th
33.	0.9684	————	————	————
34.	0.7211	————	————	————
35.	0.5925	————	————	————
36.	0.2459	————	————	————
37.	0.1183	————	————	————
38.	0.6720	————	————	————
39.	0.0477	————	————	————
40.	0.0656	————	————	————
41.	0.4239	————	————	————
42.	0.6874	————	————	————

Use the calculator to change the following millimeters to decimal parts of an inch and to the nearest 32nd of an inch.

	Millimeters	Inches	Nearest 32nd
43.	1	0.03937	$\frac{1}{32}$
44.	2	————	————
45.	3	————	————
46.	4	————	————
47.	5	————	————
48.	6	————	————
49.	7	————	————
50.	8	————	————
51.	9	————	————
52.	10	————	————
53.	11	————	————
54.	12	————	————
55.	13	————	————
56.	14	————	————
57.	15	————	————

	Millimeters	Inches	Nearest 32nd
58.	16	_____	_____
59.	17	_____	_____
60.	18	_____	_____
61.	19	_____	_____
62.	20	_____	_____

Solve the following problems with a calculator. Round off the answers to three places.

63. What is the product of 46.887 and 63.921?
64. What is the sum of 65.34, 56.38, 788.45, and 28.447?
65. Subtract 345.967 from 9876.455.
66. Divide 78.431 by 7.43.
67. Divide 0.056 by 0.543333.
68. Multiply 458.998 by 397.596.
69. Add 56.45, 365.65, 387.54, 2987.87, 9832.77, and 45.75.
70. Subtract 85.357 from 3578.911.
71. Multiply 5678.9 by 387.56.
72. Divide 65.929 by 196.45.
73. Gold weighs 1,206 lb per cubic foot. Find the weight of a cubic inch of gold if there are 1,728 cubic inches in a cubic foot.
74. How many cubic feet are there in a ton of aluminum if one cubic foot of aluminum weights 160 lb? (There are 2,000 lb in a ton.)
75. There are 0.4536 kilograms in one pound. How many kilograms are there in ten tons?
76. There are 231 cubic inches in one gallon. How many cubic inches are there in 20 gallons?
77. How many seconds are there in a 24-hour day?
78. How many seconds have you lived?
79. Light travels 186,000 miles per second. How many seconds does it take for light to reach the earth from our sun if the Earth is 93,000,000 miles from the sun?
80. How many dollars would you have if you took a penny the first day, two cents the second day, four cents the third day, and kept doubling in this manner for thirty days?
81. The volume of a sphere (ball) is found by multiplying, using this formula: $\frac{4}{3} \times 3.1416 \times$ radius $\times$ radius $\times$ radius. What is the volume of the Earth in cubic miles if the radius of the Earth is 4,000 miles?
82. A car travels 394 miles on one tankful of gasoline. If the gasoline tank holds 21 gallons, how many miles per gallon does this car get?

Self-Test

Do the following problems. All answers are to be in decimal form. Check your answers with the answers in the back of the book.

1. $0.377 + \frac{1}{4}$

2. $1.427 + 0.675$

3. $5.329 - \frac{5}{8}$

4. $7.62 - 1.81$

5. $3.2 \times \frac{3}{4}$

6. 4.7×4.3

7. $5.3 \div \frac{1}{2}$

8. $4.32 \div 2.35$

9. $\dfrac{3.2 + 1.7}{4.6}$

10. $\dfrac{7.35 \times 6.25}{\frac{5}{8}}$

Chapter Test

Do the following problems. All answers are to be in decimal form. Round off all answers to three decimal places.

Add:

1. $0.513 + \frac{1}{8}$

2. $\frac{3}{4} + 1.007$

3. $1.06 + 0.391$

4. $6\frac{1}{2} + 0.762$

5. $0.83 + 4.205$

6. $1.07 + 3.6121$

7. $0.006 + 7$

8. $3.217 + 15.3928$

9. $8 + 0.312$

10. $4.37 + 5$

Subtract:

11. $1\frac{5}{16} - 0.3125$

12. $0.362 - \frac{1}{8}$

13. $1.71 - 0.007$

14. $8.62 - 1.317$

15. $15.73 - 9.635$

16. $8.6 - 0.082$

17. $97 - 9.708$

18. $0.438 - 0.099$

19. $14.293 - 8$

20. $63 - 12.62$

Multiply:

21. $6.2 \times \frac{1}{16}$

22. $1\frac{1}{2} \times 0.079$

23. 0.03×0.07

24. 8.6×0.43

25. $1\frac{2}{3} \times 6.21$

26. 1.27×6.33

27. 4.81×0.0162

28. $0.625 \times \frac{1}{16}$

29. 4×6.357

30. $\$46.20 \times 6$

Divide:

31. $1 \div 16$

32. $3.2 \div \frac{1}{8}$

33. $0.01 \div 0.7$

34. $8.6 \div 3.21$

35. $1.2 \div 0.63$

36. $10 \div 1.63$

37. $100 \div 0.51$

38. $0.6 \div 0.006$

39. $\$5.00 \div 4$

40. $\$655.60 \div 8$

Miscellaneous Problems in Decimal Fractions

NOTE. Where the number of decimal places desired in the answer is not stated, decide what degree of accuracy is required in the answer to any given problem.

1. Cast iron weighs 450 lb per cubic foot; wrought iron, 480 lb per cubic foot; and steel, 490 lb per cubic foot. Find the weight of each per cubic inch (1 cu ft = 1728 cu in.). State the answers as decimal fractions of three places.

2. What will be the weight of a wrought iron rod that contains 62.25 cu in.?

3. What part of a cubic foot is contained in a steel shaft weighing 215.75 lb? State the answer as a decimal fraction to three places.
4. What will be the weight of 312 iron castings if each contains 86.5 cu in.?
5. The actual inside diameter of a one-inch pipe is 1.04"; the actual outside diameter is 1.315". Find the thickness of the pipe (Fig. 3-1).

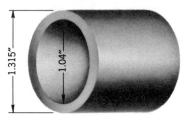

Fig. 3-1

6. A ⅞" bolt (without nut) weighs 0.664 lb. How many such bolts are there in a keg of 250 lb?
7. 3 × 3 × ⅜ angle iron weighs 7.2 lb per foot. Find the weight of a piece 12'3" long.
8. Find the difference in diameters of the piece of tapered work in Fig. 3-2.

Fig. 3-2

9. The circumference of a circle is found by multiplying its diameter by 3.1416. Find the circumference of a circle whose diameter is 2.35". State the answer to two decimal places.
10. To find the diameter of a circle when the circumference is known, divide the circumference by 3.1416. Find the diameter of a circle having a circumference of 8.7865". State the answer to one decimal place.
11. Find the whole depth of the gear tooth shown in Fig. 3-3. State the answer to the nearest ten-thousandths of an inch.

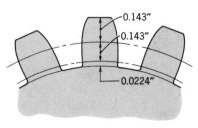

Fig. 3-3

12. How many pounds of water are there in a tank that contains 904.7 cu ft? (A cubic foot of water weighs 62.5 lb.)
13. How many gallons of water will the tank in Problem 12 contain if there are 231 cu in. in a gallon?
14. Find: (*a*) the weight of a cubic inch of water; and (*b*) the weight of a gallon of water.
15. With gasoline at $1.159 per gallon, what will it cost to fill a tank whose capacity is 1.75 cu ft? (One cu ft is approximately 7½ gallons.)
16. Find the diameter at *A* of the tapered shank in Fig. 3-4 if the difference between the small diameter and the diameter at *A* is 0.392″.

Fig. 3-4

17. A machinist's helper gets $7.75 per hour. How much will he earn in a week of 40 hours?
18. A welder gets $12.75 per hour for a 40-hour week. During the week she works 6 hours overtime at time-and-a-half. How much does she earn?
19. A sheet metal worker works a 40-hour week at $15 per hour. How much does he earn?
20. If the basic week is 40 hours at $7.35 per hour and a woman works 49 hours, how much does she earn during the week, allowing time-and-a-half for overtime?
21. The rim or cutting speed on cylindrical work is found by multiplying the circumference of the work by the number of revolutions per minute. Find the cutting speed on a piece of work whose circumference is 4.273″ if the work is making 120 revolutions per minute. Express the answer in feet per minute.
22. At 32 cents per pound, find the cost of 125 castings each weighing 8.68 lb.
23. Compute the depth of tooth in Fig. 3.5.

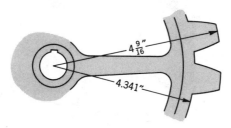

Fig. 3-5

24. Compute the root diameter of the thread in Fig. 3-6.

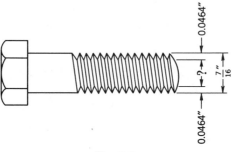

Fig. 3-6

25. Find the corner measurement for laying out an octagonal end on the square bar in Fig. 3-7.

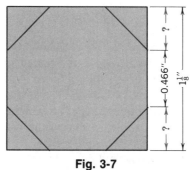

Fig. 3-7

26. Find the depth of cut required to mill a square end on the round shaft in Fig. 3-8.

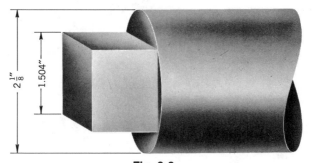

Fig. 3-8

27. What is the thickness of metal in the cross section of pipe in Fig. 3-9?

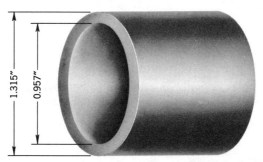

Fig. 3-9

28. (a) Find the diameter at the large end of the tapered portion in Fig. 3-10. (b) Find the difference between the diameters at the large and small ends of the tapered part. (c) Compute the missing dimension.

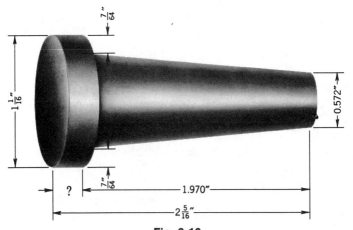

Fig. 3-10

29. A shaft that should be $1\frac{3}{16}''$ in diameter measures 1.317". How much is it oversize?

30. A round piece of work is $1\frac{5}{8}''$ in diameter. How deep a cut should be taken to bring the diameter down to 1.594"?

31. Figure 3-11 shows an American Standard thread of $\frac{1}{9}''$ pitch. Find the flat at top and bottom if the flat is $\frac{1}{8}$ of the pitch. State the answer to the nearest ten-thousandth of an inch.

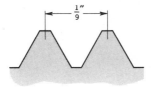

Fig. 3-11

32. Find the value of the following expressions. State the answers as deci-
    mals of three places.

    (a) $2.689 + 1\frac{7}{12}$ (b) $1\frac{2}{7} - 0.875$ (c) $3.256 \times 4\frac{1}{8}$

    (d) $4.586 \div \frac{9}{16}$ (e) $\dfrac{\frac{1}{3} \text{ of } \frac{7}{8}}{\frac{1}{2} \text{ of } \frac{3}{4}}$ (f) $\dfrac{\frac{3}{4} \div 2\frac{1}{2}}{\frac{5}{8} \times 4\frac{1}{4}}$

33. Fourteen holes, equally spaced, are to be drilled as shown in Fig. 3-12.
    Find the distance center to center of adjacent holes. Give the answer to
    the nearest sixty-fourth of an inch.

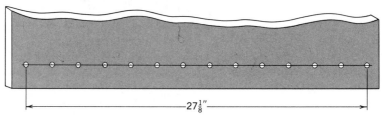

Fig. 3-12

34. A No. 10 wire has a diameter of 0.1019 in. and a No. 12 wire has a
    diameter of 0.0808 in. What is the difference between their diameters?
    (American wire gage sizes)
35. One 100-watt bulb draws 0.87 amperes of current. How many amperes
    will 17 such bulbs draw?
36. If roofing shingles cost $5.95 per bundle and three bundles make a
    square (100 sq ft), how much would 20 such squares cost?
37. The yearly rainfall in Cherrapunji, India is 449.8 inches. What is its
    average rainfall per month?
38. Suppose the population of the United States is 241,078,000 and the
    annual increase in population is estimated at 1.2 times the present popu-
    lation. What will the population be next year?
39. Mt. McKinley in Alaska is 20,320 feet high and is the highest mountain
    in the United States. How many miles high is Mt. McKinley? (5,280
    feet = 1 mile). (Source: U.S. Geological Survey.)
40. The distance from Albuquerque, New Mexico, to Philadelphia, Pennsyl-
    vania, is 2,124 miles. If you want to make a trip in three days, how many
    miles per day must you travel?

4

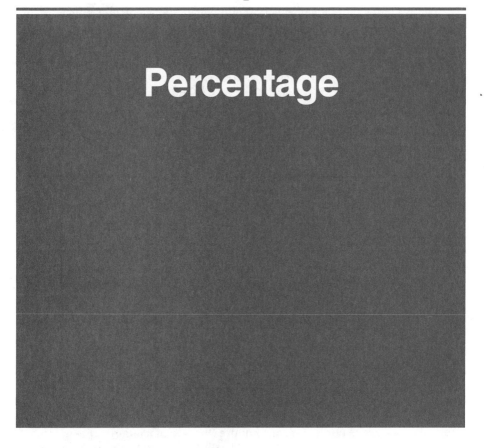

Percentage

4-1 Definitions

One **percent** of a number means one-*hundredth* of a number, six percent means six-*hundredths,* and so on.

The symbol % stands for the word percent. Thus 4 percent is written 4%, 6 percent is written 6%. The symbol % does the work of two decimal places. For example, 6% = 0.06; 25% = 0.25; 110% = 1.10; 0.5% = 0.005; $2\frac{1}{4}\%$ = 0.0225 or $0.02\frac{1}{4}$; and 5.7% = 0.057.

Since percent means hundredths, the whole of any number contains 100% of itself. Thus 100% of 75 is 75.

Every problem in percentage contains three elements: the **base,** the **rate,** and the **percentage.** When we say 6% of $100 is $6, the base is $100, the rate is 6% or 0.06, and the percentage is $6. If any two of these three elements are known, the third may be found.

The following form will help you solve percentage problems.

$$\text{Percentage}\searrow \qquad\qquad \swarrow\text{Rate}$$
$$\tfrac{5}{8} = 0.625 = 62.5\%$$
$$\text{Base}\nearrow \qquad\qquad \nwarrow\text{Decimal equivalent}$$

The student should memorize the following table, which gives the percentage equivalents of the common fractions most frequently used in practice.

PERCENTAGE EQUIVALENTS OF COMMON FRACTIONS			
$\frac{1}{8} = 12\frac{1}{2}\%$	$\frac{5}{8} = 62\frac{1}{2}\%$	$\frac{1}{6} = 16\frac{2}{3}\%$	$\frac{1}{5} = 20\%$
$\frac{1}{4} = 25\%$	$\frac{3}{4} = 75\%$	$\frac{1}{3} = 33\frac{1}{3}\%$	$\frac{2}{5} = 40\%$
$\frac{3}{8} = 37\frac{1}{2}\%$	$\frac{7}{8} = 87\frac{1}{2}\%$	$\frac{2}{3} = 66\frac{2}{3}\%$	$\frac{3}{5} = 60\%$
$\frac{1}{2} = 50\%$		$\frac{5}{6} = 83\frac{1}{3}\%$	$\frac{4}{5} = 80\%$

4-2   Finding the Percentage, Given the Base and the Rate

To find the percentage when the base and the rate are known, multiply the rate by the base. Remember to change the rate to its decimal equivalent, that is, $6\% = 0.06$, and so on.

$$\text{Decimal equivalent of rate} \times \text{Base} = \text{Percentage}$$

Example
Find 4% of 683.

SOLUTION $4\% = 0.04$ $0.04 \times 683 = 27.32$.

EXPLANATION   Since percent means hundredths, 4% of a number means 0.04 of a number. To obtain 4% of 683, multiply 683 by 0.04 and get the percentage 27.32.

RULE _____
To find the percentage, multiply the base by the rate.

Problems 4-2

Find the following percentages.

1. 10% of 50	2. 10% of 500	3. 10% of 5000
4. 1% of 500	5. 1% of 5000	6. 20% of 50
7. 20% of 500	8. 20% of 5000	9. 50% of 15
10. 50% of 500	11. 50% of 5000	12. 12% of 30
13. 15% of 75	14. 18% of 28	15. 5% of 87
16. 12½% of 88	17. 17% of 300	18. 83⅓% of 660
19. ½% of 100	20. ½% of 50	

21. White metal is made up of 3.7% copper, 88.8% tin, and 7.5% antimony. How many pounds of each metal are there in 465 lb of the alloy?

22. A machinist earning $405 per week got a wage increase of 15%. What did the machinist earn after the increase?

23. If an individual is allowed a discount of 2% from a bill amounting to $242.85, what must the person pay?

24. A dealer in electrical supplies allows discounts of 15% and 5% from the list price, and a 2% discount for cash. How much cash must be paid for a bill of goods whose list price amounts to $162.50? (Deduct 15% from the original amount, then 5% from the remainder, and 2% from the second remainder.)

25. Muntz metal is 59.5% copper, 39.9% zinc, and 0.6% lead. How many pounds of each are there in 432 lb of the alloy?

26. U.S. Navy specifications for phosphor bronze call for 85% copper, 7% tin, 0.06% iron, 0.2% lead, 0.3% phosphorus, and the remainder of zinc. How many pounds of each element are required to make 500 lb of phosphor bronze?

27. On a $15,000 contract, 72% was paid for labor and materials, 6% for supervision, and the remainder for profit. How much was paid for each item?

28. The manufacturer of a certain automobile estimates the following costs: materials, $38\frac{1}{2}$%; labor, $41\frac{1}{4}$%; overhead, $6\frac{1}{2}$%; and profit, $13\frac{3}{4}$%. Find the cost of each item in an automobile that sells for $7,589.

29. A standard formula for type metal calls for $77\frac{1}{2}$% lead, $6\frac{1}{2}$% tin, and 16% antimony. How many pounds of each metal are there in 250 lb of the type metal?

30. A certain iron ore yields $4\frac{1}{4}$% iron. How many pounds of iron are there in a ton of this ore?

31. A man's weekly pay is $277.30. At the rate of 6.13%, how much does he pay in social security taxes per week?

32. A woman works a 40-hour week at the rate of $9.75 per hour. How much should she find in her pay envelope at the end of the week if 18% is deducted for income tax and 6.13% for social security tax?

33. If sales tax is 6%, how much was the tax on a used car that sold for $1,495?

34. A customer buys an electric drill for $8.98, a disc sander for $36.75, and an electric chain saw for $39.99. How much sales tax must be paid if the sales tax is 6%?

35. Three suits are purchased at $89.99 each. The sales tax is 6% and is added to the bill. How much is the total bill?

36. Shirts are advertised at $14.75 each, less 25%. The sales tax is 6%. What will the bill be for five shirts?

37. What is $5\frac{1}{4}$% of $1,500?

38. What is $5\frac{1}{2}$% of $100?

39. If a certain bill is paid within 20 days of the billing date, a discount of 2% is allowed. How much must be paid if the original bill was for $596.50 and it is paid within the 20-day limit?

40. The interest charge per month on a charge card is $1\frac{1}{2}$% of the balance due. What is the month's interest charge on a $2,067.58 balance?

4-3 Finding the Rate, Given the Base and the Percentage

Suppose a company makes 5,000 articles and 150 are found to be defective. This relationship between articles made and articles rejected is often expressed as a rate, that is, 3% of the articles made were rejected. The problem can be stated as: What percent of 5,000 is 150?

SOLUTION 150 parts out of 5,000 is $\frac{150}{5000} = 0.03$, and $0.03 \times 100 = 3\%$.

To find the rate use the following rule:

RULE _____

To find the rate write a fraction with the percentage as the numerator and the base as the denominator and change this fraction to a decimal. Multiply the quotient by 100 to get the final percent.

$$\frac{\text{Percentage}}{\text{Base}} = \text{Decimal equivalent of rate}$$

Example 1
What percent of 8 is 7?

SOLUTION 7 is $\frac{7}{8}$ of 8, and $\frac{7}{8} = 0.875 = 87.5\%$

Example 2
What percent of 64.8 is 16.5?

SOLUTION $\dfrac{16.5}{64.8} = 0.2546 = 25.46\%$

Problems 4-3

Find the following rates.

1. What percent of 10 is 5?
2. 12 is what percent of 100?
3. What percent of 80 is 20?
4. What percent of 30 is $7\frac{1}{2}$?
5. $12\frac{1}{2}$ is what percent of 50?
6. What percent of 75 is 25?
7. What percent of 20 is 4?
8. 3 is what percent of 15?
9. What percent of 36 is 12?
10. 25 is what percent of 125?
11. What percent of 58 is 29?
12. What percent of 208 is 52?
13. 10 is what percent of 61?

14. What percent of 51 is 17?
15. $3\frac{1}{4}$ is what percent of $5\frac{1}{2}$?
16. What percent of $2\frac{1}{2}$ is $\frac{1}{2}$?
17. What percent of 1.32 is 0.6?
18. 0.25 is what percent of 8.75?
19. What percent of 0.625 is 0.25?
20. What percent of 0.025 is 0.005?
21. A motor receiving 8 hp delivers 6.8 hp of work. What percent of the input is the output?
22. A ton of ore yields 80 lb of iron. What percent of the ore is iron?
23. A man pays $5.75 for an article and sells it for $6.50. What percent profit does he make?
24. The usual allowance made for shrinkage when casting iron pipe is $\frac{1}{8}$" per foot. What percent is this?
25. In making 95 lb of solder, 38 lb of lead and 57 lb of tin were used. What percent of each was used?
26. The indicated horsepower of a steam engine is 9.4 and the effective horsepower is 8.1. What percent of the indicated horsepower is the effective horsepower?
27. What percent of a foot is $\frac{5}{8}$ of an inch?
28. What percent of a mile is 100 feet?
29. Out of a total production of 2,715 ball bearings manufactured during a day, 107 were rejected by the inspectors as imperfect. What percent of the total was rejected?
30. An alloy of common yellow brass is made of the following ingredients: copper, 170.5 lb; lead, 7.7 lb; tin, 0.55 lb; and zinc, 96.25 lb. What percent of the entire alloy does each of the metals represent?
31. A machine shop job required 42 hours on the lathe, $7\frac{1}{2}$ hours on the milling machine, and $11\frac{1}{4}$ hours on the planar. What percent of the total time should be charged to each machine?
32. A company made 1,000 automobile doors and 20 were found to be defective. What percent of the output was found to be defective?
33. A company projects that if no more than 1% of its output is defective, it will make a profit. If the company averages 98 rejected articles out of every 10,000 articles made, does it meet the projection for making a profit?

Fill out the following Weekly Production Chart.

	Day	Articles Made	Articles Rejected	Rate of Rejections
34.	Monday	4,020	43	_____
35.	Tuesday	4,070	37	_____
36.	Wednesday	4,100	40	_____
37.	Thursday	4,075	29	_____
38.	Friday	4,125	32	_____
39.	Saturday	3,760	30	_____
40.	Totals	_____	_____	_____

4-4 Finding the Base, Given the Percentage and the Rate

If a certain ore contained 5% iron and a company required a production schedule of 100 tons of iron per week, then it would have to process a certain amount of ore to produce the necessary 100 tons of iron. This is an example of knowing the percentage and the rate and using them to find the base.

RULE

To find the base, divide the percentage by the rate expressed as hundredths.

$$\frac{\text{Percentage}}{\text{Decimal equivalent of rate}} = \text{Base}$$

Example 1
1,022 is 28% of what number?

SOLUTION

$$
\begin{array}{r}
36\ 50 \\
28.\overline{)1022\,00.} \\
84 \\
\overline{182} \\
168 \\
\overline{140} \\
140
\end{array}
$$

EXPLANATION   Dividing by 28 gives 1% of the number. To get 100%, that is, the whole number, multiply by 100. The result is obtained in one step by dividing 1,022 by 0.28.

Example 2
300 is 20% more than what number?

SOLUTION

$$
\begin{array}{r}
2\ 50 \\
1\,20.\overline{)300\,00.} \\
240 \\
\overline{600} \\
600
\end{array}
$$

EXPLANATION   The problem states that 300 is more than the original number by 20%. Therefore, part of the 300 is 100% of the original number and part of the 300 is 20% of the original number. The 300 then represents 120% of the original number. Since 300 is 120% of the number, divide by 120 to get 1% of the number. To get 100%, or the whole number, multiply by 100. The result is obtained in one step by dividing 300 by 1.20.

Example 3

210 is 30% less than what number?

SOLUTION

$$\begin{array}{r} 3\ 00 \\ {}_x70.\overline{)210_x00.} \\ \underline{210} \end{array}$$

EXPLANATION   The problem states that 210 is really 70% of the original number, since 100% would equal the original number, and 30% less than 100% is 70%. Since 210 is 70% of the number, divide by 70 to get 1% of the number. To get 100%, or the whole number, multiply by 100. The result is obtained in one step by dividing 210 by 0.70.

Problems 4-4

In each of the following problems find the base. Remember the rule:

RULE

To find the base, divide the percentage by the rate expressed as hundredths.

1. 50% of what number is 8?
2. 70 is 10% of what number?
3. 20 is 100% of what number?
4. 116 is 16⅔% of what number?
5. 83⅓% of what number is 125?
6. 20% of what number is 67?
7. 28½ is 40% of what number?
8. 49 is 70% of what number?
9. 198 is 90% of what number?
10. 2% of what number is 5?
11. 5 is 25% of what number?
12. ⅔ is 75% of what number?
13. 80% of what number is 8½?
14. 30 is 10% more than what number?
15. 10% more than what number is 15?
16. 10% more than what number is 5.5?
17. 6⅞ is 1% more than what number?
18. 48 is 20% less than what number?
19. 15% less than what number is 5.8?
20. 3⅓ is 6% less than what number?
21. 75% of what number is 3?
22. 22.4 is 65% of what number?
23. 25% of what number is ⅔?
24. 30 is 16⅔% more than what number?
25. 18.2% less than what number is 48?

26. After a man's wages were reduced 15%, he got $262.40. How much did he get before the reduction?

27. A dealer sells coal at $97.75 per ton. If her profit is 12%, what does the coal cost her?

28. The efficiency of a motor is 90%, that is, the output is 90% of the input. If the motor delivers 8 hp, what is the input?

29. A person sells an article for $3.60, thereby losing 10%. What did the article originally cost?

30. A certain ore yields 5% of iron. How many tons of ore are required to produce $2\frac{1}{2}$ tons of iron?

31. An article loses $3\frac{1}{2}$% of its weight by drying. If it weighs $8\frac{1}{2}$ lb when dry, what was its weight before drying?

32. The inspectors in a factory rejected 33 pieces as imperfect. This represented $1\frac{1}{2}$% of the daily production. How many pieces were produced?

33. A motor whose efficiency is 86% delivers $10\frac{3}{4}$ hp. What horsepower does it receive?

34. What must be the length of a pattern for an $18\frac{1}{2}$"-long casting if the allowance for shrinkage is $\frac{3}{4}$ of 1%?

35. A mechanic gets an increase of $20.60 per week, which represents an increase of 4%. What is the mechanic's new weekly salary?

4-5   Using the Calculator to Find the Percentage, Given the Base and the Rate

The calculator problems in this chapter may be worked on a calculator like the one shown here. To find the percentage when the base and rate are known, multiply the base by the rate expressed in hundredths.

Example
What is 6% of $35.65?

SOLUTION   The display shows

Turn on the calculator $0.$
Enter 6 $6.$
Press ÷ $6.$
Enter 100 $100.$
Press × 0.06
Enter 35.65 35.65
Press = 2.139

Since this is a problem involving money, round the answer to $2.14.

Problems 4-5

Use the calculator to find the following percentages. Round your answers to fit the problems, that is, a money problem should be rounded to the nearest cent, and so on.

1. A car was bought for $14,095 and the state sales tax was 6%. What was the cost of the car including the sales tax?
2. A company making angle brackets expects a 1.5% reject rate daily. If 5,973 angle brackets are made daily, how many rejected brackets can be expected each day?
3. 15% of a class of 35 students got a grade of A. How many students got an A in this class?
4. A firm gives successive discounts of 10% and 2% for their preferred customers who pay their bills promptly. What amount should a preferred customer pay if the original bill was $1,258.00?
5. A baseball player is hitting at .247, which means that he has averaged a 24.7% chance of hitting the ball for each time he is up to bat. If he was up to bat 490 times, how many hits did he get?
6. What is the total social security tax on a $20,000 yearly income if the social security tax is 6.13%?
7. A cottage is sold for $56,500. If 72% of the selling price is for labor and materials and 8% is for supervision, what is the profit made on the cottage?
8. A certain gold ore contains 0.02% gold. How many ounces of gold will there be in one ton of this ore?
9. The surface area of a sphere (ball) is found by using this formula: 4 × 3.1416 × radius × radius. If the radius of the Earth is 4,000 miles and 70% of the Earth's surface is covered by water, then how many square miles of water are there on the Earth's surface?
10. In 1978 there were 10,760,521 persons receiving aid from the government; 70.2% of this total were children under 18 years of age. How many

such children were receiving aid in 1978? (Source: U.S. Department of Health, Education, and Welfare.)

4-6   Using the Calculator to Find the Rate, Given the Base and the Percentage

To find the rate, divide the percentage by the base and multiply the quotient by 100.

Example

What is the rate of death by heart disease if 723,878 people in the United States died from heart disease in 1977? The population at that time was 216,332,000.

SOLUTION	The display shows
Turn on the calculator	0.
Enter 723878	723878.
Press ÷	723878.
Enter 216332000	216332000.
Press ×	0.003346143
Enter 100	100.
Press =	0.334614389

Round this answer to 0.33% as the rate of death by heart disease in 1977. (Source: U.S. Health Service.)

Problems 4-6

Use the calculator to solve the following problems. Round the answer according to the nature of problem.

1. In a recent year there were 134,036,000 licensed drivers in the United States and 6,938,000 licensed drivers in Pennsylvania. What percentage of the licensed drivers lived in Pennsylvania? (Source: Federal Highway Administration.)
2. A ton of ore yields 36 lb of copper. What percent of the ore is copper?
3. What percent of the Earth's surface is Asia if Asia's area is 17,129,000 square miles? (Refer to Problem 9 of Problem Set 4-5.)
4. The diameter of the sun is 864,900 miles and the diameter of the Earth is 8,000 miles. What percent of the sun's diameter is the Earth's diameter?
5. Out of 1,399,000 college graduates, 322,000 received Master's degrees. What percent of these college graduates received Master's degrees?
6. Out of a total output of 380 castings per day, 27 were rejected. What is the percentage of bad castings?

7. If a person borrows $5,000 and repays $5,650, what is the rate of interest expressed in percent?
8. By installing new machinery, a manufacturer can make 10,000 special bolts where only 7,500 bolts could be made before. What is the gain in output of the new process over the old expressed in percent?
9. In 1968 gasoline was 54.9 cents per gallon, and in 1987 gasoline was 96.9 cents per gallon. What was the increase in cost per gallon expressed in percent?
10. In 1980 there were 41,323 black personnel in the U.S. Navy. If there were 522,000 personnel in the Navy at that time, what percentage were black? (Source: U.S. Department of Defense.)

4-7   Using the Calculator to Find the Base, Given the Percentage and the Rate

To find the base, divide the percentage by the rate expressed in hundredths.

Example
A certain iron ore yields 4.7% of iron. How many tons are required to produce 150 tons of iron? Round your answer to hundreds.

SOLUTION   The display shows

Turn on the calculator	$0.$
Enter 150	$150.$
Press ÷	$150.$
Enter 4.7	4.7
Press ×	$31.91489362.$
Enter 100	$100.$
Press =	$3191.489362.$

Round this answer to 3,200 tons.

Problems 4-7

Use the calculator to solve the following problems. Round the answer according to the nature of problem.

1. A certain iron ore yields 3.6% of iron. How many tons are required to produce 100 tons of iron?
2. Inspectors rejected 137 parts as not usable. This represented 1.2% of the total output. What was the total output?
3. A casting is to be 1 foot $6\frac{1}{2}$ inches long. What must be the length of the pattern if the shrinkage allowance is 0.0075%?
4. 65 is 11.5% of what number?
5. 60 is 12% more than what number?

6. 2.75 is 5.5% less than what number?
7. A motor has an efficiency rating of 87.5% and it delivers 17.5 hp. What horsepower does it receive?
8. There are 5,971 black officers in the U.S. Army. This is 6.1% of the total number of officers in the Army. How many officers are there in the Army? (Source: U.S. Department of Defense, 1980.)
9. In 1890 there were 43,731 graduates from high school. This represents 3.4% of those who could have graduated based on age. How many could have graduated in 1890 if school attendance was mandatory?
10. If approximately 9.8% of the population of the United States lives in California and the population of California is 23,667,565, then what is the population of the United States? (Source: Bureau of Statistics.)

4-8   The P-R-B Triangle

Sometimes the following device is used to help remember the rules involving percentage problems. Draw a triangle and place P (meaning *percentage*), R (meaning *rate*), and B (meaning *base*) as shown in Fig. 4-1.

Fig. 4-1

When the percentage is required, cover the P with your finger, and the device shows that the base (B) must be multiplied by the rate (R). If the rate is required, cover the R with your finger, and the device shows that the percentage (P) must be divided by the base (B). If the base is required, cover the B with your finger, and the device shows that the percentage (P) must be divided by the rate (R). You might want to try the *P-R-B triangle* to solve the percentage problems in the following tests.

Self-Test

Do the following problems. Check your answers with the answers in the back of the book.

1. $\frac{1}{2} =$ _____%
2. $\frac{1}{3} =$ _____%
3. 10% of 75 is _____.
4. What percent of 25 is 5?
5. 40 is 20% of what number?
6. $8\frac{1}{4}\%$ of $32,000 is _____.
7. 1% of 80 is _____.
8. $\frac{1}{2}\%$ of 200 is _____.
9. The inspectors rejected 15 pieces of a 500-piece lot. What was the percentage of rejections?

10. If an article was priced at $230.50 and is marked down 20%, what is its new price?

Chapter Test

Do the following problems involving percentages. Make sure that your answer is reasonable by estimating the answer first.

1. $\frac{1}{8}$ = _____%
2. $\frac{1}{4}$ = _____%
3. $\frac{3}{4}$ = _____%
4. $\frac{2}{3}$ = _____%
5. $\frac{1}{6}$ = _____%
6. $\frac{5}{8}$ = _____%
7. 10% of 250 is _____.
8. 18% of 508 is _____.
9. 45% of 127 is _____.
10. 0.75% of 30 is _____.
11. What percent of 15 is 5?
12. What percent of 35 is $17\frac{1}{2}$?
13. What percent of 40 is 6?
14. What percent of 75 is 5?
15. What percent of 1 is 16?
16. 4 is 75% of what number?
17. 36 is 20% of what number?
18. 72 is 20% less than what number?
19. 57 is $16\frac{2}{3}$% of what number?
20. 100% of 3.1416 is _____.

5

Ratio and Proportion

5-1 Ratio

A *ratio* is the comparison of two like quantities. It is expressed by the quotient obtained when the first quantity is divided by the second. For example, the relation between 10 and 5 can be stated in one of four ways: the ratio of 10 to 5, or $10 \div 5$, or $10:5$, or $\frac{10}{5}$. In each instance the expression represents the ratio of 10 to 5, and the value of the ratio is obtained by dividing 10 by 5, giving 2.

Every common fraction may be regarded as a ratio. The fraction $\frac{2}{3}$ is the ratio of 2 to 3. The two numbers compared are the *terms* of the ratio. The *inverse ratio* is the reciprocal of the original ratio. The ratio of 3 to 2 is the inverse of the ratio of 2 to 3, and vice versa.

5-2 Reduction of Ratios to Lowest Terms

Working with ratios is much like working with common fractions, and the same rules apply. The ratio of 6 to 30 is the same as the ratio of 2 to 10 and the same as the ratio of 1 to 5, and so forth. To reduce a ratio to lowest terms, proceed as in reducing a common fraction to lowest terms; that is, find the highest common factor of both terms and divide both terms using the highest common factor as their divisor.

Example 1

Express in lowest terms the ratio of 6 to 30.

SOLUTION $\dfrac{6}{30} = \dfrac{6 \div 6}{30 \div 6} = \dfrac{1}{5}$ or 1 to 5

EXPLANATION   Express the ratio as a fraction $\frac{6}{30}$ and divide both terms by the highest common divisor, 6, giving the fraction $\frac{1}{5}$. The relation between 6 and 30 is the same as between 1 and 5. The ratio of 6 to 30 is in lowest terms when expressed as the ratio of 1 to 5.

Example 2

Express in lowest terms the ratio $\frac{2}{3}$ to $\frac{4}{5}$.

SOLUTION   Ratio of $\frac{2}{3}$ to $\frac{4}{5} = \frac{2}{3} \div \frac{4}{5} = \frac{2}{3} \times \frac{5}{4} = \frac{5}{6}$

EXPLANATION   Since the value of a ratio is the quotient obtained by dividing the first term by the second, divide $\frac{2}{3}$ by $\frac{4}{5}$, obtaining the quotient $\frac{5}{6}$. That is, the ratio of $\frac{2}{3}$ to $\frac{4}{5}$ is the same as the ratio of 5 to 6.

Example 3

Divide $28 between two people in the ratio of 2 to 5.

SOLUTION $\frac{2}{7}$ of $28 = $8
$\frac{5}{7}$ of $28 = $20

EXPLANATION   To divide in the ratio of 2 to 5 means that for every $2 given to one person, $5 must be given to the other person. In other words, out of every $7, one gets $\frac{2}{7}$ or $2 and the other gets $\frac{5}{7}$ or $5. In general, one person gets $\frac{2}{7}$ of the total amount and the other gets $\frac{5}{7}$.

Problems 5-2

Express the following ratios in lowest terms.

1. 12 to 3	2. 3 to 12	3. 6 to 5	4. $\frac{3}{4}$ to $\frac{9}{16}$
5. $\frac{5}{8}$ to $\frac{7}{8}$	6. $1\frac{1}{2}$ to 3	7. $2\frac{1}{2}$ to $3\frac{1}{4}$	8. $3\frac{1}{2}$ to $16\frac{2}{3}$

Write the inverse of the following ratios and express in lowest terms.

9. 3 to 7	10. 5 to 8	11. $1\frac{1}{2}$ to 3	12. $2\frac{1}{2}$ to $3\frac{1}{2}$
13. $5\frac{1}{8}$ to 6	14. 3 to $2\frac{1}{2}$	15. 5.5 to 0.5	16. 0.25 to 0.75

17. Express the following ratios in lowest terms: 15 to 3; 8 to 16; 12 to 18; $\frac{7}{8}$ to $\frac{9}{16}$; $\frac{1}{2}$ to $\frac{1}{3}$; $2\frac{1}{2}$ to 10; $3\frac{1}{3}$ to $16\frac{2}{3}$.

18. Find the inverse of the following ratios: $5:3$; $7:8$; $4\frac{1}{2}:15$; $2\frac{2}{3}:3\frac{1}{4}$; $\frac{6}{7}:\frac{9}{10}$.

19. The smaller of two belted pulleys makes 240 revolutions per minute and the larger one makes 80. What is the ratio of their speeds?

20. Of two gears in mesh, the smaller gear makes 75 rpm and the larger gear makes 50 rpm. What is the ratio of their speeds?

21. A train runs at the rate of 60 mph and an airplane flies at the rate of 640 mph. Find the ratio of their speeds.

22. Tool steel may be worked at a cutting speed of 20′ per minute in a lathe and cast iron may be worked at a cutting speed of 45′ per minute. Find the ratio of the cutting speeds.

23. A high-speed drill $\frac{3}{4}''$ in diameter drills through 240 castings in the same time a carbon steel drill of the same diameter drills only 65 castings. Find the ratio of their speeds.

24. A foot of copper wire 0.001″ in diameter has a resistance of 10.4 ohms, whereas a foot of aluminum wire of the same diameter has a resistance of 18.7 ohms. What is the ratio of the two resistances?

25. One teenager earns $72 per week and another earns $64. What is the ratio of their earnings?

26. Bell metal is made of 4 parts of copper and 1 part of tin. Find the amount of each in a bell weighing 8.5 lb.

27. Divide $40 between two persons in the ratio of 3 to 5.

28. Britannia metal consists of 2 parts antimony, 1 part bismuth, and 1 part tin. How many pounds of each are there in a casting of Britannia metal weighing 24 lb?

29. White pine weighs 25 lb per cubic foot; steel, 490 lb per cubic foot. Find the ratio of their weights.

30. The circumference of a $2\frac{3}{4}''$ circle is 8.64″. Find the ratio between the circumference and the diameter. State the answer as a decimal of two places.

31. Refer to Table 7 (Weights of Materials) in the back of this book and find the ratio of the weight of aluminum to the weight of brass.

32. Refer to Table 7 and the column headed *Average Weight in Grams per* CM^3 and find the ratio of the weight of gold to the weight of silver.

33. Compare the weight of the lightest metal in Table 7 to the weight of the heaviest metal in Table 7. What metals are they?

34. Compare the weight of the lightest wood in Table 7 to the weight of the heaviest wood in Table 7. What woods are they?

5-3 Proportion

A *proportion* is an equality between two ratios. Study the following problems carefully and see what a powerful idea proportion is! When it is shown that one ratio is equal to another ratio, the expression is a proportion; for example, the ratio of 2 to 5 is equal to the ratio of 4 to 10. This proportion can be written $2:5 = 4:10$, or as $\frac{2}{5} = \frac{4}{10}$. Read this proportion: "Two is to five as four is to ten."

In the proportion $2:5 = 4:10$, the *outside* terms 2 and 10 are called the **extremes,** and the *inside* terms 4 and 5 are called the **means.** In the proportion $2:5 = 4:10$, notice the following:

$$5 \times 4 = 2 \times 10 \qquad (1)$$

The product of the means is equal to the product of the extremes.

$$2 = \frac{5 \times 4}{10} \qquad \text{and} \qquad 10 = \frac{5 \times 4}{2} \qquad (2)$$

Either extreme can be found by multiplying the means and dividing the product by the other extreme.

$$5 = \frac{2 \times 10}{4} \qquad \text{and} \qquad 4 = \frac{2 \times 10}{5} \qquad (3)$$

Either mean can be found by multiplying the extremes and dividing by the other mean.

Example 1

In 15 minutes a worker can machine 12 studs. How long will it take to machine 250 studs?

SOLUTION   Let x be the number of minutes it will take to machine 250 studs.

$$15:12 = x:250 \text{ (min : studs = min : studs)}$$

$$\therefore x = \frac{15 \times 250}{12} = 312\tfrac{1}{2} \text{ min} = 5 \text{ hours } 12\tfrac{1}{2} \text{ min}$$

(The symbol $\therefore$ means *therefore*.)

EXPLANATION   Let x stand for the number of minutes it will take to machine 250 studs. The ratio of the 15 min to the 12 studs machined in those 15 min is the same as the ratio of the time (x minutes) to the 250 studs to be machined in x minutes. The proportion then would be

$$15 \text{ min} : 12 \text{ studs} = x \text{ min} : 250 \text{ studs}$$

Then the mean (x minutes) can be found by multiplying the extremes (15 and 250) and dividing the product by the other mean (12); we get $x = 5$ hours $12\tfrac{1}{2}$ min.

Example 2

An 18-in. gear meshes with a 6-in. gear. If the large gear has 72 teeth, how many teeth will the small gear have?

SOLUTION $18:6 = 72:x$ $\therefore x = \dfrac{72 \times 6}{18} = 24$ teeth

EXPLANATION   Let x represent the number of teeth on the smaller gear. The ratio of the size of the larger gear to the size of the smaller gear (18 in. to 6 in.) is the same as the ratio of the number of teeth on the larger gear to the number of teeth on the smaller gear (72 teeth to x teeth). The proportion would be

$$18 \text{ in.} : 6 \text{ in.} = 72 \text{ teeth} : x \text{ teeth}$$

The extreme (x teeth) can be found by multiplying the means (6 and 72) and dividing by the other extreme (18); we get $x = 24$ teeth.

Example 3

If a young man earns \$285 per week, how long must he work to earn \$3,420?

SOLUTION $285 : 3{,}420 = 1 : x \quad \therefore x = \dfrac{3{,}420 \times 1}{285} = 12$ weeks

EXPLANATION   Since the same relation exists between the lengths of time as between the amounts earned, the ratio $285 : 3{,}420$ is equal to the ratio $1 : x$, where x is the length of time he must work to earn \$3,420.

If two pulleys are connected together by a belt and are rotating, the speed at the rim of each pulley must be the same. A larger pulley must rotate more slowly than a smaller pulley for their rim speeds to be equal. The ratio of the sizes of the pulleys are inversely proportional to the ratio of their revolutions per minute (rpm).

Example 4

A 2-in. pulley on a 3,450-rpm motor drives a 3-in. pulley. What are the revolutions per minute of the larger pulley?

SOLUTION $2 \text{ in.} : 3 \text{ in.} = x \text{ rpm} : 3{,}450 \text{ rpm}$

$$\therefore x = \frac{2 \times 3{,}450}{3} = 2{,}300 \text{ rpm}$$

EXPLANATION   The larger pulley must rotate slower than the smaller pulley so that the sizes are inversely proportional to the speeds. Let x be the rpm of the larger pulley and compare their sizes inversely as in the example.

Problems 5-3

Solve for x in the following proportions.

1. $3 : 4 = 9 : x$
2. $4 : 7 = 8 : x$
3. $15 : 9 = x : 3$
4. $3 : 6 = x : 24$
5. $24 : x = 8 : 12$
6. $7 : x = 28 : 84$

7. $x:6 = 4:12$ 8. $x:18 = 24:6$
9. $4.5:10 = 9:x$ 10. $2.6:x = 9.1:1.75$
11. $3:10.5 = x:5.25$ 12. $x:6.5 = 5.52:15$

Solve the following problems involving proportion.

13. If a 14-lb casting costs $1.04, what would a 30-lb casting cost?
14. A pattern made of white pine weighs 3.74 lb. What will a brass casting made from this pattern weigh if white pine weighs 25 lb per cubic foot and brass weighs 520 lb per cubic foot?
15. The lengths of the two rectangles shown in Fig. 5-1 are proportional to their widths. What is the length of the smaller rectangle?

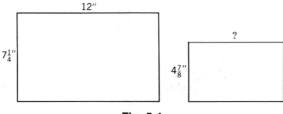

Fig. 5-1

16. A copper wire 400′ long has a resistance of 1.084 ohms. What is the resistance of 2500′ of copper wire?
17. Cast iron weighs 450 lb per cubic foot; white pine, 25 lb per cubic foot. If a certain pattern made of white pine weighs $2\frac{1}{4}$ lb, what will the casting weigh?
18. A 24″ pulley running at 180 rpm drives a 14″ pulley. How many revolutions per minute will the smaller pulley make?

NOTE. The ratio of their speeds is the inverse of the ratio of their sizes.

19. A 14″ pulley makes 240 rpm and drives a larger pulley making 210 rpm. What is the diameter of the larger pulley?
20. A 16″ grinding wheel makes 1,000 rpm and the driving pulley is 10″ in diameter. If the driven pulley is 6″ in diameter, how many revolutions per minute does the driving pulley make?
21. A 15″ shaft is found to have a taper of 0.204″ in 4″. Find the taper in the entire length of the shaft.
22. If the corresponding sides of the two triangles in Fig. 5-2 are to be proportional, find the missing dimensions of the small triangle.

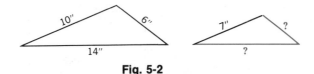

Fig. 5-2

23. In the two triangles shown in Fig. 5-3, the heights are proportional to the bases. Find the height of the smaller triangle.

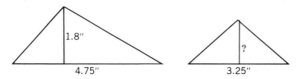

Fig. 5-3

24. The circumferences of circles are proportional to their diameters. Find the circumference of a 7″ circle if the circumference of a 3″ circle is 9.42″.

25. The areas of circles are proportional to the squares of their diameters. Find the area of a 9″ circle if the area of a 5″ circle is 19.635 sq in.

26. The pressure of water increases with the depth. If the pressure is 6.51 lb per square inch at a depth of 15′, what is the pressure at a depth of 80′?

27. The power of a gas engine increases with the area of the piston. If an engine with a piston area of 8.30 sq in. develops 25.5 hp, how many horsepower will be developed by an engine with a piston area of 7.07 sq in.?

5-4 Using the Calculator to Solve Proportion Problems

Proportion problems are one ratio (fraction) equated to another ratio (fraction). The solution to a proportion problem is that of finding the missing term of one of the ratios. To find the missing term, multiplication and division are

usually required. A calculator, such as the one shown here, is an excellent tool to solve these problems. The operator must be careful to multiply those parts that should be multiplied and divide those parts that should be divided.

RULE _____

The product of the extremes is equal to the product of the means.

Example

If a worker can stamp out 36 receptacle plates in 15 minutes, how many could he stamp out in an 8-hour shift?

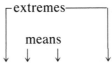

SOLUTION   The problem becomes $36:15 = x:(8 \times 60)$. Stampings are compared to minutes as unknown stampings (x) are compared to minutes (i.e., 8 hours $\times$ 60 minutes = 480 minutes).

	The display shows
Turn on the calculator	0.
Enter 36	36.
Press $\times$	36.
Enter 480	480.
Press $\div$	17280.
Enter 15	15.
Press $=$	1152. (answer)

The answer is the number of stampings the worker could make in an 8-hour shift by working at the rate of 36 stampings for each 15 minutes.

Problems 5-4

Use the calculator to solve the following proportions. Analyze each answer to determine if the answer is reasonable for the problem that was solved.

1. Copper wire has an average weight of 550 lb per cubic foot. What will one cubic inch of copper weigh? There are 1,728 cubic inches in one cubic foot.
2. There are 16.387 cubic centimeters in one cubic inch. How many cubic centimeters are there in one cubic foot?
3. A cubic foot of water weighs 62.5 lb. A gallon contains 231 cubic inches. What would 1,000 gallons of water weigh?
4. There are 3.58267 inches in 91 millimeters. How many inches are there in 25 millimeters?
5. Black walnut wood weighs 38 lb per cubic foot when dry. Refer to Problem 2 and give the weight of dry black walnut in grams per cubic centimeter. (1 lb = 452.6 grams)

6. A 1.25″ disc has a small hole in its center as shown in Fig. 5-4. To measure the hole, the disc is projected through a film strip projector; the shadow of the disc measures $6\frac{7}{16}$″ and the image of the hole measures 2.5 mm. How large is the hole in the disc, to the nearest 0.1 mm?

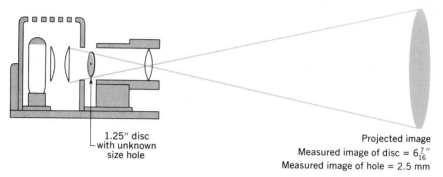

1.25″ disc
with unknown
size hole

Projected image
Measured image of disc = $6\frac{7}{16}$″
Measured image of hole = 2.5 mm

Fig. 5-4

7. The material in a 15.75-lb casting costs $4.76. What would a 67.5-lb casting of the same material cost?

8. Silver weighs 10.51 grams per cubic centimeter. If a silver charm weighs 15.5 grams, then how many cubic centimeters does it contain?

9. Spruce weighs 25 lb per cubic foot. What would be the weight of a brass casting (brass weighs 512 lb per cu ft) made from a spruce pattern weighing 16.8 lb?

10. A 2″ pulley running at 3,400 rpm drives a 2.5″ pulley. What is the rate at which the driven pulley is turning? (Hint: Does the larger pulley turn faster or slower?)

11. Fill out the following chart. The water pressure is proportional to the depth.

Depth (ft)		10	20	30	40	50	100	200	300
Water Pressure (lb)	4.34								

12. The areas of circles are proportional to the square of their diameters. Find the area of an 8.5-inch circle if a 10-inch circle has an area of 78.53975 square inches.

5-5   Averages

To find the *average,* or *mean,* grade for five tests whose scores are 90, 85, 75, 80, and 100, the five scores would be added and the sum divided by 5. In this example, 90 + 85 + 75 + 80 + 100 = 430. Dividing 430 by 5 equals 86, which is the average grade for the five tests. This is the procedure to be followed for all average problems: Add together all the items and divide by the number of items.

RULE _____

To find the average of two or more quantities, divide their sum by the number of quantities.

Example
Find the average length of four rods whose respective lengths are 5 ft, 7 ft, 12 ft, and 16 ft.

SOLUTION   Total length = 5 + 7 + 12 + 16 = 40 ft
            Average length = $\frac{40}{4}$ = 10 ft

EXPLANATION   Divide the total length of the rods by the number of rods.

Problems 5-5
_____

Answer the following questions involving averages.
1. Find the average weight of the following five castings: 8 lb, 10 lb, 14 lb, 9 lb, 19 lb.
2. Find the average length of the following rods: 6'8", 5'4", 3'10", 4'6", 5'2", 4'9".
3. One of three apprentices turns out 184 cotter pins and the other two turn out 206 and 195, respectively. What is the average output per apprentice?
4. A machinist works $8\frac{1}{2}$ hours on Monday, $8\frac{1}{4}$ hours on Tuesday, 8 hours on Wednesday, $9\frac{1}{4}$ hours on Thursday, 9 hours on Friday, and $4\frac{1}{4}$ hours on Saturday. What is the average number of hours worked per day?
5. What would you consider the correct diameter of a ball bearing if you measure it three times with a vernier caliper and find a different reading each time as follows: 0.214", 0.212", 0.213"?
6. The measurements of the diameter of a piece of work are found to be as follows: 0.4206", 0.4203", 0.4209", 0.4204". What is the most probable diameter?
7. Find the mean of the following dimensions: 1.6435", 1.6440", 1.6438", 1.6429", 1.6432", and 1.6426".
8. Which of the dimensions in Problem 7 differed most from the mean and by how much?
9. The average weight of 24 castings is 16.5 lb. What does the lot weigh?
10. Four mechanics (*A, B, C,* and *D*) in a shop require different amounts of time for a certain job. *A* can do it in 3 hours 20 minutes, *B* in 2 hours 50 minutes, *C* in 2 hours 45 minutes, and *D* in 3 hours 35 minutes. What is the average amount of time required in the shop for that job?
11. In a shop where a careful inspection is maintained, the following number of pieces were rejected during a certain week: Monday 140, Tuesday 166, Wednesday 161, Thursday 171, Friday 155, and Saturday 93. What was the average number of rejects per day?

12. Several samples of steel wire taken from the same lot are tested for tensile strength with the following results: 278 lb, 276 lb, 285 lb, 270 lb, 281 lb, and 275 lb. What is the average strength of the wire?

13. Here is a trick question you can use on your friends. (It is impossible to do.) A man drives up a hill at 30 mph. How fast must he drive down the hill to average 60 mph?

5-6   Using the Calculator to Find the Average of a Set of Numbers

The calculator solves average problems by adding the individual items and dividing by the number of items. Suppose the average daily temperature at noon in Philadelphia during January was desired. It would be necessary to know the temperature to be used for each day. Suppose the temperature at noon was as follows for each day of January:

Date	1	2	3	4	5	6	7	8	9	10	11	12	13	14
Temp. at noon	21°	25°	35°	40°	42°	41°	37°	35°	20°	15°	12°	10°	17°	19°

Date	15	16	17	18	19	20	21	22	23	24	25	26	27	28
Temp. at noon	15°	17°	21°	25°	35°	40°	45°	50°	52°	50°	40°	35°	30°	21°

Date	29	30	31
Temp. at noon	25°	27°	12°

To use the calculator to solve this problem, each temperature is entered and added to the previous sum of temperatures; when the last temperature is added, the total sum is divided by 31 (the number of days in January). For repetitive operations, a flow chart (Fig. 5-5) is often helpful to show the order of operations.

Problems 5-6

Use the calculator to find the following averages.

1. Use the temperature chart for January and the flow chart in Fig. 5-5 to find the average temperature in January.

2. The linemen of the football team of Eureca Prep School weighed as follows: right end, 165 lb; left end, 185 lb; right guard, 210 lb; left guard, 192 lb; center, 217 lb. What was the average weight of the Eureca Prep line?

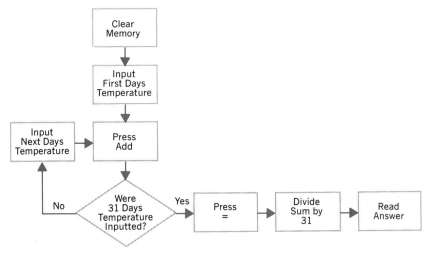

Fig. 5-5

3. Using a ten-thousandth micrometer, a mechanic made five measurements of a steel ball. They were 2.0156″, 2.0155″, 2.0157″, 2.0156″, and 2.0155″. What measurement should be accepted as probably correct?

4. In one year 14,033 ships used the Panama Canal and transported 147,907,000 metric tons of cargo. What is the average tonnage (metric tons) per ship?

5. John, Kenny, and Malcolm went fishing on 11 successive days last June. A tally of their total catch was as follows:

Date	11	12	13	14	15	16	17	18	19	20	21
Fish caught	15	22	5	33	8	17	12	8	17	6	10
Weight in pounds of catch	39	55	11	75	19	43	29	14	36	14	24

What was the average number of fish caught per person? What was the average weight per fish?

Self-Test

Do the following problems. Check your answers with the answers in the back of the book.

1. Express the ratio 8 to 6 in lowest terms.
2. Express the ratio 4 to 16 in lowest terms.
3. Write the inverse of the ratio 3 to 4.
4. Write the inverse of the ratio 6 to 5.
5. Express the ratio $4\frac{1}{4}$ to $8\frac{1}{2}$ in lowest terms.

6. If 100 ft of copper wire has a resistance of 0.54 ohm, what is the resistance of 200 ft of copper wire?
7. Two gears in mesh have 36 teeth and 18 teeth, respectively. What is the ratio of their speeds?
8. Divide $100 between two people in the ratio 3 to 2.
9. If five castings weigh 8 lb, 7 lb, 8 lb, 6 lb, and 6 lb, respectively, what is their average weight?
10. If your average earnings are $150 a week, what is your yearly income?

Chapter Test

Do the following examples involving ratios, proportions, and averages. Make sure that all your answers are in lowest terms.

Express the following ratios in lowest terms.

1. 12 to 4   2. $\frac{3}{4}$ to $\frac{5}{8}$ 3. $1\frac{1}{2}$ to 6   4. $2\frac{1}{2}$ to $5\frac{1}{2}$

Write the inverse of the following ratios.

5. 3 to 8   6. $\frac{1}{4}$ to $\frac{3}{8}$ 7. $3\frac{1}{2}$ to 7   8. $5\frac{1}{4}$ to $6\frac{1}{8}$

Find the following ratios.

9. Of two gears in mesh, the smaller gear makes 66 rpm and the larger gear makes 22 rpm. What is the ratio of their speeds?
10. Divide $60 between two people in the ratio of 3 to 2.
11. One worker earns $60 a week and another earns $50. What is the ratio of their earnings?
12. What is the ratio of one week to one year?

Solve the following proportions.

13. A copper wire 200' long has a resistance of 1.084 ohms. What is the resistance of 1000'?
14. An 18" shaft is found to have a taper of 0.375" in 6". Find the taper in the entire length of the shaft.
15. A 10" pulley makes 150 rpm and drives a larger pulley 75 rpm. What is the diameter of the larger pulley?

Find the following averages.

16. Find the average weight of the following five castings: 10 lb, 12 lb, 16 lb, 20 lb, and 22 lb.
17. Find the average length of the following rods: 5'6", 3'2", 6'5", and 2'7".
18. The average weight of 19 castings is 12 lb. What is the weight of the entire lot?
19. If you earn $300 a week for three weeks and $280 a week for the following five weeks, what are your average weekly earnings?
20. If your average earnings are $305.50 a week, what is your yearly income?

6

Practical Algebra

6-1   Use of Letters

Algebra uses letters to represent numbers. This is of great advantage because rules and formulas can be expressed in concise form. For example, to find the area of a rectangle, the length is multiplied by the width. This rule can be expressed in algebraic form.

If: A = area l = length w = width
Then: $A = l \times w$ or $A = lw$

The times sign ($\times$) is understood when letters are written next to each other, with no sign or space between them. Thus, $lw = l \times w$.

6-2   Negative Numbers

Negative numbers are numbers less than zero. Numbers greater than zero are *positive numbers.* We are familiar with the negative numbers used in winter weather reports when the temperature is below zero degrees. On a February morning in Maine, it might be minus 10 degrees ($^-10°$) with a wind chill factor

109

of minus 30 degrees (⁻30°). You may also have heard of a business having a negative cash flow, which means that there is more money being spent than being earned.

Negative numbers are distinguished from positive numbers by a minus sign. Thus ⁻3 means 3 below zero, while ⁺3 or 3 means 3 above zero. In Fig. 6-1, positive numbers are represented by spaces *above* zero and negative numbers by spaces *below* zero. To add ⁺3 to ⁺1, start at ⁺1, go *up* three spaces, and arrive at ⁺4; that is, ⁺1 + ⁺3 = ⁺4. To add ⁺5 to ⁻2, start at ⁻2, go up five spaces, and arrive at ⁺3; that is, ⁻2 + ⁺5 = ⁺3.

To subtract ⁺3 from ⁺5, start at ⁺5, go down three spaces, and arrive at ⁺2; that is, ⁺5 − ⁺3 = ⁺2. To subtract ⁺5 from ⁺2, start at ⁺2, go down five spaces, and arrive at ⁻3; that is, ⁺2 − ⁺5 = ⁻3. To subtract ⁺3 from ⁻2, start at ⁻2, go down three spaces, and arrive at ⁻5; that is, ⁻2 − ⁺3 = ⁻5.

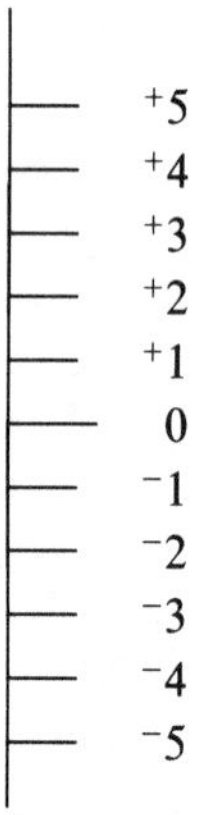

⁺5
⁺4
⁺3
⁺2
⁺1
0
⁻1
⁻2
⁻3
⁻4
⁻5

Fig. 6-1

Problems 6-2

Fill in the following charts using the diagram of positive and negative numbers (Fig. 6-1). (Note that a positive number does not need a plus sign to indicate that it is positive.)

ADDING

	Starting number	Add *n* (Go up *n*)	Arrive at
1.	1	3	_____
2.	3	2	_____
3.	5	3	_____
4.	⁻2	4	_____
5.	⁻5	3	_____

SUBTRACTING

	Starting number	Subtract *n* (Go down *n*)	Arrive at
6.	5	3	_____
7.	3	1	_____
8.	2	5	_____
9.	⁻1	2	_____
10.	⁻3	3	_____

6-3   Definitions

Absolute Value

The *absolute value* of a number is its value without reference to its sign. Thus the absolute value of $^+6$ is 6; the absolute value of $^-6$ is also 6. In algebra, the symbol used to indicate that the absolute value of a number is to be used rather than the number itself is two vertical bars, one on either side of the number, such as $|^-12| = 12$. When distance is discussed, the signs can be used to indicate direction; that is, away from a particular point can be considered positive, and toward the point can be considered negative. Distance, however, can only be positive. You couldn't drive $^-20$ miles, for instance.

Problems 6-3a

State the value of the following numbers.

1. $|+7|$ 2. $|-7|$ 3. $|85|$ 4. $|-2.6|$ 5. $|-1010|$

Factors

Numbers that can be multiplied together to produce a given number are *factors* of the given number. When the given number is a prime number, only the prime number and 1 are considered to be the factors. If the number is non-prime, then look for all the prime numbers that can be factors. Thus, the factors of 5 are 5 and 1, because 5 is a prime number. The factors of 6 are 2 and 3. Since 6 is a non-prime number, we are interested in the prime numbers that are factors of 6, but we are not interested in using 1 as a factor.

Algebra uses numbers and letters together. An algebraic expression such as $10xy$ means 10 times x times y. We may not know what values x and y represent, so we cannot factor them other than claiming $x = x$ times 1 and $y = y$ times 1. We can, however, factor the expression $10xy$ as 5 times 2 times x times y. The factor 1 is generally not necessary for operations such as this.

Problems 6-3b

Fill in the blanks with factors for the number or expression as indicated.

6. $10 = $ _____ $\times$ _____
7. $27 = $ _____ $\times$ _____ $\times$ _____
8. $6x = $ _____ $\times$ _____ $\times$ _____
9. $17xyz = $ _____ $\times$ _____ $\times$ _____ $\times$ _____
10. $39mn = $ _____ $\times$ _____ $\times$ _____ $\times$ _____

Power

A *power* is the product of two or more identical factors. Thus, *aa*, which means *a* times *a*, is the second power of *a* and is written a^2; *aaa*, which means *a* times *a* times *a*, is the third power of *a* and is written a^3; and *aaaa*, which means *a* times *a* times *a* times *a*, is the fourth power of *a* and is written a^4; and so on.

Problems 6-3c

State the power to which each of the following numbers is to be raised.

11. 2^2 12. a^4 13. 7^9 14. b^3 15. y^5

Exponent

An *exponent* is a number written above and to the right of a number or letter and shows how many times the number or letter is to be used as a factor. In the expression a^4, 4 is the exponent of *a* and shows that *a* is to be used as a factor four times; that is, $a^4 = aaaa$. When no exponent is written over a number or letter, the exponent 1 is understood; that is, $a^1 = a$; $2^1 = 2$, and so forth.

Problems 6-3d

Write the following expressions using exponents.

16. *aaaa*   17. *2xx*   18. *xxyyy*   19. *mmmnn*
20. *aaabbbcccddd*

Coefficients

In a product any factor or the product of two or more factors is the *coefficient* of the remaining factor or of the product of the remaining factors. Thus in the expression 3*a*, 3 is the coefficient of *a*; in the expression 5*abx*, 5 is the coefficient of *abx*, 5*a* is the coefficient of *bx*, and 5*ab* is the coefficient of *x*. In elementary algebra, the coefficient is the numerical part of the term. Where no coefficient is shown, the coefficient 1 is understood; thus *a* means 1*a*; *xy* means 1*xy*.

The student should learn to distinguish between the significance of exponents and that of coefficients. Thus $5^3 = 5 \times 5 \times 5 = 125$, but $3 \times 5 = 15$; likewise $a^3 = aaa$, but 3*a* = 3 times *a*.

Answer the following questions involving coefficients.

21. In the expression $6x$ the coefficient of the x is what?
22. In the expression $5ax$ the coefficient of the x is what?
23. The coefficient of xyz in the expression $3xyz$ is what?
24. The coefficient of bc in the expression $7abc$ is what?
25. In the expression $25axy$ the coefficient of axy is what?

Roots

A **root** is one of the equal factors of a number. Since $5 \times 5 = 25$, 5 is the square root of 25; since $2 \times 2 \times 2 = 8$, 2 is the cube root of 8. The root of a number is indicated by the symbol $\sqrt{}$. Thus the square root of 25 is written $\sqrt{25}$; the cube root of 8 is written $\sqrt[3]{8}$.

The **index** of the root is the number that tells what root is to be taken. It is written to the left and above the root sign. When no index appears, the square root is understood. Thus $\sqrt{16}$ denotes the square root of 16, which is 4 since $4 \times 4 = 16$; but $\sqrt[4]{16}$ denotes the fourth root of 16, which is 2, since $2 \times 2 \times 2 \times 2 = 16$.

Give a root of the following.

26. $\sqrt{25}$ 27. $\sqrt{225}$ 28. $\sqrt{x^2}$ 29. $\sqrt[3]{27}$ 30. $\sqrt[4]{625}$

6-4 Substitution

Study the examples shown below. If the numbers the letters represent are known, then the numerical value of the expression can be found by following the rules of arithmetic. Remember that an expression such as $4a$ means *4 times the value of a*. Parentheses indicate that the expression inside the parentheses is to be treated as a unit. Example 4 shows how to solve a problem that includes parentheses.

Evaluate the expression inside the parentheses first. Roots and powers are done before multiplication and division, which is done before addition and subtraction.

$$\left\{\begin{array}{c}\text{Parentheses} \\ \text{First}\end{array}\right\} \rightarrow \left\{\begin{array}{c}\text{Roots} \\ \text{or} \\ \text{Powers}\end{array}\right\} \rightarrow \left\{\begin{array}{c}\text{Multiplication} \\ \text{or} \\ \text{Division}\end{array}\right\} \rightarrow \left\{\begin{array}{c}\text{Addition} \\ \text{or} \\ \text{Subtraction}\end{array}\right\}$$

Example 1
If $a = 3$ and $x = 5$, find the value of $a + x$.

SOLUTION $a + x = 3 + 5 = 8$

Example 2
 If $a = 3$ and $x = 5$, find the value of $4a + 7x$.

SOLUTION $4a + 7x = 4 \times 3 + 7 \times 5$
$$= 12 + 35 = 47$$

Example 3
 If $a = 3$ and $x = 5$, find the value of $6a^2 + 2x^3$.

SOLUTION $6a^2 + 2x^3 = 6 \times 3^2 + 2 \times 5^3$
$$= 6 \times 9 + 2 \times 125 = 54 + 250$$
$$= 304$$

Example 4
 If $a = 4$, $b = 6$, and $y = 9$, find the value of $5(3a + 2b^2)\sqrt{y}$.

SOLUTION $5(3a + 2b^2)\sqrt{y} = 5(3 \times 4 + 2 \times 6^2)\sqrt{9}$
$$= 5(12 + 72)3 = 5 \times 84 \times 3$$
$$= 1{,}260$$

Problems 6-4

Find the numerical values of the following expressions if $a = 2$, $b = 5$, $c = 4$, $x = 3$, and $y = \frac{1}{2}$.

1. $5c$	2. $4x$	3. $2by$
4. $4a^2$	5. $5b^2$	6. $6abcx$
7. $3a^2b^2c^2$	8. $4(b + x^2)$	9. $2a(cy + bc)$
10. $4y + 3b$	11. $8ac + 5ax$	12. $2(a^2 + b^2 + x)$
13. $\sqrt{c}$	14. $\sqrt{x^2}$	15. $\sqrt[3]{2c}$
16. $\sqrt[3]{b^3}$	17. $a\sqrt{b^2}$	18. $(a + c)\sqrt{10ab}$
19. $c(b - a)$	20. $(c + x)^2$	21. $a^2 - y^2$
22. $\dfrac{c}{a} + \dfrac{b}{y}$	23. $\dfrac{1}{a} + \dfrac{1}{c}$	24. $\dfrac{1}{x} + \dfrac{1}{y}$
25. $4bc - 3ax$	26. $2c^2 - x^2$	27. $x^2 + y^2$
28. $(x + y)^2$	29. $(x + y)^3$	30. $(x - y)^2$

6-5   Addition

If an expression is written as the sum of several quantities, each of these quantities is called a **term** of the expression. Thus x, $5cb$, $3ab^2 \times 5cd$, $\dfrac{2a^3}{3c}$ are one term each. The expression $2a - 3bc + d$ consists of the three terms $2a$, ^-3bc, and d.

Similar or *like terms* are terms having the same letters raised to the same powers; thus $3a^2bc$ and $^-5a^2bc$ are like terms.

Dissimilar or *unlike terms* are terms that are not similar; $3a^2bc$ and $4ab^2c^2$ are unlike terms because the same letters are not raised to the same power; $2ax$ and $3dy$ are also dissimilar terms because they have different letters.

RULE

To add similar terms having the same sign, add the absolute values of the numerical coefficients, annex the common letters, and prefix the common sign to the result.

Example 1

Add $3a$, $5a$, and $4a$.

SOLUTION

$$\begin{array}{r} 3a \\ 5a \\ + 4a \\ \hline 12a \end{array}$$

EXPLANATION   Adding the numerical coefficients, we get $3 + 5 + 4 = 12$. Annex to this sum the common letter a; and since all terms are positive, their sum is positive. Hence the answer is $12a$.

Example 2

Add ^-6ab, ^-2ab, ^-9ab.

SOLUTION

$$\begin{array}{r} ^-6ab \\ ^-2ab \\ + ^-9ab \\ \hline ^-17ab \end{array}$$

EXPLANATION   The sum of the absolute values of the numerical coefficients is $6 + 2 + 9 = 17$. Annexing the common letters ab gives $17ab$; and since the common sign is negative, the sign of the sum is negative. Hence the answer is ^-17ab.

RULE

If the terms to be added have unlike signs, proceed as follows:

1. *Add the positive coefficients.*
2. *Add the negative coefficients.*
3. *Evaluate the absolute value of the coefficients of both sums and subtract the smaller from the larger. Use the sign of the larger as the sign of the result.*
4. *Annex the common letters.*

Example 1
  Add $4x$, ^-7x, ^-3x, $6x$, and ^-5x.

SOLUTION Add positive
          coefficients
          (Step 1)

Add negative
coefficients
(Step 2)

Follow steps
3 and 4

$$4x$$
$$+6x$$
$$\overline{10x}$$

$$^-7x$$
$$^-3x$$
$$+^-5x$$
$$\overline{-15x}$$

$$^-15x$$
$$+\ 10x$$
$$\overline{-5x}$$

EXPLANATION Evaluate the absolute value of both sums.

Absolute value of $^-15 = 15$
Absolute value of $\ 10 = 10$

Subtract the smaller from the larger, use the sign of the larger and annex the common letters.

Example 2
  Add $7a - 3b$, $4a + 6b$, $^-5a + 4b$.

SOLUTION $$7a - 3b$$
$$4a + 6b$$
$$+^-5a + 4b$$
$$\overline{6a + 7b}$$

EXPLANATION Write like terms under one another, then add each column separately as in Example 1.

Problems 6-5

Perform the following additions.

1. $7a$
    $5a$
    $4a$
    $+3a$

2. $16bx$
    $12bx$
    $+\ 8bx$

3. ^-5abc
    ^-8abc
    $+^-10abc$

4. $^-2a^2x$
    $^-9a^2x$
    $+^-4a^2x$

5. ^+25by
    ^+18by
    $+^-12by$

6. ^-16xy
    ^-14xy
    $+^+15xy$

7. $3b$
    ^-5b
    $+\ 8b$

8. ^-mn
    $5mn$
    $+^-12mn$

9. $^-15ax^2$
    $^-18ax^2$
    $22ax^2$
    $+\ 12ax^2$

10. $11b^2y$
    $^-6b^2y$
    $^-7b^2y$
    $+\ 4b^2y$

11. $6ab$
    $5ab$
    ^-8ab
    $+^-13ab$

12. $^-14bc^2$
    $6bc^2$
    $^-18bc^2$
    $+\ 9bc^2$

13.	$4a - 6c$	14.	$^-7x + 4y$	15.	$8b + 3d$
	$^-5a - 5c$		$^-9x - 6y$		$^-3b - 5d$
$+$	$^-8a + 4c$	$+$	$5x + 7y$	$+$	$^-12b - 10d$

6-6   Subtraction

RULE

To subtract, change the sign of the number to be subtracted (subtrahend) and then combine the two numbers as in addition.

Example 1
Subtract $5x$ from $12x$.

SOLUTION
$$\begin{array}{r} 12x \\ +\,^-5x \\ \hline 7x \end{array}$$

EXPLANATION   Changing the sign of $5x$ makes it ^-5x. Adding $12x$ and ^-5x gives $7x$.

Example 2
Subtract $5x$ from ^-12x.

SOLUTION
$$\begin{array}{r} ^-12x \\ +\,^-5x \\ \hline ^-17x \end{array}$$

EXPLANATION   Changing the sign of $5x$ makes it ^-5x. Adding ^-12x and ^-5x gives ^-17x.

Example 3
Subtract ^-5x from $12x$.

SOLUTION
$$\begin{array}{r} 12x \\ +5x \\ \hline 17x \end{array}$$

EXPLANATION   Changing the sign of ^-5x makes it $5x$. Adding $12x$ and $5x$ gives $17x$.

Example 4
Subtract ^-5x from ^-12x.

SOLUTION
$$\begin{array}{r} 12x \\ +\,^-5x \\ \hline ^-7x \end{array}$$

EXPLANATION   Changing the sign of ^-5x makes it $5x$. Adding ^-12x and $5x$ gives ^-7x.

Example 5

From $5x - 2y$ take $8x + 12y$.

SOLUTION
$$
\begin{array}{r}
5x + {}^-2y \\
-\ \ 8x + \ \ 12y \\
\hline
{}^-3x + {}^-14y = {}^-3x - 14y
\end{array}
$$

EXPLANATION   Write like terms under each other and proceed with each pair of like terms as explained in the preceding examples.

Problems 6-6

Do the following subtraction problems as indicated.

1. Subtract $12ax$ from $18ax$.
2. Subtract $14b^2y^2$ from $^-5b^2y^2$.
3. Subtract $^-16a^2b^2$ from $^-22a^2b^2$.
4. Subtract ^-10mm from $8mn$.
5. Subtract $^-36x^2y^2$ from $^-24x^2y^2$.
6. Subtract $21ac$ from ^-9ac.
7. Subtract $7ay$ from $3ay$.
8. Subtract ^-15abc from ^-18abc.
9. Subtract ^-24xy from $32xy$.
10. Subtract $9ac^2$ from $^-6ac^2$.
11. Subtract $48z + 14y$ from $36z - 12y$.
12. Subtract $^-30abc - 10bx$ from $20abc - 18bx$.
13. Subtract $8a + 9b$ from $6a + 4b$.
14. From $^-21x - 25y$ take $33x + 28y$.
15. From $14x + 16y$ take $^-7x - 8y$.

6-7   Symbols of Grouping

The symbols of grouping are the parentheses ( ), the brackets [ ], and the braces { }. The method of removing signs of grouping and expressing the resulting quantity in its simplest form will be clear from a study of the following examples.

Example 1

$5 + (3 + 4)$

SOLUTION $5 + (3 + 4) = 5 + 7 = 12$

The same result would be obtained if the parentheses were dropped first and then the numbers added; thus $5 + (3 + 4) = 5 + 3 + 4 = 12$.

RULE _____

Parentheses preceded by a positive sign (⁺) *may be dropped without chang-ing the sign of any term.*

Example 2
   $15 - (6 + 2)$

SOLUTION $15 - (6 + 2) = 15 - 8 = 7$
   The negative sign before the parentheses means that the value of the numbers within the parentheses is to be subtracted from 15. The rule for subtraction is: Change the sign in the subtrahend and combine with the minuend as in addition.

RULE _____

Parentheses preceded by a negative sign (⁻) *may be dropped if the sign of every term within the parentheses is changed. When one pair of parentheses appears within another, remove one pair at a time, beginning with the innermost and proceeding outward.*

Example 3
   $6a - [7a - (3a + 2b)]$

SOLUTION $6a - [7a - (3a + 2b)]$ (1)
   $= 6a - [7a - 3a - 2b]$ (2)
   $= 6a - 7a + 3a + 2b$ (3)
   $= 2a + 2b$ (4)

EXPLANATION   Removing the parentheses around $3a + 2b$ gives line (2); next remove the brackets and obtain line (3). Combining like terms gives line (4), the answer.

Problems 6-7

Simplify the following expressions by removing the parentheses and combin-ing like terms.

1. $12 + (5 + 2)$ 2. $12 + (5 - 2)$
3. $12 - (5 + 2)$ 4. $12 - (5 - 2)$
5. $a + (b + a)$ 6. $a + (6 - 2a)$
7. $a - (b - 2a)$ 8. $4a + (b - a)$
9. $5a - (a - b)$ 10. $9 - (2 - 4)$
11. $10 - (6 + 3)$ 12. $(3a - 3b) - (b + c)$
13. $8ab - (2ab - 2a^2)$ 14. $a - [-(-a)]$
15. $a + [b - (a - b)]$

6-8   Multiplication

In the multiplication of algebraic quantities, two important laws must be followed.

The Law of Signs

The product of two numbers having like signs is positive; the product of two numbers having unlike signs is negative. Thus,

$$(^+3) \times (^+5) = {}^+15$$
$$(^-3) \times (^-5) = {}^+15$$
$$(^+3) \times (^-5) = {}^-15$$
$$(^-3) \times (^+5) = {}^-15$$

The Law of Exponents

The exponent of a letter in the product is found by adding the exponents of that letter in the factors. Thus,

$$a^2 \times a^3 = a^{2+3} = a^5; \qquad ab^2c \times a^2b^3c^2 = a^3b^5c^3$$

RULE

To multiply one term by another, find the product of their numerical coefficients, prefix the sign according to the law of signs, annex the letters that appear in the factors, and obtain the exponent for each letter in the product by adding the exponents of that letter in the factors.

Example 1

Multiply $6a^2x$ by $^-4ax^2y$.

SOLUTION
$$\begin{array}{r} 6a^2x \\ \times\ {}^-4ax^2y \\ \hline {}^-24a^3x^3y \end{array}$$

EXPLANATION   Multiplying 6 by 4 gives 24. Since the signs are unlike, prefix the negative sign to the product. Annex all the letters that appear in both factors, writing each letter but once. Since the exponent of a in the multiplicand is 2 and in the multiplier 1, the exponent of a in the product is 2 + 1, or 3. Similarly the exponent of x is 1 + 2, or 3. The exponent of y is 1 because it appears in the multiplier to the first power and does not appear at all in the multiplicand.

Example 2

Simplify $^-3a(2x - 7y)$.

SOLUTION $^-3a(2x - 7y) = {}^-6ax {}^- 21ay = {}^-6ax + 21ay$

EXPLANATION   To simplify the given expression means to perform the indicated operations. Each term within the parentheses is multiplied by the multiplier ^-3a. To obtain the first term in the product we have $^-3a \times 2x = ^-6ax$. The second term is equal to $^-3a \times ^-7y = 21ay$. The complete answer is $^-6ax + 21ay$.

Example 3
Multiply $8a - 3b$ by $2a - 5b$.

SOLUTION

$$\begin{array}{r} 8a - 3b \\ \times 2a - 5b \\ \hline 16a^2 - 6ab \\ - 40ab + 15b^2 \\ \hline 16a^2 - 46ab + 15b^2 \end{array}$$

EXPLANATION First multiply each term of the multiplicand by the first term of the multiplier $(2a)$, giving the partial product $16a^2 - 6ab$. Next multiply each term of the multiplicand by the second term of the multiplier (^-5b), giving the partial product, $^-40ab + 15b^2$. The term ^-40ab is written under the similar term ^-6ab in the first partial product. Adding the two partial products gives the final result, $16a^2 - 46ab + 15b^2$.

Problems 6-8

Perform the following multiplications as indicated.

1. $a \times a$
2. $c^2 \times c$
3. $a^2 \times a^4$
4. $2a^2 \times 3a^3$
5. $^-5b \times ^-3b^2$
6. $^-b^2 \times ^-b^3$
7. $4a \times ^-3a^2$
8. $^-3c \times 8c^3$
9. $2a \times 3a^2 \times 5a$
10. $6a^2 \times ^-4a \times 3a$
11. $^-x^2 \times ^-3x \times ^-2x^3$
12. $2ab(a + b + c)$
13. $3a^2c(2a - 3b - c)$
14. $^-5a(2x^2 - 2y^2)$
15. $(3a - 4b)(5a - 6b)$
16. $(a + b)(a - b)$
17. $(2a + 3b)(4a + 5b)$
18. $(8a - 2b)(2b + 4a)$
19. $(2x + 3y)(2x - 3y)$
20. $(2x + 3y)(2x + 2y)$

6-9 Division

To perform a division, reverse the process of multiplication.

RULE
To divide one term by another, divide the numerical coefficient of the dividend by the numerical coefficient of the divisor; the result is the numerical coefficient of the quotient. Then follow the law of signs and the law of exponents to determine the sign and exponent of the quotient.

The Law of Signs
The quotient is positive if the signs are the same; the quotient is negative if the signs are opposite.

The Law of Exponents
Subtract the exponent of a letter in the divisor from the exponent of the same letter in the dividend; the result is the exponent of the same letter in the quotient.

Example 1
Divide $40a^2x^4$ by $5ax^2$.

SOLUTION $\dfrac{40a^2x^4}{5ax^2} = \dfrac{40}{5} \times \dfrac{a^2}{a} \times \dfrac{x^4}{x^2} = 8 \times a^{2-1} \times x^{4-2} = 8ax^2$

EXPLANATION   Dividing 40 by 5 gives 8. Since both signs are alike, the sign of the quotient is positive. Write after the numerical quotient 8 the letters that appear in both dividend and divisor, namely a and x. To obtain the exponent of a in the quotient, subtract the exponent of a in the divisor from the exponent of a in the dividend; thus $2 - 1 = 1$, which is not written after the a in the quotient. Similarly the exponent of x in the quotient is found by subtracting the exponent of x in the divisor from the exponent of x in the dividend; thus $4 - 2 = 2$.

Example 2
Divide $5a^2x^4$ by $5a^2x$.

SOLUTION $\dfrac{5a^2x^4}{5a^2x} = x^3$

EXPLANATION   The 5 in the dividend contains the 5 in the divisor once; the a^2 in the dividend contains the a^2 in the divisor once. Hence the given example is equivalent to $1 \times 1 \times \dfrac{x^4}{x} = 1 \times 1 \times x^3 = x^3$.

Example 3
Divide $^-12a^3x^2y^5$ by $3ay^2$.

SOLUTION $\dfrac{^-12a^3x^2y^5}{3ay^2} = {}^-4a^2x^2y^3$

EXPLANATION $12 \div 3 = 4$; the signs being unlike, the sign of the quotient is negative. Write in the quotient the letters that appear in both dividend and divisor. The exponent of a is $3 - 1 = 2$; the exponent of x is $2 - 0 = 2$; the exponent of y is $5 - 2 = 3$. The answer is $^-4a^2x^2y^3$.

Example 4
  Divide $20a^2b^2 - 16a^3b$ by $4ab$.

SOLUTION $\quad \dfrac{20a^2b^2 - 16a^3b}{4ab} = \dfrac{20a^2b^2}{4ab} - \dfrac{16a^3b}{4ab} = 5ab - 4a^2$

EXPLANATION $\quad$ Dividing $4ab$ into the first term of the dividend $20a^2b^2$ gives $5ab$; dividing $4ab$ into the second term of the dividend gives $^-4a^2$. The quotient is $5ab - 4a^2$.

Example 5
  Divide $a^2 - 8ab + 15b^2$ by $a - 3b$.

SOLUTION
$$
\begin{array}{r}
a - 5b \\
a - 3b\overline{)a^2 - 8ab + 15b^2} \\
\underline{a^2 - 3ab} \\
-5ab + 15b^2 \\
\underline{-5ab + 15b^2}
\end{array}
$$

EXPLANATION $\quad$ Divide the first term of the divisor into the first term of the dividend to obtain the first term of the quotient; thus $a^2 \div a = a$. Multiply each term of the divisor by the first term of the quotient; thus $a(a - 3b) = a^2 - 3ab$. Subtract this product from the dividend, leaving $^-5ab + 15b^2$. Divide the first term of the divisor into the first term of the remainder to obtain the second term of the quotient; thus $^-5ab \div a = ^-5b$. Multiply the divisor by the second term of the quotient; thus $^-5b(a - 3b) = ^-5ab + 15b^2$. Subtracting this product from the remainder previously obtained leaves 0. Since the remainder is exactly 0, the quotient is exactly $a - 5b$.

We know from our experience in division problems that when the remainder is 0, the divisor and the quotient are factors of the dividend. When we are called upon to factor an algebraic expression, we look for a divisor that will give a quotient without a remainder.

Problems 6-9

Perform the following divisions as indicated.

1. $18x^3 \div 6x$ 2. $24x^4y^3 \div 3x^2y^2$
3. $^-16x^3y^4 \div ^-8xy$ 4. $^-12abc \div ^-2ab$
5. $32a^2x^3 \div ^-4a^2x$ 6. $27b^5c^4 \div ^-9bc$
7. $^-36x^3y^3 \div 4xy^2$ 8. $^-40a^3d^3 \div 5a^2d$
9. $(24a^2b^2 + 18a^3b^3) \div 2ab$ 10. $(16a^3x^2 - 12a^2x^3) \div 4ax$
11. $(21b^4y^5 + 18b^5y^6) \div ^-3b^2y^2$ 12. $(14a^4y^4 - 7a^2y^2) \div ^-7ay$
13. $(x^2 + 5x + 6) \div (x + 2)$ 14. $(x^2 - 8x + 15) \div (x - 3)$

15. $(x^2 - y^2) \div (x + y)$

16. $(a^2 + 2ab + b^2) \div (a + b)$

17. $(x^2 - 2xy + y^2) \div (x - y)$

18. $\dfrac{48m^7n^7p}{{}^-6mp}$

19. $\dfrac{{}^-14a^2b^3c^4}{{}^-14a^2b^3c^4}$

20. $\dfrac{{}^-12xy^3}{{}^-12y^3}$

6-10 Equations

An *equation* is a statement that two quantities are equal. For example, the equation $3x = 15$ states that $3x$ and 15 are equal to each other.

Members

That part of an equation which is to the left of the equality sign ($=$) is called the *first member, left side,* or *left-hand member.* That part of an equation which is on the right of the equality sign is called the *second member, right side,* or *right-hand member.* In the equation $3x = 15$, $3x$ is the first or left-hand member, and 15 is the second or right-hand member.

An equation is used to find the value of an unknown number from its relation to known numbers. The unknown number is usually represented by some letter, as x, y, or z. The first letters of the alphabet, a, b, or c, usually represent known numbers.

The solution of equations depends upon the following axioms (statements or principles that are accepted as true).

AXIOMS

1. *Equal numbers added to equal numbers give equal sums.*
   If $x = y$, then $x + 2 = y + 2$.
2. *Equal numbers subtracted from equal numbers give equal remainders.*
   If $x = y$, then $x - 2 = y - 2$.
3. *Equal numbers multiplied by equal numbers give equal products.*
   If $x = y$, then $2x = 2y$.
4. *Equal numbers divided by equal numbers give equal quotients.*
   If $x = y$, then $\dfrac{x}{2} = \dfrac{y}{2}$.

Transposition of Terms

To *transpose* a term is to transfer it from one side of the equation to the other. A term may be transposed from one member of an equation to the other provided the sign of the term is changed.

Proof 1

Given the equation

$$x - a = b$$

Add a to each member,

$$x - a + a = b + a$$

But $^-a + a = 0$; therefore, $x = b + a$.

The result is the same as if the ^-a from the left side were transposed to the right and its sign changed.

Proof 2

Given the equation

$$x + a = b$$

Subtract a from each member,

$$x + a - a = b - a$$

But $a - a = 0$; therefore, $x = b - a$.

In this case the same result is obtained by transposing a from the left side to the right side and changing the sign.

The sign of every term of an equation may be changed without affecting the equality.

Proof 3

If in the equation

$$a - x = b + c \tag{1}$$

each term is multiplied by $^-1$ (Axiom 3), the result is

$$^-a + x = ^-b - c \tag{2}$$

Equation (2) is the same as Equation (1) with the sign of each term changed.

Solution of Equations

To solve an equation, it is usual to transpose all the unknown values to the left side of the equation and all the known values to the right side of the equation. Then by combining and simplifying, it is possible to solve for the unknown value. Study the examples that follow.

Example 1

Solve the equation $6x - 4 = 4x + 6$.

SOLUTION   Transposing $4x$ to the left side and $^-4$ to the right side,

$$6x - 4x = 6 + 4$$

Combining like terms,

$$2x = 10$$

Dividing both members by the coefficient of x,

$$x = 5$$

CHECK   Substitute 5 for x in both members of the equation,

$$6 \times 5 - 4 = 4 \times 5 + 6$$
$$26 = 26$$

Therefore the answer, $x = 5$, is correct.

Example 2

Solve the equation $\frac{2}{3}(x + 3) = \frac{1}{2}(x + 8)$.

SOLUTION   Multiply by 6 to clear fractions.

Simplifying	$4(x + 3) = 3(x + 8)$
Simplifying	$4x + 12 = 3x + 24$
Transposing	$4x - 3x = 24 - 12$
Combining	$x = 12$

CHECK   Substitute 12 for x in the given equation,

$$\tfrac{2}{3}(12 + 3) = \tfrac{1}{2}(12 + 8)$$
$$\tfrac{2}{3} \times 15 = \tfrac{1}{2} \times 20$$
$$10 = 10$$

Example 3

Solve the equation for x.

$$\frac{2x}{a} = b + c$$

SOLUTION   Multiplying both members by a,

$$2x = a(b + c)$$

Dividing by the coefficient of x,

$$x = \frac{a(b + c)}{2}$$

Example 4

Solve the equation for x.

$$ax + 2b = bx + c$$

SOLUTION   Transposing,

$$ax - bx = c - 2b$$

Factoring the left side gives

$$(a - b)x = c - 2b$$

Dividing by the coefficient of x,

$$x = \frac{c - 2b}{a - b}$$

From the preceding examples, we formulate the following rule for solving a simple equation.

RULE

Transpose all unknown terms to the left side and all known terms to the right side. Combine like terms and divide both members by the coefficient of the unknown term.

Example 5
$$4x^2 - 6 = 2x^2 + 12$$

SOLUTION   Transposing,

$$4x^2 - 2x^2 = 12 + 6$$

Combining like terms,

$$2x^2 = 18$$

Dividing by the coefficient of x^2,

$$x^2 = 9$$

Taking the square root of both sides,
$$\sqrt{x^2} = \sqrt{9}$$
$$x = 3 \text{ or } ^-3$$

Since $^+3^2 = 3 \times 3 = 9$, and $^-3^2 = ^-3 \times ^-3 = 9$, the square root of 9 is either 3 or $^-3$. From the conditions of the problem, it would be possible to tell which of the two values of x is the correct one.

Example 6
$$D = \frac{N + 2}{P}. \text{ Solve for } P.$$

SOLUTION   Multiplying by P,

$$PD = N + 2$$

Dividing by D,

$$P = \frac{N + 2}{D}$$

Solve the following equations for x or as indicated.

1. $5x = 10 + 4x$

2. $12x - 4 = 7x + 6$

3. $3x + 3 = 18$

4. $4x - 5 = 3x + 2$

5. $ax = b$

6. $ax + b = c$

7. $ax + b = cx + d$

8. $6(x + 2) = 5(x + 4)$

9. $4(x - 2) = 2(x + 6)$

10. $\dfrac{3x}{b} = a - c$

11. $\dfrac{5x}{a} = b + c$

12. $3x^2 + 3 = 15$

13. $2x^2 - 12 = x^2 + 24$

14. If $C = 2\pi r$, solve for r.

15. $V = lwh$. Solve for h.

16. $N = PD$. Solve for D.

17. $E = IR$. Solve for I and R.

18. Horsepower $= \dfrac{PLAN}{33,000}$. Solve for P.

19. Solve formula in Problem 18 for L.

20. Solve formula in Problem 18 for A.

21. $R = \dfrac{10.4 \times L}{CM}$. Solve for L and for CM.

22. $D = \dfrac{N + 2}{P}$. Solve for N.

23. $A = \pi r^2$. Solve for r.

24. $D = M - 3W + 1.732P$. Solve for P.

25. $\dfrac{a}{b} = \dfrac{c}{d}$. Solve for b and c.

26. $V^2 = 2gh$. Solve for g and h.

27. $C = \frac{5}{9}(F - 32)$. Solve for F.

28. $a = \dfrac{V - v}{t}$. Solve for V and v and t.

29. $M = \frac{1}{6}bh^2$. Solve for b and h.

30. $A = \pi ab$. Solve for a and b.

31. $S = 4\pi r^2$. Solve for r.

32. Horsepower $= 0.4D^2N$. Solve for D.

33. $OD = \dfrac{(N + 2)P_c}{\pi}$. Solve for N.

34. $A = lw$. Solve for l and w.

35. $V = \frac{4}{3}\pi r^3$. Solve for r.

36. $P = \dfrac{\pi}{P_c}$. Solve for P_c.

37. $A = 0.7854d^2$. Solve for d.

38. $N = \dfrac{\pi d}{P_c}$. Solve for d and P_c.

39. Power factor $= \dfrac{\text{Effective power}}{\text{Apparent power}}$. Solve for effective power and for apparent power.

40. Efficiency $= \dfrac{\text{Output}}{\text{Input}}$. Solve for output and for input.

41. $BHP = \dfrac{2\pi LNW}{33,000}$. Solve for L and N and W.

42. Cutting speed $=$ Circumference $\times$ rpm. Solve for circumference and for rpm.

43. Watts $=$ Amperes $\times$ Volts. Solve for amperes and for volts.

44. Area $= \dfrac{ab}{2}$. Solve for a and for b.

45. $A = 0.866s$. Solve for s.
46. $s = 0.707d$. Solve for d.
47. $f = 1.732s$. Solve for s.
48. $C = 1.083f$. Solve for f.
49. $A = 0.433s^2$. Solve for s.
50. $A = 0.866f^2$. Solve for f.

Self-Test

Do the following problems. Check your answers with the answers in the back of the book.

1. Add 3 to $^-2$.
2. What is the absolute value of $^-11$?
3. What is the exponent of the letter x in the term $5x^3y^2$?
4. What is the coefficient $15x^2yz^3$?
5. If $a = 3$, $b = {}^-2$, and $c = 7$, what is the value of $4ab^2c$?
6. Add $5ab^2$, $^-2ab^2$, $10ab^2$, $^-7ab^2$, and $3ab^2$.
7. Subtract $^-3x^2y$ from $5x^2y$.
8. Multiply: $(3x + 2y)(2x - 5y)$.
9. Divide $x^2 + 2xy + y^2$ by $x + y$.
10. Solve for a if $V = \frac{1}{3}\pi r^2a$.

Chapter Test

Do the following problems as indicated.

1. Add $^-4$, $^+7$, and $^-2$.
2. What is the absolute value of $^-7x + 3x$?
3. If $x = 7$, $y = 3$, and $z = {}^-3$, what is the value of $2x^2yz^3 - 4xy^2$?

4. What is the sum of $3mn^2$, $^-2mn^2$, mn^2, and $12mn^2$?

5. Subtract $^-2xy^3$ from $^-5xy^3$.

6. Multiply: $(4a^2 - b)(4a^2 + b)$.

7. Divide $8x^3y^2z$ by $2x^2yz$.

8. Solve for r if $A = \pi r^2$.

9. Solve for b if $A = \frac{1}{2}ab$.

10. $I = \dfrac{E}{R}$. Solve for R.

11. $\dfrac{2}{x} = \dfrac{x}{8}$. Solve for x.

12. $\frac{9}{5}(C) + 32 = F$. Solve for C.

13. $E = mc^2$. Solve for c.

14. $V = \dfrac{\pi d^3}{6}$. Solve for d.

15. $P = I^2R$. Solve for R and I.

7

Rectangles and Triangles

7-1 Area of Surfaces, Units of Area

Common units of area are the square inch, the square foot, the square yard, the acre, and the square mile. It is interesting that acre is automatically a square measure; there is no need to use the word "square" when referring to land measure in acres. In metric measure the common units are the square millimeter, the square centimeter, the square meter, the are, and the square kilometer. The metric "are" is automatically a square measure and is used for land measure (see Chapter 10). An are is much smaller than an acre.

A *square inch* is the area contained in a square 1 in. on each side (Fig. 7-1).

A *square foot* is the area contained in a square 1 ft on each side (Fig. 7-2).

A *square yard* is the area contained in a square 1 yd on each side.

Since there are 12 in. in a foot, a square foot contains 12 in. × 12 in., or 144 sq in.

Similarly, since there are 3 ft in a yard, a square yard contains 3 ft × 3 ft, or 9 sq ft (Fig. 7-3).

A figure frequently encountered is the rectangle. A *rectangle* is a four-sided figure whose angles are right angles (90°) and whose opposite sides are equal in length (Fig. 7-4).

131

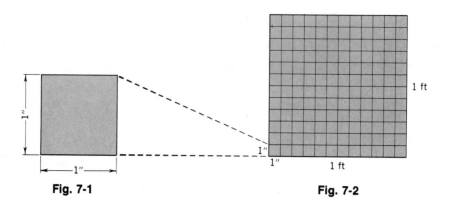

Fig. 7-1   **Fig. 7-2**

RULE
The area of a rectangle is found by multiplying the length by the width.

If l = length and w = width, the rule may be expressed by the formula:

$$\text{Area} = l \times w$$

Example 1
Find the area of a rectangle whose length is 5 in. and whose width is 3 in.

SOLUTION   Area = $l \times w$
= 5 in. $\times$ 3 in. = 15 sq in.

EXPLANATION   Since the length of the rectangle is 5 in. and the width 3 in., to get the area, multiply the length 5 in. by the width 3 in., obtaining the area 15 sq in.

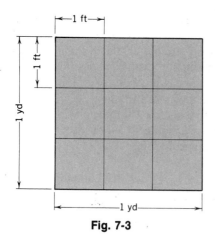

Fig. 7-3

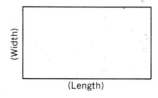

Fig. 7-4

Example 2

Find the area in square feet of a floor that is 18 ft 5 in. long and 12 ft 9 in. wide.

SOLUTION   Change 18 ft 5 in. and 12 ft 9 in. to inches.

$$18 \text{ ft } 5 \text{ in. } = (18 \times 12) + 5 = 216 \text{ in. } + 5 \text{ in. } = 221 \text{ in.}$$
$$12 \text{ ft } 9 \text{ in. } = (12 \times 12) + 9 = 144 \text{ in. } + 9 \text{ in. } = 153 \text{ in.}$$
$$\text{Area} = l \times w = 221 \times 153 = 33{,}813 \text{ sq in.}$$
$$33{,}813 \div 144 = 234.8125 \text{ sq ft} \approx 234.8 \text{ sq ft}$$

ALTERNATE SOLUTION   Change 18 ft 5 in. and 12 ft 9 in. to feet before applying the formula for area.

$$18 \text{ ft } 5 \text{ in. } = 18\tfrac{5}{12} \text{ ft} = 18.417 \text{ ft}$$
$$12 \text{ ft } 9 \text{ in. } = 12\tfrac{9}{12} \text{ ft} = 12.75 \text{ ft}$$
$$A = l \times w = 18.417 \text{ ft} \times 12.75 \text{ ft} = 234.81675 \text{ sq ft} \approx 234.8 \text{ sq ft}$$

EXPLANATION   Floor area is generally expressed in square feet. Divide the 33,813 sq in. by 144 (the number of sq in. in a sq ft). Since one decimal place is sufficiently accurate for this type of work, state the answer as 234.8 sq ft. In the alternate solution, both dimensions are changed to feet before the formula for area is used. It is not necessary to divide by 144 because the units are already in feet.

The wise mechanic estimates the answer before solving a problem. This practice avoids errors in computation. To estimate the answer of Example 2, round the two dimensions to the nearest foot. If both the dimensions were in feet, then it would be a simple matter of just multiplying the two. 18 ft 5 in. is a little over 18 ft, and 12 ft 9 in. is almost 13 ft. An approximate answer to the original problem would therefore be the product of 18 ft and 13 ft. Therefore the estimated area would be

$$\text{Area}_{est} = 18 \text{ ft} \times 13 \text{ ft} = 234 \text{ sq ft (estimated answer)}$$

A comparison of the estimated answer with the actual answer shows that the computed answer is probably correct. In general, the estimated answer will be fairly accurate if one number can be rounded up and the other number can be rounded down. Common sense must be used when comparing the computed answer with the estimated answer. If there is a great difference between the two, then the processes used to find the answers must be reviewed to discover where a mistake might lie. Mistakes are costly and should always be avoided.

Fill in the areas (in square feet) in the following chart. Estimate the answer before doing the necessary computation. Use one decimal place in the answer.

RECTANGLES

	Length	Width	Estimated Area	Computed Area
1.	5′	4′	20 sq ft	_____
2.	5′3″	4′0″	20 sq ft	_____
3.	5′3″	4′9″	25 sq ft	_____
4.	6′9″	4′4″	_____	_____
5.	8′9″	6′6″	_____	_____
6.	12′3″	10′9″	_____	_____
7.	7′4″	5′3″	_____	_____
8.	9′6″	7′9″	_____	_____
9.	5′8″	4′4″	_____	_____
10.	3′4″	2′6″	_____	_____
11.	6′9″	3′3″	_____	_____
12.	12′10″	8′2″	_____	_____

Find the following areas. Use units in the answers that would be reasonable for each particular problem. Estimate the answers before doing the problems.

13. Find the area of a page of a book $7\frac{1}{2}$″ long and 5″ wide.
14. Find the area of the floor of a room 16′4″ long and 10′8″ wide. Give the answer in square feet to one decimal place.
15. How many square feet of pavement are there in a sidewalk 198′ long and 18′ wide?
16. Find the area of the walls of a room 9′2″ high if the room is 19′4″ long and 12′3″ wide. (Make no allowance for doors or windows.)
17. How many square feet are there in the top of a desk 52″ long and 28″ wide? (144 sq in. = 1 sq ft)
18. How many square inches are there in a pane of glass $15\frac{1}{2}$″ × $19\frac{1}{2}$″?
19. A parking lot is 186′ × 91′. How many square yards does it contain? Find the cost of paving this area with concrete at $16 per square yard.
20. How many square feet are there in a rectangular lot 88′4″ deep with a frontage of 74′2″?
21. Allowing 25 sq ft per pupil, how many children may safely play in a playground 200′ long and 140′ wide?
22. How many square inches are there in the surface of a closed cubical box that measures 8″ on a side?

23. How many square feet are there in the six sides of a closed box $8\frac{1}{2}'' \times 6''$ $\times 3\frac{1}{2}''$?

24. How many square feet of corrugated iron are required for the walls of a garage $15'6'' \times 8'9''$ and $8'3''$ high? (Subtract 64 sq ft as the door and window allowance.)

Find the dimensions of the parts marked (?) in the following composite rectangular figures. Remember that a property of the rectangle is that its opposite sides are equal.

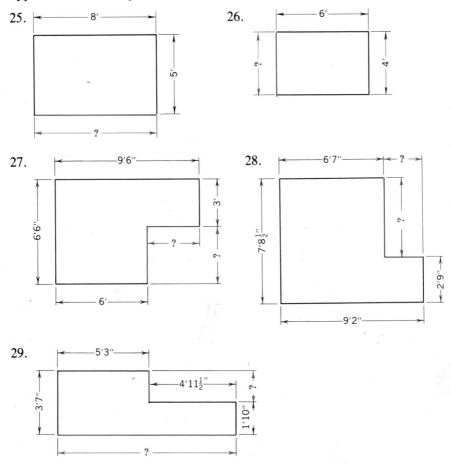

25.

26.

27.

28.

29.

Challenge Find the areas of the following figures, which are composite rectangles. Divide each figure into rectangles, and the area of the figure will be the sum of the areas of the rectangles into which the figure can be divided. Try estimating the areas before doing the actual computation. Area = length times width.

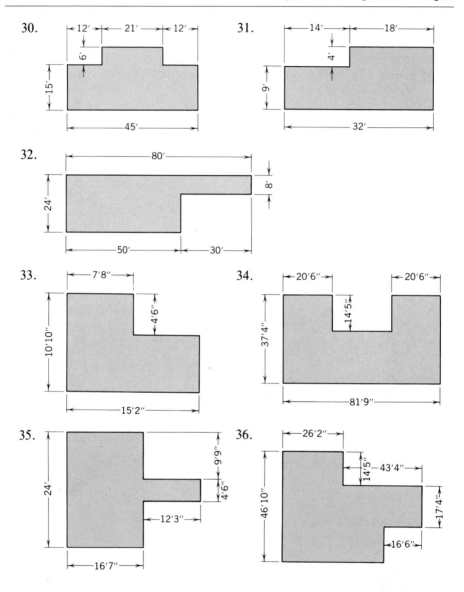

30.

31.

32.

33.

34.

35.

36.

37. Find the cost of covering the floor of a room 13'9" × 21'6" with carpet costing $12.95 a square yard.
38. How many 9"-by-9" vinyl floor tiles will be needed to cover a floor of a room 15'9" by 18'9"?
39. If the tiles in Problem 38 cost 64 cents each, what would be the cost of covering the floor in that problem?
40. If carpeting costs $13.50 a square yard and 9"-by-9" cork tiles cost 85 cents each, which floor covering would be less expensive to buy for the floor in Problem 38, and what would be the difference in cost between the two floor coverings?

7-2   The Perimeter of a Rectangle

The *perimeter* of a rectangle is the distance around the outside of the rectangle. Since a property of the rectangle is that its opposite sides are equal, the dimensions given for a particular rectangle need only be a given length and a given width. The perimeter of a rectangle then would be twice the sum of the width and length.

$$p = 2(l + w)$$

Example

Find the perimeter of a rectangle which is 5 ft long and 3 ft wide.

SOLUTION $p = 2(l + w)$
$= 2(5 \text{ ft} + 3 \text{ ft})$
$= 2(8 \text{ ft})$
$= 16 \text{ ft}$

   NOTE.   Perimeter is length or linear measure.

Problems 7-2

Fill in the perimeters in the following chart. Determine the units that best fit the information given.

RECTANGLES

	Length	Width	Perimeter
1.	15′	10′	_____
2.	12′	6′6″	_____
3.	10′	6′8″	_____
4.	12′6″	5′6″	_____
5.	9′9″	5′4″	_____
6.	3′2″	1′8″	_____
7.	2′6½″	3′7″	_____
8.	5′3¼″	3′2¾″	_____

7-3   Finding the Width or Length of a Rectangle

If the area and the length of a rectangle are known, divide the area by the length to find the width.

$$\text{width} = \frac{\text{area}}{\text{length}}$$

Example 1

The area of a rectangle is 72 sq in. and the length is 12 in. Find the width.

SOLUTION   width $= \dfrac{\text{area}}{\text{length}}$

$= \dfrac{72 \text{ sq in.}}{12 \text{ in.}}$

$= 6$ in.

If the area and the width of a rectangle are known, divide the area by the width to find the length.

$$\text{length} = \dfrac{\text{area}}{\text{width}}$$

Example 2

Find the length of a rectangle whose area is 135 sq in. and whose width is 7.5 in.

SOLUTION   length $= \dfrac{\text{area}}{\text{width}}$

$= \dfrac{135 \text{ sq in.}}{7.5 \text{ in.}}$

$= 18$ in.

Problems 7-3

Find the missing dimensions indicated in the following chart. Include the units of measure in your answers.

RECTANGLES

	Width	Length	Area
1.	5′	_____	15 sq ft
2.	_____	10′6″	126 sq ft
3.	4′8″	_____	$30\frac{1}{3}$ sq ft
4.	5′6″	_____	$47\frac{2}{3}$ sq ft
5.	_____	3′4″	15 sq ft
6.	2′6″	_____	$9\frac{1}{6}$ sq ft
7.	_____	2′9″	$9\frac{5}{8}$ sq ft
8.	3′6″	_____	$14\frac{7}{8}$ sq ft
9.	_____	12′8″	$107\frac{2}{3}$ sq ft
10.	11′6″	_____	$140\frac{7}{8}$ sq ft

Solve the following problems. Make sure that the units of measure are reasonable for each problem.

11. A floor contains 360 sq ft and is 24' long. How wide is it?
12. What is the length of a parking space 29,100 sq ft in area if it is 145'6" wide?
13. A building lot has an area of 3,125 sq ft. If it is 25'0" wide, how deep is the lot?
14. The floor of a room has an area of 308 sq ft. If the room is 22' long, how wide is it?
15. A rug 14 ft 6 in. long has an area of 152.25 sq ft. How wide is the rug?
16. A playground covers an area of 11,250 sq ft. If it is 62 ft 6 in. wide, how long is it?
17. Find the length of a factory floor 49 ft 6 in. wide if its area is 6,162.75 sq ft.
18. What is the width of a loft 82 ft 9 in. long if its area is 2,027.38 sq ft?
19. A rectangle 2 ft 4 in. wide has an area of 1,456 sq in. How long is it?
20. The floor area of a garage 65 ft by 85 ft is to be increased by 1,000 sq ft. If it is made 6 ft longer, how much will the width have to be increased?

Challenge

21. The area of a square with 6-in. sides is how many times greater than the area of a square with 3-in. sides?
22. The area of a square with 15-in. sides is how many times greater than the area of a square with 3-in. sides?
23. If you double the length of the sides of a square, then how many times is the area increased?
24. The area of a square with 8-in. sides is how many times greater than the area of a square with 2-in. sides?
25. The area of a square with 7-in. sides is how many times greater than the area of a square with 3-in. sides?
26. The area of an 8 ft by 4 ft rectangle is how many times greater than the area of a 4 ft by 2 ft rectangle?
27. The area of a 9 in. by 9 in. tile is how much smaller than the area of a 1 ft by 1 ft tile?
28. The area of a rectangle whose length is 3 ft 6 in. and whose width is 1 ft 6 in. is how many times smaller than the area of a rectangle whose length is 4 ft 6 in. and whose width is 2 ft 3 in.?
29. If the length and width of a rectangle are doubled, then its area is _____ .
30. If only the length or only the width of a rectangle is doubled, then its area is _____ .

7-4 Squares and Square Roots

In Fig. 7-5 the 1-in. square *a* contains 1 sq in.; the 2-in. square *b* contains 4 sq in.; the 5-in. square *c* contains 25 sq in. In each instance the area of the square is obtained by multiplying the length of a side by itself.

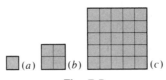

Fig. 7-5

(a) area of 1-in. square = 1 in. × 1 in. = 1 sq in.
(b) area of 2-in. square = 2 in. × 2 in. = 4 sq in.
(c) area of 5-in. square = 5 in. × 5 in. = 25 sq in.

There are short ways of saying and writing certain expressions in mathematics. For example, to show that 3 multiplied by itself (3 × 3) equals 9, you may write $3^2 = 9$. The small number 2 used in this way is called an *exponent* and indicates how many times the number is used as a factor. When the exponent of a number is 2, the resulting product is called the *square* of the number.

The square of 3 is 9, or $3^2 = 9$.
The square of 7 is 49, or $7^2 = 49$.
The square of 5 is 25, or $5^2 = 25$.

To find the length of the side of a square that contains a certain number of square inches, or the number that must be multiplied by itself to obtain a certain number, it is necessary to reverse the operation. For instance, what number multiplied by itself equals 9? Since 3 × 3 = 9, 3 is the *square root* of 9. The sign $\sqrt{\ }$ takes the place of the four words "the square root of," so that $\sqrt{9} = 3$ means "the square root of 9 equals 3."

The following table of squares and square roots should be memorized.

Squares	Square Roots
$1^2 = 1$	$\sqrt{1} = 1$
$2^2 = 4$	$\sqrt{4} = 2$
$3^2 = 9$	$\sqrt{9} = 3$
$4^2 = 16$	$\sqrt{16} = 4$
$5^2 = 25$	$\sqrt{25} = 5$
$6^2 = 36$	$\sqrt{36} = 6$
$7^2 = 49$	$\sqrt{49} = 7$
$8^2 = 64$	$\sqrt{64} = 8$
$9^2 = 81$	$\sqrt{81} = 9$
$10^2 = 100$	$\sqrt{100} = 10$

Perform the following indicated operations.

1. $11^2 =$ _____	2. $12^2 =$ _____	3. $\sqrt{169} =$ _____
4. $100^2 =$ _____	5. $\sqrt{121} =$ _____	6. $\sqrt{144} =$ _____
7. $\sqrt{10,000} =$ _____	8. $50^2 =$ _____	9. $60^2 =$ _____
10. $\sqrt{2,500} =$ _____	11. $\sqrt{3,600} =$ _____	12. $\sqrt{4,900} =$ _____

7-5 Finding the Square Root of a Whole Number

When a number is an exact square, there is generally no difficulty in finding its square root. The mechanic usually has a reference book or a calculator available to find the square root of a number. The following method may be used to find the square root of a number when other methods are not available.

Example
Find the square root of 5,625.

SOLUTION

$$
\begin{array}{r}
7\ \ 5 \\
\sqrt{56\ 25} \\
49 \\
\hline
145)\ 7\ 25 \\
7\ 25 \\
\hline
0
\end{array}
$$

EXPLANATION   Write the number and group it into pairs of figures from right to left, $\overgroup{56}\ \overgroup{25}$. The first pair of figures is 56. The largest square equal to or less than 56 is 49. The square root of 49 is 7. Write the 49 under the 56 and the 7 over the 56. Subtracting 49 from 56 leaves 7. Bring down the next pair of figures, 25, with the 7, making 725. Now double the answer already obtained: $2 \times 7 = 14$. Put this before the remainder 725 as a *trial divisor*. Disregard the last figure of the 725 and divide 14 into 72. This gives 5 for the second figure of the answer. Write the new figure 5 above the second pair, 25, and after the trial divisor, 14, making the complete divisor 145. Multiply the divisor 145 by the new figure 5; $5 \times 145 = 725$. Subtracting 725 from the divisor leaves no remainder. This shows that the exact square root of 5,625 is 75.

CHECK $75 \times 75 = 5,625.$

Find the square root of the following whole numbers.

1. $\sqrt{7225}$	2. $\sqrt{6400}$	3. $\sqrt{8836}$	4. $\sqrt{1849}$
5. $\sqrt{2025}$	6. $\sqrt{576}$	7. $\sqrt{4225}$	8. $\sqrt{784}$
9. $\sqrt{3969}$	10. $\sqrt{1024}$	11. $\sqrt{1225}$	12. $\sqrt{2809}$

7-6   The Square Root of Mixed Numbers

Numbers like 16 and 81 are called perfect squares because their square roots have no remainder. Numbers that occur in the trades usually are not perfect squares, and their square roots do have a remainder. The mechanic must decide how accurate the square root of a number should be before using the procedure to find the square root.

To obtain the square root of 56.25, or $56\frac{1}{4}$, proceed in the same manner. Group the number 56.25 into pairs of figures *to the right and left of the decimal point* and place the decimal point of the square root directly above the decimal point of the square.

$$\frac{7.5}{\sqrt{56.25}} \qquad \sqrt{56.25} = 7.5 \quad \text{or} \quad 7\tfrac{1}{2}$$

The square root of 731, for example, has a remainder after getting the second figure. Suppose the square root of 731 is to be accurate to two decimal places. Four figures after the decimal point are needed in the square to give two decimal places in the square root.

RULE _____

Two decimal places are required in the square for each decimal place in the root.

Therefore, to find the square root of 731 to two decimal places, write 731.0000 and mark off the figures into pairs to the right and left of the decimal point: $\overline{7}\ \overline{31}.\overline{00}\ \overline{00}$.

Example
Find the square root of 731 to two decimal places.

SOLUTION
$$
\begin{array}{r}
2\ 7\ .\ 0\ \ 3 \\
\sqrt{7\ 31.00\ 00} \\
4 \\
\hline
47)\overline{3\ 31} \\
3\ 29 \\
\hline
5403)\quad 2\ 00\ 00 \\
1\ 62\ 09 \\
\hline
37\ 91
\end{array}
$$

EXPLANATION   The largest square less than 7 is 4, of which the square root is 2. Write the 2 over the 7 and the 4 under the 7. Subtracting 4 from 7 leaves a remainder of 3. Bring down the next pair, 31. Double the root already obtained; $2 \times 2 = 4$, the trial divisor. 4 divides into 33 eight times, but if we try 8, we will find it too large; so we try a smaller number, 7. Write the 7 over the 31 and after the trial divisor, making the complete divisor 47. Multiply the divisor 47 by the new root figure 7; $7 \times 47 = 329$. Subtract 329 from 331, leaving the remainder 2.

Bring down the next pair of figures, 00. Double the root already obtained, 27; $2 \times 27 = 54$, the new trial divisor. 54 divides into 20 zero times. Write the 0 over the 00 and after the 54. Bring down the next pair of figures, 00, making the dividend 20,000. Dividing the trial divisor, 540, into 2,000 gives the figure 3 as the fourth figure of the root. Multiply the complete divisor 5403 by the new root figure 3; $3 \times 5,403 = 16,209$. Subtracting 16,209 from 20,000 leaves a remainder of 3,791. Because the fraction $\frac{3791}{5403}$ is more than $\frac{1}{2}$, we add 1 to the last figure of the root, making the answer 27.04. The square root of 731, accurate to two decimal places, equals 27.04. We see that after the operation of obtaining the first figure of the root, all the other figures are obtained by exactly the same process.

The important things to bear in mind are:

1. Begin at the decimal point and group the number into pairs of figures on each side of the decimal point.
2. Decide the number of decimal places the work requires.
3. Arrange your work for that number of decimal places.
4. Place the decimal point in the root directly over the decimal point in the square.

Problems 7-6

The following examples illustrate how to arrange the work for finding the square roots of various numbers.

1. Find $\sqrt{3859}$ to two decimal places. $\sqrt{\overline{38}\ \overline{59}.\overline{00}\ \overline{00}}$

2. Find $\sqrt{2}$ to three decimal places. $\sqrt{\overline{2}.\overline{00}\ \overline{00}\ \overline{00}}$

3. Find $\sqrt{713.801}$ to three decimal places. $\sqrt{\overline{7}\ \overline{13}.\overline{80}\ \overline{10}\ \overline{00}}$

4. Find $\sqrt{8.2733}$ to three decimal places. $\sqrt{\overline{8}.\overline{27}\ \overline{33}\ \overline{00}}$

5. Find $\sqrt{2733816}$ to no decimal places. $\sqrt{\overline{2}\ \overline{73}\ \overline{38}\ \overline{16}.}$

6. Find $\sqrt{0.273}$ to three decimal places. $\sqrt{\overline{0}.\overline{27}\ \overline{30}\ \overline{00}}$

7. Find $\sqrt{0.08601}$ to three decimal places. $\sqrt{\overline{0}.\overline{08}\ \overline{60}\ \overline{10}}$

8. Find $\sqrt{3\frac{5}{8}}$ to three decimal places. $\sqrt{\overline{3}.\overline{62}\ \overline{50}\ \overline{00}}$

9. Find $\sqrt{5\frac{2}{3}}$ to two decimal places. $\sqrt{5.66\ 67}$

10. Find $\sqrt{6\frac{1}{3}}$ to two decimal places. $\sqrt{6.33\ 33}$

7-7 Finding the Square Root of a Fraction

The simplest method of finding the square root of a fraction is to convert the fraction into a decimal of twice as many places as are required in the root. For instance, to get the square root of $\frac{2}{3}$ to three decimal places, convert the fraction $\frac{2}{3}$ to the decimal 0.666667; the square root of 0.666667 is 0.816; hence $\sqrt{\frac{2}{3}} = 0.816$.

To obtain the square root of $\frac{191}{3.14}$ to two decimal places, divide 191 by 3.14, carrying the work to four decimal places: $\frac{191}{3.14} = 60.8280$. The square root of 60.8280 is 7.80; hence $\sqrt{\frac{191}{3.14}} = 7.80$.

Problems 7-7

Find the square root of the following numbers.

1. $\sqrt{2\ 25}$ 2. $\sqrt{6\ 25}$ 3. $\sqrt{12\ 25}$ 4. $\sqrt{121}$
5. $\sqrt{256}$ 6. $\sqrt{169}$ 7. $\sqrt{289}$ 8. $\sqrt{361}$
9. $\sqrt{324}$ 10. $\sqrt{400}$ 11. $\sqrt{1.44}$ 12. $\sqrt{72.25}$

Try estimating the answers to the following problems before performing the operation.

13. Find the square root of 4,624.
14. Find the square root of 54,756.
15. Find the square root of 161.29.
16. Find the square root of 2,316 to two decimal places.
17. Find the square root of 8.609 to three decimal places.
18. Find the square root of 75.3 to three decimal places.
19. Find the length of the side of a square which contains 1,000 sq in.
20. Find the length of the side of a square which contains 2 sq in.
21. Find the length of the side of a square which contains 5 sq in.
22. Find the length of the side of a square which contains 10 sq ft.
23. Find the length of the side of a square which contains 35 sq ft.
24. Find the length of the side of a square which contains 50 sq ft.

7-8 Using the Calculator to Solve Square Root Problems

Many calculators, like the one shown here, have a square root function key. On these calculators, you can compute the square root of the number in the display by pressing the square root function key.

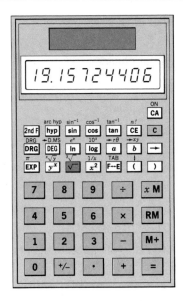

Example 1

Find the square root of 367.

SOLUTION   Use the calculator in the following manner.

The display shows

Turn on the calculator	$0.$
Enter 367	$367.$
Press $\sqrt{}$	19.15724406

Round off the solution to the number of decimal places desired. The square of 367 to two decimal places would be 19.16.

Problems 7-8

Use the calculator to solve the following problems.

1. $\sqrt{0.426}$ to three decimal places
2. $\sqrt{0.4891}$ to four decimal places
3. $\sqrt{0.0893}$ to three decimal places
4. $\sqrt{0.00524}$ to four decimal places
5. $\sqrt{0.000618}$ to four decimal places
6. $\sqrt{\frac{1}{3}}$ to three decimal places
7. $\sqrt{9\frac{3}{4}}$ to three decimal places
8. $\sqrt{2\frac{7}{16}}$ to three decimal places
9. $\sqrt{1.02}$ to three decimal places
10. $\sqrt{\frac{2}{7}}$ to three decimal places
11. $\sqrt{\frac{5}{12}}$ to three decimal places

12. $\sqrt{\frac{23}{32}}$ to three decimal places

13. $\sqrt{\dfrac{1.33}{3.14}}$ to four decimal places

14. $\sqrt{\dfrac{2763}{0.83}}$ to one decimal place

15. $\sqrt{\dfrac{11\frac{1}{8}}{96\frac{3}{4}}}$ to four decimal places

16. $\sqrt{0.4}$ to one decimal place

7-9   Applications of Square Root

A common and most useful application of square root is in finding the third side of a right triangle when the lengths of two of the sides are known. A *right triangle* is a triangle that has a right angle (90°). Of the six triangles shown in Fig. 7-6, only (*a*) and (*b*) are right triangles, because they each have a right angle.

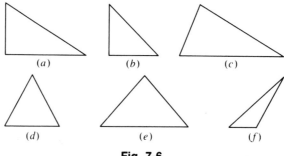

(*a*)   (*b*)   (*c*)

(*d*)   (*e*)   (*f*)

Fig. 7-6

Shown in Fig. 7-7 are the names given to the three sides of a right triangle: *base, altitude,* and *hypotenuse.*

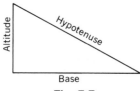

Fig. 7-7

To find the relation between the three sides, construct a right triangle with a 4-in. base and a 3-in. altitude (Fig. 7-8). Measure the hypotenuse. It should be 5 in. The area of the square built on the hypotenuse is equal to the sum of the areas of the squares built on the base and the altitude.

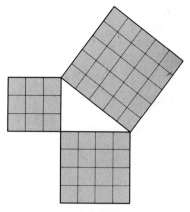

Fig. 7-8

$$\text{Base}^2 = (4 \text{ in.})^2 = 16 \text{ sq in.}$$
$$\underline{\text{Altitude}^2 = (3 \text{ in.})^2 = 9 \text{ sq in.}}$$
$$\text{Sum of squares} = 25 \text{ sq in.}$$

The square root of 25 sq in. is 5 in., the length of the hypotenuse. From this is derived the rule for the relation between the three sides of a right triangle.

RULE
_____

The square of the hypotenuse is equal to the sum of the squares of the base and the altitude.

If we call the base b, the altitude a, and the hypotenuse c, we can express the rule thus: $c^2 = a^2 + b^2$. To find c, the hypotenuse, we must square a, square b, add the squares, and get the square root of the sum. This may be expressed by the formula

$$c = \sqrt{a^2 + b^2}$$

which says that the hypotenuse c is equal to the square root of the sum of a^2 and b^2. This rule is called the ***Pythagorean Theorem.***

Example 1
Find the hypotenuse of the right triangle in Fig. 7-9.

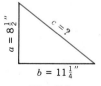

Fig. 7-9

SOLUTION $c = \sqrt{a^2 + b^2}$
$$= \sqrt{(8\tfrac{1}{2})^2 + (11\tfrac{1}{4})^2}$$
$$= \sqrt{72.25 + 126.5625}$$
$$= \sqrt{198.8125}$$
$$= 14.10$$

The length of the hypotenuse is 14.10″.

If the hypotenuse and one of the sides are known, then the other side can be found by subtracting the squares instead of adding them.
To find the altitude, use the formula

$$a = \sqrt{c^2 - b^2}$$

To find the base, use the formula

$$b = \sqrt{c^2 - a^2}$$

Example 2
Find the altitude of the right triangle in Fig. 7-10.

Fig. 7-10

SOLUTION $a = \sqrt{c^2 - b^2}$
$$= \sqrt{20^2 - 16^2}$$
$$= \sqrt{400 - 256}$$
$$= \sqrt{144} = 12$$

The altitude is 12″.

Example 3
Find the base of the right triangle in Fig. 7-11.

Fig. 7-11

SOLUTION $b = \sqrt{c^2 - a^2}$
$$= \sqrt{20^2 - 12^2}$$
$$= \sqrt{400 - 144}$$
$$= \sqrt{256} = 16$$

The base is 16″.

Summary

To find the hypotenuse of a right triangle: Square both sides, add the squares, and take the square root of the sum.

To find the base or the altitude of a right triangle: Square the hypotenuse and the given side, subtract the square of the side from the square of the hypotenuse, and take the square root of the difference.

Problems 7-9

Find the missing sides of the right triangles in the following chart. The letter *a* indicates the altitude, *b* indicates the base, and *c* indicates the hypotenuse.

RIGHT TRIANGLES

	a	b	c
1.	5″	12″	_____
2.	8″	15″	_____
3.	7″	24″	_____
4.	2′6″	6′0″	_____
5.	1′4″	_____	2′10″
6.	6″	_____	10″
7.	1′2″	_____	4′2″
8.	9″	_____	15″
9.	_____	2′0″	2′2″
10.	_____	3′9″	4′3″
11.	_____	3′0″	3′1½″
12.	_____	1′10½″	2′1½″

Solve the following problems.

13. A common problem is that of computing the diagonal of a square. How long is the diagonal of a square with 1-in. sides? A square with 2-in. sides? A square with 3-in. sides?

14. Compute the center-to-center distance between the bolt holes in Fig. 7-12.

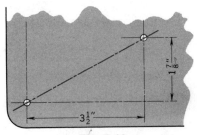

Fig. 7-12

15. In printing, good proportion calls for a page whose diagonal is twice the width. What should be the length of a page that is $3\frac{7}{8}$ in. wide?

16. Compute the overall length along the center line of the piece of work shown in Fig. 7-13.

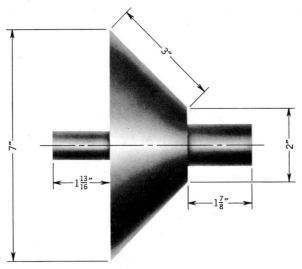

Fig. 7-13

17. Compute the length of the stair stringer shown in Fig. 7-14.

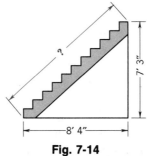

Fig. 7-14

18. Figure 7-15 shows a roof truss. Would a 20′ rafter be long enough to build this truss? (The overhang is a part of the rafter.)

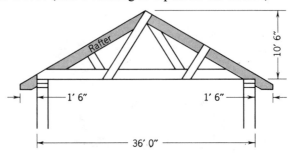

Fig. 7-15

19. In Fig. 7-16, pulley *A* is 18′4″ above and 20′8″ to the right of pulley *B*. Find the distance center to center of pulleys.

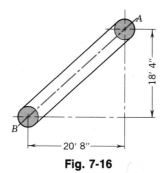

Fig. 7-16

20. In building a pier, the stringers were fastened to the piles with square-headed bolts 2″ on a side. If the heads are to be countersunk, what size holes should be bored in the stringers, allowing ½″ clearance around the bolt heads? See Fig. 7-17.

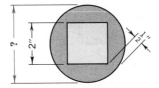

Fig. 7-17

7-10 Triangles

We shall deal with the following types of triangles:

A *right triangle* is a triangle that has one right angle (Fig. 7-18).

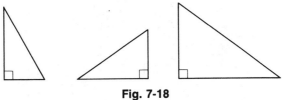

Fig. 7-18

An *isosceles triangle* is a triangle that has two equal sides and two equal angles (Fig. 7-19).

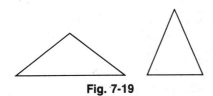

Fig. 7-19

An *equilateral triangle* is a triangle in which the three sides are equal and the three angles are equal (Fig. 7-20).

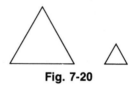

Fig. 7-20

A *scalene triangle* is a triangle that has no two sides equal (Fig. 7-21).

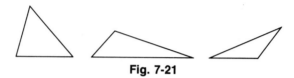

Fig. 7-21

Areas of Triangles

In the rectangle (Fig. 7-22*a*), the area is found by multiplying the altitude by the base. Area = 10 in. × 16 in. = 160 sq in. Drawing the diagonal *CF* divides the rectangle into two equal right triangles, *CEF* and *CDF*.

The area of each right triangle is ½ of 160 sq in., or 80 sq in.; ½ × 16 in. × 10 in. = 80 sq in. The area of the right triangle is therefore equal to ½ the base times the altitude. Calling the altitude *a* and the base *b*, we find:

Area of right triangle = ½ × *a* × *b*

A study of Fig. 7-22*b* shows that the same rule applies to that triangle. In fact, the following is a general formula for finding the area of any triangle when the base and the altitude are known or can be computed:

Area = ½ × *a* × *b*

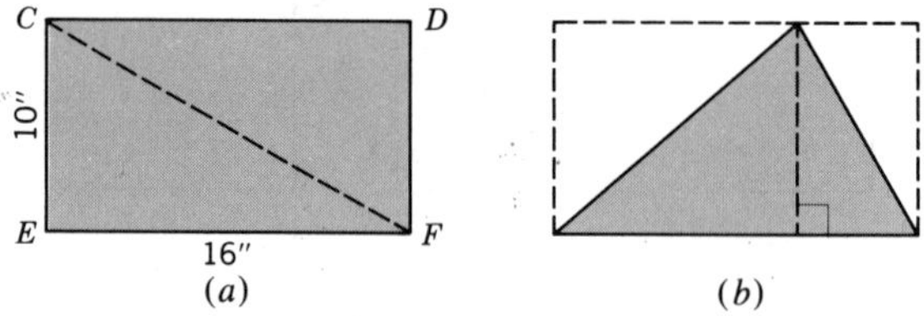

$$C \quad\quad\quad\quad\quad D$$

(a)   (b)

Fig. 7-22

It is customary to omit the sign of multiplication in mathematical formulas such as the preceding. (The expression *ab* always means *a times b*.) The formula is then written:

Area = ½*ab*

Example 1
Find the area of the right triangle shown in Fig. 7-23.

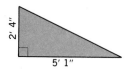

Fig. 7-23

SOLUTION $2'4'' = 28''$, the altitude
$5'1'' = 61''$, the base

$$\text{Area} = \tfrac{1}{2}ab$$
$$= \tfrac{1}{2} \times 28 \text{ in.} \times 61 \text{ in.}$$
$$= 854 \text{ sq in.}$$
$$= 5.93 \text{ sq ft}$$

Example 2
Find the area of the right triangle shown in Fig. 7-24.

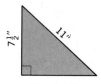

Fig. 7-24

SOLUTION   First find the length of the base and then proceed as in Example 1.

$$b = \sqrt{11^2 - (7\tfrac{1}{2})^2} = \sqrt{121 - 56.25} = \sqrt{64.75} = 8.05 \text{ in.}$$

$$\text{Area} = \tfrac{1}{2}ab$$
$$= \tfrac{1}{2} \times 7\tfrac{1}{2} \text{ in.} \times 8.05 \text{ in.}$$
$$= 30.19 \text{ sq in.}$$

Problems 7-10

Find the areas of the right triangles in the following chart. Use the appropriate formula for the relation between the sides of a right triangle to determine the missing side: $c = \sqrt{a^2 + b^2}$; $a = \sqrt{c^2 - b^2}$; $b = \sqrt{c^2 - a^2}$.

RIGHT TRIANGLES

	Side a, Altitude	Side b, Base	Side c, Hypotenuse	Area
1.	5"	11"	———	———
2.	2'0"	2'6"	———	———
3.	$3\frac{1}{2}$"	$5\frac{1}{2}$"	———	———
4.	1'0"	———	15"	———
5.	8"	———	1'5"	———
6.	$7\frac{1}{2}$"	———	$1'7\frac{1}{2}$"	———
7.	$1'4\frac{1}{2}$"	$2'1\frac{1}{4}$"	———	———
8.	———	2'0"	2'1"	———
9.	———	$10\frac{1}{2}$"	$3'1\frac{1}{2}$"	———
10.	2'6"	2'6"	———	———
11.	1.732"	———	2.000"	———
12.	$3\frac{1}{2}$"	———	7"	———

7-11 Areas of Isosceles Triangles

In isosceles and equilateral triangles, computations are easily made because a perpendicular dropped from the **vertex**, that is, the point opposite the base, divides the triangles into two equal right triangles. In Fig. 7-25 the perpendicular divides the isosceles triangle into two equal right triangles; the base of each is 16 in. and the hypotenuse of each is 20 in. The altitude is found by the usual method for finding the altitude of a right triangle:

$$\text{altitude} = \sqrt{20^2 - 16^2} = \sqrt{400 - 256} = \sqrt{144} = 12 \text{ in.}$$

The area of the isosceles triangle is found by the formula:

$$\begin{aligned}
\text{Area} &= \tfrac{1}{2}ab \\
&= \tfrac{1}{2} \times 12 \text{ in.} \times 32 \text{ in.} \\
&= 192 \text{ sq in.}
\end{aligned}$$

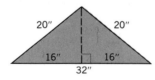

Fig. 7-25

Problems 7-11

Find the areas of the isosceles triangles in the following chart. The base and one of the equal sides is given. Use your knowledge of the relationship between the sides of a right triangle to find the altitude.

ISOSCELES TRIANGLES

	Base	One of the Equal Sides	Altitude to the Base	Area
1.	8"	5"	_____	_____
2.	2'6"	1'5"	_____	_____
3.	1'2"	2'1"	_____	_____
4.	3'0"	2'6"	_____	_____
5.	2'0"	2'0"	_____	_____
6.	1'4"	10"	_____	_____
7.	2'1"	2'6½"	_____	_____
8.	8'0"	5'0"	_____	_____
9.	19.2"	2'0"	_____	_____
10.	7.8"	7.8"	_____	_____
11.	38.4"	20.8"	_____	_____
12.	5'6"	5'6"	_____	_____

7-12 Areas of Scalene Triangles Using Hero's Formula

In a scalene triangle, the area is found by the same formula

$$\text{Area} = \tfrac{1}{2}ab$$

provided the base and altitude are known. If the altitude is not known but the three sides are known, then the area can be found by using a special method called *Hero's Formula,* shown here.

$$\text{Area of a triangle} = \sqrt{s(s - a)(s - b)(s - c)}$$

Denote the three sides by the letters a, b, and c; s denotes half the sum of the sides (semi-perimeter). The formula reads as follows: Area = the square root of the product of s, $(s - a)$, $(s - b)$, and $(s - c)$. In Fig. 7-26 call the 20-in. side a, the 14-in. side b, and the 12-in side c. Since the triangle is not known to be a right triangle, the formula $c^2 = a^2 + b^2$ cannot be used.

Example
Find the area of the scalene triangle in Fig. 7-26.

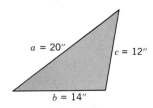

Fig. 7-26

SOLUTION   20 + 14 + 12 = 46 $s = \frac{1}{2}$ of 46 = 23

Now subtract each side in turn from this half sum (s):

$$s - a = 23 - 20 = 3$$
$$s - b = 23 - 14 = 9$$
$$s - c = 23 - 12 = 11$$

Then multiply together the half sum (s) and the three quantities obtained by taking each side in turn from the half sum: 23 × 3 × 9 × 11 = 6831. The square root of this product is the area of the triangle: $\sqrt{6,831}$ = 82.6. The area is 82.6 sq in.

$$s = \tfrac{1}{2} \text{ sum of sides} = \frac{20 + 14 + 12}{2} = \frac{46}{2} = 23$$

$$\text{Area} = \sqrt{s(s - a)(s - b)(s - c)}$$
$$= \sqrt{23(23 - 20)(23 - 14)(23 - 12)}$$
$$= \sqrt{23 \times 3 \times 9 \times 11}$$
$$= \sqrt{6,831}$$
$$= 82.6$$

Problems 7-12

Find the area of the following triangles using Hero's Formula.

1.

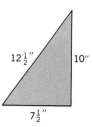

$12\frac{1}{2}''$ $10''$

$7\frac{1}{2}''$

2.

$6''$ $6''$

$6''$

3.

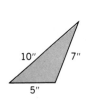

$10''$ $7''$

$5''$

4.

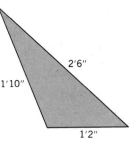

$2'6''$

$1'10''$

$1'2''$

5.

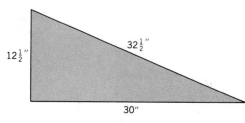

$12\frac{1}{2}''$

$32\frac{1}{2}''$

$30''$

6.

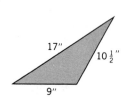

$17''$

$10\frac{1}{2}''$

$9''$

7-13 Using the Calculator to Solve Triangle Problems

The calculator demonstration given here depends on your calculator having a square key (x^2), a square root key ($\sqrt{\ }$), a memory key ($x \rightarrow M$), a memory addition key (M+), and a memory recall key (RM). Read your instruction manual if your calculator does not have these keys.

Example 1

Using a calculator, find the hypotenuse of the right triangle in Fig. 7-27. Round the answer to two decimal places.

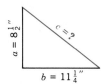

Fig. 7-27

SOLUTION   To solve this problem, use the Pythagorean Theorem and solve for the hypotenuse c. $c = \sqrt{a^2 + b^2}$

	The display shows
Turn on the calculator	$0.$
Enter 8	$8.$
Press +	$8.$
Enter 1	$1.$
Press ÷	$1.$
Enter 2	$2.$
Press =	8.5
Press x^2	72.25
Press $x \rightarrow M$	72.25 The 72.25 is now in the calculator's memory.
Enter 11	$11.$
Press +	$11.$
Enter 1	$1.$
Press ÷	$1.$
Enter 4	$4.$
Press =	11.25
Press x^2	126.5625
Press M+	126.5625 This has now been added to the 72.25 in the memory.
Press RM	198.8125 The new amount in the memory.
Press $\sqrt{\ }$	14.10008865 (answer)

Round off the length of the hypotenuse to 14.10 in.

Example 2

Using a calculator, find the altitude of the right triangle in Fig. 7-28.

Fig. 7-28

SOLUTION To solve this problem, use the Pythagorean Theorem and solve for the altitude a. $a = \sqrt{c^2 - b^2}$

The display shows

Turn on the calculator	0.
Enter 20	20.
Press x^2	400.
Press $x \rightarrow$ M	400.
Enter 16	16.
Press x^2	256.
Press +/−	−256.
Press M+	−256.
Press RM	144.
Press $\sqrt{}$	12. (answer)

The length of the altitude is 12 in.

Example 3

Using a calculator, find the area of the right triangle shown in Fig. 7-29. Round the answer to two decimal places.

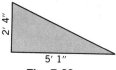

Fig. 7-29

SOLUTION In solving for the area of a triangle when the base and altitude are known or can be found, use the formula $A = .5 \times a \times b$. Convert 2 ft 4 in. to 28 in. and 5 ft 1 in. to 61 in. before solving this problem.

The display shows

Turn on the calculator	0.
Enter .5	0.5
Press $\times$	0.5
Enter 28	28.
Press $\times$	14.

Enter 61	$61.$
Press =	$854.$
Press ÷	$854.$
Enter 144	$144.$
Press =	5.930555556 (answer)

Round off the area to 5.93 sq ft.

EXPLANATION Because the numbers entered were converted to inches, the 854 means 854 sq in. It usually makes sense to reduce numbers to more manageable amounts whenever possible, so converting the answer to square feet is a good idea here. In every square foot there are 12 in. by 12 in., or 144 sq in. Dividing 854 by 144 will give the proper answer in square feet.

Example 4

Using a calculator, find the area of the right triangle shown in Fig. 7-30. Round the answer to two decimal places.

Fig. 7-30

SOLUTION Use the formula for the area of a triangle, $A = .5 \times a \times b$. To first solve for the unknown base, use the Pythagorean Theorem, $b = \sqrt{c^2 - a^2}$.

	The display shows
Turn on the calculator	$0.$
Enter 11	$11.$
Press x^2	$121.$
Press −	$121.$
Enter 7.5	7.5
Press x^2	56.25
Press =	64.75
Press $\sqrt{\ }$	8.04673847
Press $x \rightarrow M$	8.04673847
Enter .5	0.5
Press ×	0.5
Enter 7.5	7.5
Press ×	3.75
Press RM	8.04673847
Press =	30.17526926 (answer)

Round off the area to 30.18 in.

Example 5

Using a calculator, find the area of the scalene triangle in Fig. 7-31. Round the answer to one decimal place.

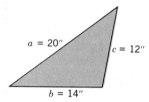

Fig. 7-31

SOLUTION   Use Hero's Formula to solve this problem.

$$\text{Area} = \sqrt{s(s - a)(s - b)(s - c)}$$

$$s = .5(a + b + c)$$

This problem requires a lot of keying in on the calculator, and the memory should be utilized to hold subtotals while the problem progresses. It is not difficult to do if the problem solution is thoroughly understood. (You may find that the work is easier to key in if some of the problem is worked out on paper and the calculator is used to do the more difficult operations.)

The display shows

Turn on the calculator	$0.$
Enter 20	$20.$ a
Press +	$20.$
Enter 14	$14.$ b
Press +	$34.$
Enter 12	$12.$ c
Press =	$46.$ $a + b + c$
Press ×	$46.$
Enter .5	0.5
Press =	$23.$ $.5(a + b + c) = s$
Press $x \rightarrow$ M	$23.$ s is now in memory.
Press −	$23.$
Enter 20	$20.$
Press =	$3.$ $s - a$
Press ×	$3.$
Press MR	$23.$
Press =	$69.$ $s(s - a)$
Press $x \rightarrow$ M	$69.$ $s(s - a)$ is now in memory, remembered as the value of s.
Enter 23	$23.$
Press −	$23.$
Enter 14	$14.$
Press =	$9.$ $s - b$

Press ×	9.
Press MR	69.
Press =	621. $s(s - a)(s - b)$
Press $x \to$ M	621. $s(s - a)(s - b)$ is now in memory, remembered as the value of s.

Enter 23	23.
Press −	23.
Enter 12	12.
Press =	11. $s - c$
Press ×	11.
Press MR	621.
Press =	6831. $s(s - a)(s - b)(s - c)$
Press $\sqrt{}$	82.64986388 (answer)

Round off the area to 82.6 sq in.

Test on Areas of Triangles

Find the areas of the following triangles. Make sure that you include the units of measure in your answer. Try to estimate the area of the triangle before doing the actual computation.

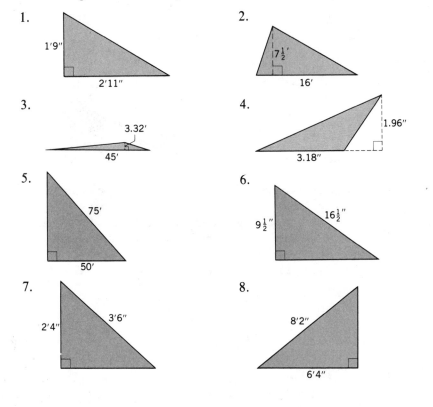

1.
1'9"
2'11"

2.
$7\frac{1}{2}$'
16'

3.
3.32'
45'

4.
1.96"
3.18"

5.
75'
50'

6.
$9\frac{1}{2}$"
$16\frac{1}{2}$"

7.
3'6"
2'4"

8.
8'2"
6'4"

9.

10.

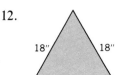

11.

12.

7-14 Angles in Triangles

Angles are defined as follows:

A *right angle* is a 90-degree angle (Fig. 7-32*a*).
An *acute angle* is an angle that is less than 90 degrees (Fig. 7-32*b*).
An *obtuse angle* is an angle that is more than 90 degrees but less than 180 degrees (Fig. 7-32*c*).

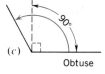

Fig. 7-32

RULE

The sum of the three angles in any triangle is equal to 180 degrees.

In a right triangle, one angle is 90 degrees. The sum of the two acute angles is $180 - 90 = 90$ degrees.

In an isosceles triangle, the angles opposite the equal sides are equal.

In an equilateral triangle, since all the sides are equal, the angles are all equal; each angle $= \frac{180}{3} = 60$ degrees.

Angles are commonly measured in *degrees, minutes,* and *seconds.*

$$1 \text{ degree} = 60 \text{ minutes}$$

$$1 \text{ minute} = 60 \text{ seconds}$$

The symbols commonly used for indicating degrees, minutes, and seconds are: degrees (°), minutes ('), seconds ("). Write 27 degrees, 18 minutes, and 30 seconds as 27°18′30″.

It is customary to label the vertices (the plural of "vertex") of a triangle A, B, and C. The angle at each vertex is then called angle A, angle B, and angle C, respectively.

Example 1
In the right triangle of Fig. 7-33, find the value of angle B.

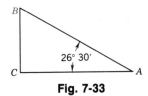

Fig. 7-33

SOLUTION
$$A + B = 90° = 89°60'$$
$$\underline{A \qquad\qquad = 26°30'}$$
$$\text{Angle } B \quad = 63°30'$$

EXPLANATION Since angles A and B together are 90°, subtract 26°30' from 90°, leaving 63°30' for angle B. In order to subtract, change 90° to 89°60'.

Example 2
In the isosceles triangle of Fig. 7-34, compute the base angles B and C.

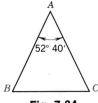

Fig. 7-34

SOLUTION
$$A + B + C = 180° = 179°60'$$
$$\underline{A \qquad\qquad\qquad = \quad 52°40'}$$
$$B + C \qquad = 127°20'$$

$$\text{Each base angle} = \frac{127°20'}{2} = \frac{126°80'}{2} = 63°40'$$

EXPLANATION Since the sum of the three angles is 180°, subtract 52°40' from 180° (changed to 179°60'), leaving 127°20' for the sum of the two base angles B and C. Each base angle is one-half of 127°20'. Changing 127°20' to 126°80' simplifies the division.

Example 3
In the scalene triangle of Fig. 7-35, angle $A = 52°40'30''$ and angle $B = 27°30'50''$. Find angle C.

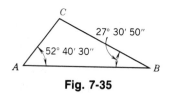

Fig. 7-35

SOLUTION $A = 52°40'30''$
                    $B = 27°30'50''$
                    _____
                    $A + B = 79°70'80''$
                          $= 80°11'20''$

Sum of three angles $= A + B + C = 180° = 179°59'60''$
                    $A + B$ $= 80°11'20''$
                    _____
                         Angle C $= 99°48'40''$

EXPLANATION   Since the sum of the three angles is 180°, add the two given angles and subtract their sum from 180°.

Example 4

In the isosceles triangle of Fig. 7-36, the base angles are 47°39′51″ each. Find the vertex angle, that is, the angle opposite the base.

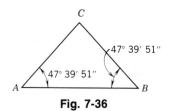

Fig. 7-36

SOLUTION   Base angle =   47°39′ 51″
                            Base angle =   47°39′ 51″
                            _____
                    Sum of base angles =   94°78′102″
                                       =   95°19′ 42″

                    Sum of three angles = 180° = 179°59′ 60″
                    Sum of base angles   =   95°19′ 42″
                    _____
                            Vertex angle =   84°40′ 18″

EXPLANATION   The base angles in an isosceles triangle are equal. To obtain the vertex angle, subtract the sum of the base angles from 180°.

Problems 7-14

Find the missing angles of the triangles in the following chart.

TRIANGLES

	Angle A	Angle B	Angle C
1.	30°	60°	_____
2.	45°	45°	_____
3.	20°	_____	80°
4.	_____	50°	40°
5.	27°10′	52°20′	_____
6.	_____	70°27′	25°15′
7.	40°45′	_____	40°45′
8.	120°30′	15°30′	_____
9.	75°50′	25°28′	_____
10.	10°37′	_____	47°29′
11.	40°27′10″	_____	56°12′20″
12.	_____	36°47′35″	72°25′42″

7-15 Using the Calculator for Angle Measure

To do problems involving degrees, minutes, and seconds on the calculator, it is necessary on many calculators to do the problems in degrees and decimal parts of a degree and then convert back to degrees, minutes, and seconds by using the → D.MS function on the calculator. There may be some slight round-off errors with the calculator, because many calculators only have 10 decimal places.

To change minutes to a decimal part of a degree, divide minutes by 60.

To change seconds to a decimal part of a degree, divide seconds by 3,600.

Example 1
In the right triangle (Fig. 7-37), find the value of angle B.

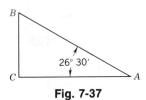

Fig. 7-37

SOLUTION   Angle B is the complement of angle A, that is, the sum of the two acute angles of a right triangle is 90 degrees (A + B = 90°).

The display shows

Turn on the calculator	0.
Enter 30	30. (minutes)
Press ÷	30.⎤ Change minutes to
Enter 60	60.⎦ decimal parts of a degree.

Press =	0.5
Press +	0.5
Enter 26	26.
Press =	26.5
Press +/−	−26.5
Press +	−26.5
Enter 90	90.
Press =	63.5
Press 2nd F	63.5
Press → D.MS	63.300000

Degrees, minutes, and seconds are read from the calculator display thus:
The numbers to the left of the decimal point indicate degrees, the first two
numbers to the right of the decimal point indicate minutes, the next two
numbers represent seconds, and the last two numbers indicate hundredths of a second.

The answer to this problem is 63 degrees 30 minutes.

Example 2

In the following isosceles triangle (Fig. 7-38), compute the base angles B
and C.

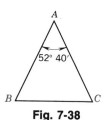

Fig. 7-38

SOLUTION   To solve this problem, we must remember that the base angles of
an isosceles triangle are equal and that the sum of the three angles of a
triangle is 180 degrees. If we subtract the vertex angle from 180 degrees,
we then have the sum of the base angles. By dividing this sum by 2 we
have the value of each base angle.

The display shows

Turn on the calculator	0.
Enter 40	40.
Press ÷	40.
Enter 60	60.
Press =	0.666666666
Press +	0.666666666
Enter 52	52.
Press =	52.66666667
Press +/−	−52.66666667
Press +	−52.66666667
Enter 180	180.

Press =	!27.3333333
Press ÷	!27.3333333
Enter 2	2.
Press =	63.66666667
Press 2nd F	63.66666667
Press → D.MS	63.400000

We read this answer as 63 degrees 40 minutes.

Example 3

In the scalene triangle (Fig. 7-39), angle A = 52°40'30" and angle B = 27°30'50". Find angle C.

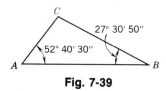

Fig. 7-39

SOLUTION   In this example, we have two angles in minutes and seconds that must be converted to decimal parts of a degree. We are going to have to use the memory of the calculator to hold one of these angles while we work on the other. To change the seconds to a decimal part of a degree, we must divide by 3600. (60 parts of a degree is a minute and 60 parts of a minute is a second.)

The sum of the three angles of the triangle is 180 degrees. We are going to add the two angles we know and subtract their sum from 180 degrees to solve our problem.

	The display shows
Turn on the calculator	0.
Enter 50	50.
Press ÷	50.
Enter 3600	3600.
Press =	0.013888888
Press $x \rightarrow$ M	0.013888888
Enter 30	30.
Press ÷	30.
Enter 60	60.
Press =	0.5
Press M+	0.5
Enter 27	27.
Press M+	27.

The angle 27 degrees 30 minutes 50 seconds is now in decimal form and is saved in memory. We will now change 52 degrees 40 minutes 30 seconds to decimal form and add it to the memory also.

Enter 30	$30.$
Press ÷	$30.$
Enter 3600	$360.$
Press =	0.008333333
Press M+	0.008333333
Enter 40	$40.$
Press ÷	$40.$
Enter 60	$60.$
Press =	0.666666666
Press M+	0.666666666
Enter 52	$52.$
Press M+	$52.$

The memory now has the sum of the two angles in decimal form. We are going to subtract the contents of memory from 180 to get the decimal form of the answer.

Enter 180	$180.$
Press −	$180.$
Press MR	80.18888889
Press =	99.81111111
Press 2nd F	99.81111111
Press →D.MS	99.484000

Read this number as 99 degrees 48 minutes 40 seconds.

Problems 7-15

Use your calculator to do Problem Set 7-14, 5–12. Check the answers you get with the D.MS key on the calculator against the results you got with pencil and paper. Practice means gaining skill!

Self-Test

Do the following problems. Check your answers with the answers in the back of the book.

1. Find the area of a rectangle whose width is 8″ and whose length is 10″.
2. Find the perimeter of the rectangle in Problem 1.
3. Find the missing dimension of the following rectangle.

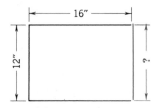

4. Find the width of a rectangle whose area is 234 sq in. and whose length is 18 in.
5. What is the square of 40?
6. What is the length of each side of a square whose area is 1,089 sq in.?
7. The legs of a right triangle are 6 in. and 8 in. What is the length of the hypotenuse?
8. One of the acute angles of a right triangle is 39°. What is the size of the other acute angle?
9. The hypotenuse of a right triangle is 12.5 in. and the altitude is 12 in., what is its area?
10. Find the area of an equilateral triangle whose sides are each 10 in.

Chapter Test

Do the following problems involving rectangles and triangles. Try to estimate the answer before doing the actual computation. Include the correct units of measure in your answer.

1. Find the area of a rectangle whose length is 15 in. and whose width is 10½ in.
2. Find the area of a rectangle whose length is 2 ft 6 in. and whose width is 1 ft 8 in.
3. Find the area of a rectangle whose length is 30½ in. and whose width is 25¾ in.
4. Find the missing dimension in the following rectangle.

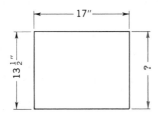

5. Find the missing dimension in the following composite rectangle.

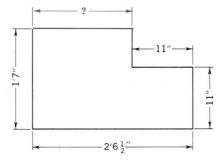

6. Find the perimeter of a rectangle whose length is 28 in. and whose width is 15 in.

7. Find the perimeter of a rectangle whose length is 2 ft $5\frac{1}{2}$ in. and whose width is 1 ft $3\frac{3}{4}$ in.

8. Find the width of a rectangle whose area is 144 sq in. and whose length is 16 in.

9. Find the length of a rectangle whose area is 256 sq in. and whose width is 2 ft 8 in.

10. Find the square of 27.

11. Find the square of $32\frac{1}{2}$.

12. Find the square root of 225.

13. Find the square root of 8,649.

14. Find the diagonal of a square 10 in. on a side.

15. Find the diagonal of a rectangle whose length is 3 ft 9 in. and whose width is 2 ft 0 in.

16. Find the hypotenuse of a right triangle whose altitude is 1 ft 4 in. and whose base is 1 ft 0 in. What is its perimeter?

17. Find the base of a right triangle whose altitude is 10 in. and whose hypotenuse is 26 in. What is its perimeter?

18. Find the altitude of a right triangle whose base is 7 in. and whose hypotenuse is 25 in.

19. Find the area of a triangle whose base is 33 in. and whose altitude is 31 in.

20. Find the area of a right triangle whose altitude is $10\frac{1}{2}$ in. and whose hypotenuse is $17\frac{1}{2}$ in.

21. Find the area of a square whose edge is $15\frac{7}{8}$ in.

22. Find the area of an isosceles triangle whose base is 40 in. and where one of the equal sides is 25 in.

23. Find the area of an equilateral triangle all of whose sides are 25 in.

24. If two angles of a triangle are 45° and 55°, how large is the remaining angle?

25. If two angles of a triangle are 50°27′ and 44°52′, how large is the third angle?

8

Regular Polygons and Circles

8-1 Definitions

A *polygon* is a closed figure in a plane. All polygons will have at least three sides. The type of polygon that occurs most often in practical work is the *regular polygon.* In a regular polygon all the sides are equal in length and all the angles formed by any two adjacent sides have the same measure. The nine shapes in Fig. 8-1 depict regular polygons.

The area of a regular polygon, its perimeter, the length of its diagonal, and the perpendicular distance between any two sides are all related to one another in length. If one is known, the others can be computed by the application of the laws for right triangles. Considerable labor is involved in this computation, but the work may be greatly shortened by means of certain ratios, called factors or constants, that have been derived for the purpose. These ratios usually appear in a *Table of Constants* such as the one that follows.

The formulas for the regular polygons that appear in the Table of Constants are accurate to 3 decimal places. Greater accuracy requires a knowledge of trigonometry.

TABLE OF CONSTANTS

Equilateral Triangle	Square

Equilateral Triangle

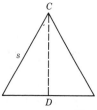

s = side

Let $CD = a$ = altitude

$$a = \frac{\sqrt{3}}{2} s \approx 0.866s$$

$$s = \frac{2}{\sqrt{3}} a \approx 1.155a$$

$$\text{Area} = \frac{\sqrt{3}}{4} s^2 \approx 0.433s^2$$

$$= \frac{1}{\sqrt{3}} a^2 \approx 0.577a^2$$

Square

s = side
d = diagonal

$$d = \sqrt{2}\, s \approx 1.414s$$

$$s = \frac{\sqrt{2}}{2} d \approx 0.707d$$

$$\text{Area} = s^2$$
$$= 0.5d^2$$

The Regular Hexagon

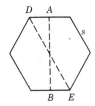

Let $AB = f$ = distance across flats

Let $DE = d$ = distance across corners (diagonal)

s = side

$$f = \frac{\sqrt{3}}{2} d \approx 0.866d \qquad f = \sqrt{3}\, s \approx 1.732s$$

$$d = 2s \qquad d = \frac{2}{\sqrt{3}} f \approx 1.155f$$

$$s = 0.5d \qquad s = \frac{1}{\sqrt{3}} f \approx 0.577f$$

$$\text{Area} = \frac{\sqrt{3}}{2} f^2 \approx 0.866f^2$$

$$= \tfrac{3}{8}\sqrt{3}\, d^2 \approx 0.650d^2$$

$$= \tfrac{3}{2}\sqrt{3}\, s^2 \approx 2.598s^2$$

The Regular Octagon

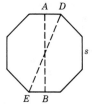

Let $AB = f$ = distance across flats

Let $DE = d$ = distance across corners (diagonal)

s = side

$$f \approx 0.924d \qquad f \approx 2.414s$$
$$d \approx 1.083f \qquad d \approx 2.613s$$
$$s \approx 0.414f \qquad s \approx 0.383d$$

$$\text{Area} \approx 0.828f^2$$

$$= \frac{\sqrt{2}}{2} d^2 \approx 0.707d^2$$

$$\approx 4.828s^2$$

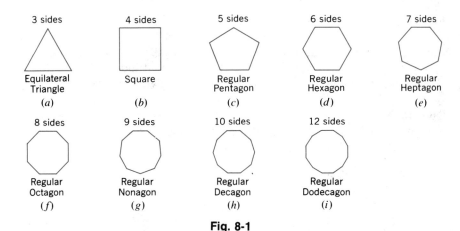

3 sides	4 sides	5 sides	6 sides	7 sides
Equilateral Triangle	Square	Regular Pentagon	Regular Hexagon	Regular Heptagon
(a)	(b)	(c)	(d)	(e)

8 sides	9 sides	10 sides	12 sides
Regular Octagon	Regular Nonagon	Regular Decagon	Regular Dodecagon
(f)	(g)	(h)	(i)

Fig. 8-1

8-2   Equilateral Triangles

An *equilateral triangle* is a regular polygon of three sides. All the sides are equal to one another. The essential parts of the equilateral triangle are the side, the altitude, the perimeter, and the area. If one of these is known, the other parts can be computed by referring to the equilateral triangle shown in the Table of Constants.

Example 1
Find the altitude of the equilateral triangle shown in Fig. 8-2.

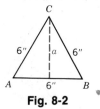

Fig. 8-2

SOLUTION $a \approx 0.866s \approx 0.866 \times 6 \approx 5.196$ in.

EXPLANATION   From the Table of Constants for the equilateral triangle, select the formula that gives the altitude, $a = 0.866s$. Substituting the value of s in this formula gives the altitude.

Example 2
Find the side of the equilateral triangle shown in Fig. 8-3.

Fig. 8-3

SOLUTION $s = 1.155a = 1.155 \times 8 = 9.24$ in.

EXPLANATION   From the Table of Constants for the equilateral triangle, select the formula that gives the side in terms of the altitude, $s = 1.155a$. Substituting the value of a (8 in.) in this formula gives the side.

Example 3
Find the area of the equilateral triangle shown in Fig. 8-4.

12″   12″

12″
Fig. 8-4

SOLUTION   Area $= 0.433s^2 = 0.433 \times 12 \times 12 = 62.352$ sq in.

EXPLANATION   From the Table of Constants for the equilateral triangle, select the formula that gives the area in terms of the side, Area $= 0.433s^2$. Substituting the value of s (12 in.) in this formula gives the area.

Example 4
Find the area of the equilateral triangle shown in Fig. 8-5.

10 cm

Fig. 8-5

SOLUTION   Area $= 0.577a^2 = 0.577 \times 10 \times 10 = 57.7$ cm^2

EXPLANATION   From the Table of Constants for the equilateral triangle, select the formula that gives the area in terms of the altitude, Area $= 0.577a^2$. Substituting the value of a (10 cm) in this formula gives the area.

Problems 8-2

Use the Table of Constants to find the missing information about equilateral triangles on the following chart. Give the answer to the nearest hundredth. Try estimating the answer before solving the problem.

EQUILATERAL TRIANGLES

	Side	Altitude	Area
1.	10″	————	————
2.	8″	————	————
3.	$4\frac{1}{2}$″	————	————
4.	5.5 cm	————	————
5.	6.25 cm	————	————
6.	————	5″	————
7.	————	11 cm	————
8.	————	$3\frac{1}{2}$″	————
9.	————	4.7 m	————
10.	————	7.75″	————

8-3 Squares

A square is a regular polygon of four sides. All the sides are equal to one another and all the angles are 90°. The essential parts of a square are its side, its perimeter, its diagonal, and its area. If one of these is known, the other parts can be computed by referring to the square shown in the Table of Constants.

Example 1

Find the diagonal of the square shown in Fig. 8-6.

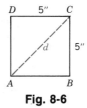

Fig. 8-6

SOLUTION $d = 1.414s = 1.414 \times 5 = 7.070$ in.

EXPLANATION   From the Table of Constants for the square, select the formula that gives the diagonal in terms of the side, $d = 1.414s$. Substituting the value of s (5 in.) in this formula gives the diagonal.

Example 2

Find the side of the square shown in Fig. 8-7.

Fig. 8-7

SOLUTION $s = 0.707d = 0.707 \times 8 = 5.656$ in.

EXPLANATION   From the Table of Constants for the square, select the formula that gives the side in terms of the diagonal, $s = 0.707d$. Substituting the value of d (8 in.) in this formula gives the side.

Example 3
Find the area of the square shown in Fig. 8-8.

6 m

6 m

Fig. 8-8

SOLUTION   Area $= s^2 = 6 \times 6 = 36$ m²

EXPLANATION   From the Table of Constants for the square, select the formula that gives the area in terms of the side, Area $= s^2$. Substituting the value of s (6 m) in this formula gives the area.

Example 4
Find the area of the square shown in Fig. 8-9.

7″

Fig. 8-9

SOLUTION   Area $= 0.5d^2 = 0.5 \times 7 \times 7 = 24.5$ sq in.

EXPLANATION   From the Table of Constants for the square, select the formula that gives the area in terms of the diagonal, Area $= 0.5d^2$. Substituting the value of d (7 in.) in this formula gives the area.

Problems 8-3

Use the Table of Constants to find the missing information about squares on the following chart. Give your answer to the nearest hundredth. Try estimating the answer before solving the problem.

SQUARES			
	Side	Diagonal	Area
1.	7 m	————	————
2.	15"	————	————
3.	2½"	————	————
4.	3¼"	————	————
5.	3.82 cm	————	————
6.	————	9 cm	————
7.	————	3½"	————
8.	————	4¼"	————
9.	————	6.5 cm	————
10.	————	3.05"	————

8-4 The Regular Hexagon

The regular hexagon has six equal sides, and all the angles formed by any two adjacent sides are equal in measure. Probably the most common use of the hexagonal shape is for bolt heads and nuts. The essential parts of a regular hexagon are its side, its diagonal, the distance across its flats, and its area. (The distance across the flats is what determines the size of an open-end wrench.)

Example 1
Find the distance across the flats (line GH or f) of the regular hexagon whose side is 5 in., as shown in Fig. 8-10.

Fig. 8-10

SOLUTION $f = 1.732s = 1.732 \times 5 = 8.66$ in.

EXPLANATION From the Table of Constants for the regular hexagon, select the formula that gives the distance across the flats (f) in terms of the side (s), $f = 1.732s$. Substituting the value of s (5 in.) in this formula gives the distance across the flats.

Example 2
Find the diagonal (line DA or d) of the regular hexagon whose side is 3 in., as shown in Fig. 8.11.

Fig. 8-11

SOLUTION $d = 2s = 2 \times 3 = 6$ in.

EXPLANATION   From the Table of Constants for the regular hexagon, select the formula that gives the diagonal (d) in terms of the side (s), $d = 2s$. Substituting the value of s (3 in.) in this formula gives the length of the diagonal.

Example 3

Find the area of the regular hexagon whose side is 4 in., as shown in Fig. 8-12.

Fig. 8-12

SOLUTION   Area $= 2.598s^2 = 2.598 \times 4 \times 4 = 41.568$ sq in.

EXPLANATION   From the Table of Constants for the regular hexagon, select the formula that gives the area in terms of the side (s), Area $= 2.598s^2$. Substituting the value of s (4 in.) in this formula gives the area of the hexagon.

Example 4

Find the diagonal (line DA) of the regular hexagon whose distance across the flats (line GH) is 3 in., as shown in Fig. 8-13.

Fig. 8-13

SOLUTION $d = 1.155f = 1.155 \times 3 = 3.465$ in.

EXPLANATION From the Table of Constants for the regular hexagon, select the formula that gives the diagonal (d) in terms of the distance across the flats (f), $d = 1.155f$. Substituting the value of f (3 in.) in this formula gives the diagonal.

Example 5
Find the area of the regular hexagon whose distance across the flats (line GH) is 6 cm as shown in Fig. 8-14.

Fig. 8-14

SOLUTION Area $= 0.866f^2 = 0.866 \times 6 \times 6 = 31.176$ cm²

EXPLANATION From the Table of Constants for the regular hexagon, select the formula that gives the area in terms of the distance across the flats (f), Area $= 0.866f^2$. Substituting the value of f (6 cm) in this formula gives the area.

Example 6
Find the area of the regular hexagon whose diagonal (line DA) is 4.5 in., as shown in Fig. 8-15.

Fig. 8-15

SOLUTION Area $= 0.650d^2 = 0.650 \times 4.5 \times 4.5 = 13.1625$ sq in.

EXPLANATION From the Table of Constants for the regular hexagon, select the formula that gives the area in terms of the diagonal (d), Area $= 0.650d^2$. Substituting the value of d (4.5 in.) in this formula gives the area.

Problems 8-4

Use the Table of Constants to find the missing information about regular hexagons on the following chart. Give your answer to the nearest hundredths. Try estimating the answer before solving the problem.

REGULAR HEXAGONS

	Side	Distance across Flats	Diagonal	Area
1.	1″	———	———	———
2.	2 cm	———	———	———
3.	2½″	———	———	———
4.	3.75″	———	———	———
5.	———	4 m	———	———
6.	———	5½″	———	———
7.	———	5.85″	———	———
8.	———	———	6 cm	———
9.	———	———	7½″	———
10.	———	———	5.25″	———

8-5   The Regular Octagon

A regular octagon has eight equal sides and all the angles formed by any two adjacent sides are equal in measure. The essential parts of a regular octagon are its side, its diagonal, the distance across its flats, and its area.

Example 1
Find the distance across the flats (line IJ) of the regular octagon whose side is 5 in., as shown in Fig. 8-16.

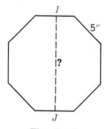

Fig. 8-16

SOLUTION $f = 2.414s = 2.414 \times 5 = 12.07$ in.

EXPLANATION   From the Table of Constants for the regular octagon, select the formula that gives the distance across the flats (f) in terms of a side, $f = 2.424s$. Substituting the value of s (5 in.) in this formula gives the distance across the flats.

Example 2
Find the length of the diagonal of a regular octagon whose side is 3 in., as shown in Fig. 8-17.

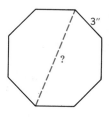

Fig. 8-17

SOLUTION $d = 2.613s = 2.613 \times 3 = 7.839$ in.

EXPLANATION   From the Table of Constants for the regular octagon, select the formula that gives the diagonal (d) in terms of the side (s), $d = 2.613s$. Substituting the value of s (3 in.) in this formula gives the length of the diagonal.

Example 3
Find the area of a regular octagon whose side is 4 in., as shown in Fig. 8-18.

SOLUTION   Area = $4.828s^2 = 4.828 \times 4 \times 4 = 77.248$ sq. in.

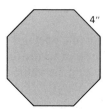

Fig. 8-18

EXPLANATION   From the Table of Constants for the regular octagon, select the formula that gives the area in terms of the side (s), Area = $4.828s^2$. Substituting the value of s (4 in.) in this formula gives the area.

Example 4
Find the length of the diagonal (line EH) of a regular octagon whose distance across the flats (line IJ) is 10 cm, as shown in Fig. 8-19.

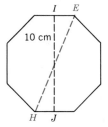

Fig. 8-19

SOLUTION $d = 1.083f = 1.083 \times 10 = 10.83$ cm

EXPLANATION   From the Table of Constants for the regular octagon, select the formula that gives the diagonal (d) in terms of the distance across the flats (f), $d = 1.083f$. Substituting the value of f (10 cm) in this formula gives the diagonal.

Example 5
Find the area of a regular octagon whose distance across the flats (line IJ) is 8 in., as shown in Fig. 8-20.

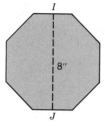

Fig. 8-20

SOLUTION   Area $= 0.828f^2 = 0.828 \times 8 \times 8 = 52.992$ sq in.

EXPLANATION   From the Table of Constants for the regular octagon, select the formula that gives the area in terms of the distance across the flats (f), Area $= 0.828f^2$. Substituting the value of f (8 in.) in this formula gives the area.

Example 6
Find the area of a regular octagon whose diagonal (line EA) is 6 in., as shown in Fig. 8-21.

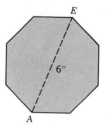

Fig. 8-21

SOLUTION   Area $= 0.707d^2 = 0.707 \times 6 \times 6 = 25.452$ sq in.

EXPLANATION   From the Table of Constants for the regular octagon, select the formula that gives the area in terms of the diagonal (d), Area $= 0.707d^2$. Substituting the value of d (6 in.) in this formula gives the area.

Problems 8-5

Find the missing parts of the regular octagons as indicated. Use the Table of Constants. Try to estimate the answer before solving each problem.

1. Find the area of a regular octagon that is 4.78″ across the flats.
2. Find the area of a regular octagon with $5\frac{1}{2}$″ sides.
3. Find the flats of a regular octagon that has 6″ sides.
4. Find the side of a regular octagon that measures 6.3″ across the flats.
5. Find the area of a regular octagon 9″ across corners.

Find the missing parts of the regular octagons in the following chart. Use the given information to find the missing parts. Try to estimate the answer before solving each problem.

REGULAR OCTAGONS

	Side	Distance across Flats	Diagonal	Area
6.	7″	_____	_____	_____
7.	$5\frac{1}{2}$″	_____	_____	_____
8.	_____	8 cm	_____	_____
9.	_____	$2\frac{1}{2}$″	_____	_____
10.	_____	_____	8 cm	_____
11.	_____	_____	$5\frac{1}{4}$″	_____
12.	3.02″	_____	_____	_____
13.	4.75″	_____	_____	_____

8-6 Practice Problems on the Use of the Table of Constants

Use the Table of Constants to find the missing parts of the regular figures as indicated in the following problems. Make a sketch of the figure and show the given part. Organize your computations and label all parts of the problem. Try to estimate the answer to each problem before solving it.

1. Find the area of an equilateral triangle with $4\frac{1}{2}$″ sides.
2. Find the area of an equilateral triangle whose altitude is $4\frac{3}{4}$″.
3. Find the diagonal of a 1.75″ square.
4. Find the side of a square whose diagonal is 3.45″.
5. Find the area of a regular hexagon that is 10″ across flats.
6. Find the area of a regular hexagon with $1\frac{3}{4}$″ sides.
7. Find the area of a regular hexagon that measures $\frac{7}{8}$″ across corners.
8. Find the side of an equilateral triangle whose altitude is $8\frac{1}{4}$″.
9. Find the altitude of an equilateral triangle with $5\frac{1}{2}$″ sides.
10. Find the area of a square whose diagonal is $7\frac{1}{2}$″.

11. Find the area of a regular hexagon that is $8\frac{1}{4}''$ across flats.
12. Find the diagonal of a regular octagon that is 6.03″ across flats.
13. Find the area of an equilateral triangle with 3.16″ sides.
14. Find the area of a regular hexagon with 1.09″ sides.
15. A sharp V-thread (Fig. 8-22) has the cross section of an equilateral triangle. If the pitch or distance AB is $\frac{1}{12}''$, find the depth CD.

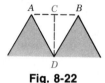

Fig. 8-22

16. What should be the diameter of a piece of round stock (Fig. 8-23) if a $1\frac{3}{4}''$ square is to be milled on the end of it?

Fig. 8-23

17. Eight bolt holes are to be drilled evenly spaced on the circumference of a 9″ circle, as in Fig. 8-24. What will be the distance center to center of holes?

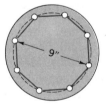

Fig. 8-24

18. A wrench (Fig. 8-25) is required to be made that will fit a hexagon nut which measures $2\frac{1}{4}''$· across corners. What must be the width of the opening between the jaws of the wrench? Allow 0.02″ for clearance.

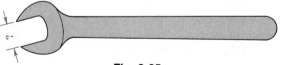

Fig. 8-25

19. How many feet of railing are required around a regular octagonal platform that measures 38′6″ across flats?

20. Find the corner measurement (Fig. 8-26) for laying out a regular octagon on a piece of 2½″ square stock.

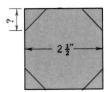

Fig. 8-26

21. A piece of 1¾″ round stock is to have a regular hexagonal end milled on it, as in Fig. 8-27. Find the distance across flats of the hexagon.

Fig. 8-27

22. Figure 8-28 represents a room with a bay window in the shape of a half of a regular hexagon. The distance across *AB* is 2.6 m. Find the distance *CD*.

23. What is the largest square that can be cut from a piece of ⅞″ round stock (Fig. 8-29)?

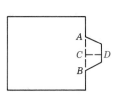
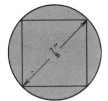

Fig. 8-28   **Fig. 8-29**

8-7   Regular Polygons and the Calculator

To solve problems involving regular polygons with the calculator, we will be using the square root key ($\sqrt{\ }$) and the reciprocal key ($1/x$). On some calculators, the reciprocal key is a second function, and you must press the 2nd F key to activate the reciprocal function key. Read your manual to see what your sequence of keys must be to use these functions.

The problems done here will have the number of decimal places to match the example problems in order to show that slight differences will occur because of the accuracy of the calculator.

Example 1

Using a calculator, find the altitude of the equilateral triangle shown in Fig. 8-30. Round off your answer to three decimal places.

Fig. 8-30

SOLUTION   Use the formula $a = \dfrac{\sqrt{3}}{2} s$.

	The display shows
Turn on the calculator	$0.$
Enter 3	$3.$
Press $\sqrt{}$	1.732050808
Press $\div$	1.732050808
Enter 2	$2.$
Press $\times$	0.866025403
Enter 6	$6.$
Press $=$	5.196152423 (answer)

Round off this answer to 5.196 in.

Example 2

Using a calculator, find the side of the equilateral triangle shown in Fig. 8-31. Round off your answer to two decimal places.

Fig. 8-31

SOLUTION   Use the formula $s = \dfrac{2}{\sqrt{3}} a$.

	The display shows
Turn on the calculator	0.
Enter 2	2.
Press ÷	2.
Enter 3	3.
Press $\sqrt{}$	1.732050808
Press ×	1.154700538
Enter 8	8.
Press =	9.237604307 (answer)

Round off this answer to 9.24 in.

Example 3

Using a calculator, find the area of the equilateral triangle shown in Fig. 8-32. Round off your answer to three decimal places.

Fig. 8-32

SOLUTION   Use the formula $A = \dfrac{\sqrt{3}}{4} s^2$.

	The display shows
Turn on the calculator	0.
Enter 3	3.
Press $\sqrt{}$	1.732050808
Press ÷	1.732050808
Enter 4	4.
Press ×	0.433012701
Enter 12	12.
Press x^2	144.
Press =	62.35382907 (answer)

Round off this answer to 62.354 sq. in.

Example 4

Using a calculator, find the area of the equilateral triangle shown in Fig. 8-33. Round off your answer to one decimal place.

Fig. 8-33

SOLUTION   Use the formula $A = \dfrac{1}{\sqrt{3}}\, a^2$.

The display shows

Turn on the calculator	0.
Enter 3	3.
Press $\sqrt{}$	1.732050808
Press 2nd F	1.732050808
Press $1/x$	0.577350269
Press $\times$	0.577350269
Enter 10	10.
Press x^2	100.
Press $=$	57.73502692 (answer)

Round off this answer to 57.7 cm².

Problems involving squares, hexagons, and octagons use similar procedures.

Problems 8-7

Do Problem Sets 8-3, 8-4, and 8-5 again using the calculator. Compare the two sets of answers.

8-8 Quadrilaterals

Figures with four sides are called *quadrilaterals.* The various types of quadrilaterals are shown in Fig. 8-34.

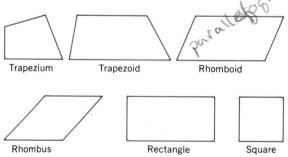

Fig. 8-34

A *parallelogram* is a quadrilateral that has two pairs of parallel sides. The types of parallelograms are:

The *rectangle,* in which all the angles are right angles.

The *square,* in which all the angles are right angles and the four sides are equal.

The *rhomboid,* which has two pairs of parallel sides but no right angles.

The *rhombus,* which has four equal sides but no right angles.

The *trapezoid* is a quadrilateral in which only two of the sides are parallel. The *isosceles trapezoid* is a trapezoid in which the two nonparallel sides are equal.

The *trapezium* is a quadrilateral that has no two sides parallel; in other words, an irregular four-sided figure.

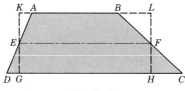

Fig. 8-35

8-9   Area of a Trapezoid

The area of the trapezoid *ABCD* (Fig. 8-35) can be found by the following method. Draw a center line parallel to and midway between the upper base *AB* and the lower base *CD*. Where this center line *EF* meets the sides, draw perpendiculars to it, forming the rectangle *KLHG*. The rectangle *KLHG* is equal in area to the trapezoid *ABCD* because the two triangles that are cut off the trapezoid in forming the rectangle are equal to the triangles that are gained. Triangle *KAE* is equal to triangle *EGD*, and triangle *BLF* is equal to triangle *FCH*.

The area of a rectangle is found by multiplying the base by the altitude. To find the area of the rectangle *KLHG*, multiply the height *LH* by the width *EF*. But the area of the trapezoid *ABCD* is equal to the area of the rectangle *KLHG*. The area of the trapezoid can be found by multiplying the height by the width along the center line *EF*. Since *EF* is half-way between the upper and the lower bases, it is equal to their average, or half their sum.

$$\text{Area of trapezoid} = \text{average width} \times \text{altitude} = \left(\frac{b_1 + b_2}{2}\right) a$$

Example 1
Find the area of the trapezoid in Fig. 8-36.

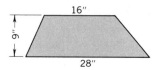

Fig. 8-36

SOLUTION Area = average width × altitude = $\left(\dfrac{b_1 + b_2}{2}\right)a$

$$\text{Average width} = \frac{16 + 28}{2} = \frac{44}{2} = 22 \text{ in.}$$

$$\text{Area} = 22 \times 9$$

$$= 198 \text{ sq in.}$$

Example 2
Find the altitude of the isosceles trapezoid in Fig. 8-37.

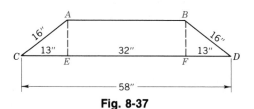

Fig. 8-37

SOLUTION $CE + FD = 58 - 32 = 26$ in.

$$CE = \tfrac{26}{2} = 13 \text{ in.}$$

$$AE = \sqrt{\text{hypotenuse}^2 - \text{base}^2}$$

$$= \sqrt{16^2 - 13^2}$$

$$= \sqrt{256 - 169} = \sqrt{87} = 9.33 \text{ in.}$$

EXPLANATION The perpendiculars AE and BF cut off equal distances CE and FD on either side of EF. But since $EF = AB = 32$ in., then

$$CE = FD = \frac{58 - 32}{2} = \frac{26}{2} = 13 \text{ in.}$$

In triangle ACE, the base is 13 in. and the hypotenuse is 16 in. Hence the altitude = $\sqrt{16^2 - 13^2} = 9.33$ in.

Problems 8-9

Find the average width and the area of each trapezoid in the following chart.

TRAPEZOIDS

	Upper Base	Lower Base	Average Width	Altitude	Area
1.	10 cm	14 cm	12	8 cm	56
2.	8″	10½″		10″	
3.	6½″	8″		7½″	
4.	5½″	6½″		4″	
5.	5.5″	7.75″		6.2″	
6.	1′2″	1′8″		8″	
7.	1′6½″	1′10″		1′3″	
8.	2′6½″	3′2½″		2′7½″	
9.	4.02″	8.07″		6.02″	
10.	5.61″	7.22″		5.05″	

Find the areas of the following trapezoids.

11.
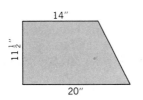

12.
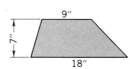

13.
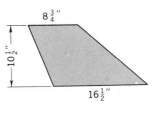

14.
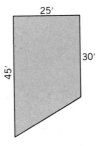

15.
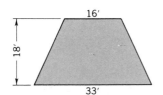

16.
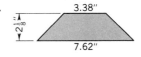

17.
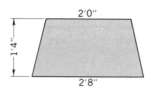

18.
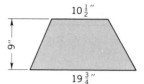

Challenge Find the altitudes and areas of the following isosceles trapezoids.

19.
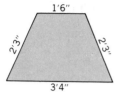

20.
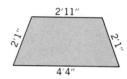

21.
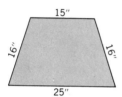

22.
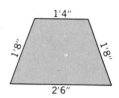

8-10 Areas of Composite Figures

It is sometimes necessary to find the area of a figure that does not belong to any of the classes of figures studied. If the figure can be divided into other shapes—the areas of which can be found by the rules already learned—then the area of the figure can be found.

Example 1

Find the area of the figure in Fig. 8-38.

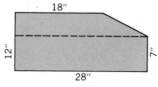

Fig. 8-38

SOLUTION This figure can be divided into a trapezoid and a rectangle by the dotted construction line as shown.

The altitude of the trapezoid = 12 − 7 = 5 in.

The average width of the trapezoid = $\dfrac{18 + 28}{2} = \dfrac{46}{2} = 23$ in.

Area of trapezoid = 23 × 5 = 115 sq in.
Area of rectangle = 28 × 7 = 196 sq in.

Total area = 311 sq in.

Example 2

Find the area of the figure in Fig. 8-39.

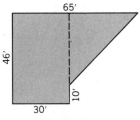

Fig. 8-39

SOLUTION   The dotted construction line divides the figure into a rectangle and a triangle.

$$\text{Altitude of triangle} = 46 - 10 = 36 \text{ ft}$$
$$\text{Base of triangle} = 65 - 30 = 35 \text{ ft}$$

$$\text{Area of triangle} = \tfrac{1}{2} \times 35 \times 36 = \quad 630 \text{ sq ft}$$
$$\text{Area of rectangle} = 30 \times 46 \qquad = \underline{1,380 \text{ sq ft}}$$
$$\text{Total area} = 2,010 \text{ sq ft}$$

Problems 8-10

Find the area of the following composite figures. Make a sketch of each figure and show how it can be divided into figures whose areas can be found. Label all parts of your computations.

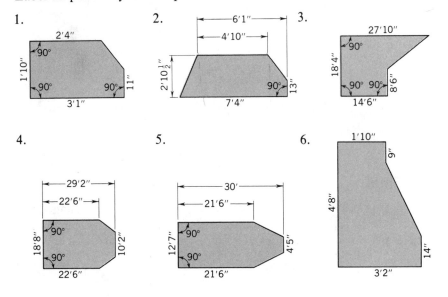

7.

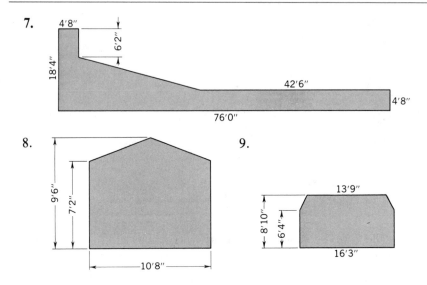

8.   **9.**

8-11 Scale

Diagrams of objects are frequently drawn smaller than the objects they represent. For example, the top of an end table 3 ft long and 1 ft 3 in. wide may be represented by a diagram 3 in. long and $1\frac{1}{4}$ in. wide (Fig. 8-40). Every inch in the diagram would therefore represent a foot length in the table. And, since the ratio of 1 in. to 1 ft is 1 to 12, the diagram is drawn to a scale of $\frac{1}{12}$, 1 : 12, or 1 in. = 1 ft. In general, scale means the ratio of any dimension on the drawing to the corresponding dimension on the object represented.

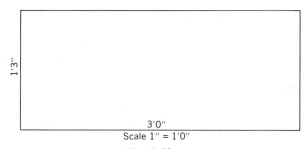

Scale 1″ = 1′0″

Fig. 8-40

8-12 Finding the Drawing Measure

If a drawing is to be made to a scale of $\frac{1}{4}$, it means that every distance on the object is to be represented by a line one-fourth as long. An object 8 ft $4\frac{1}{2}$ in. long drawn to a $\frac{1}{4}$ scale would be represented by a diagram 2 ft $1\frac{1}{8}$ in. in length, or $\frac{1}{4}$ of 8 ft $4\frac{1}{2}$ in.

To find the length of any line on the diagram, multiply the dimension on the object by the scale ratio.

Problems 8-12

Find the lengths of the following drawing measures.

1. A screwdriver 10″ long is drawn to a scale of $\frac{1}{2}$. What will be the length of the line in the drawing that represents the length of the screwdriver?

2. A wrench $10\frac{1}{2}$″ long is drawn to $\frac{1}{4}$ scale. What will be the length of the line in the drawing that represents the length of the wrench?

3. A box 5′4″ long, 4′8″ wide, and 3′0″ high is drawn to a scale of $\frac{1}{8}$. Compute the lengths of the lines in the diagram that will represent the length, width, and height of the box.

4. The length of an object is 2′4″. Find the length of the corresponding line in a diagram drawn to: (*a*) $\frac{1}{2}$ size, (*b*) $\frac{1}{4}$ size, (*c*) a scale of $\frac{3}{8}$, (*d*) a scale of $\frac{1}{8}$″ = 1′, (*e*) a scale of $\frac{1}{8}$.

5. A drawing of an object is made to a scale of 3 : 4. Find the lengths of the lines in the drawing that will represent the following measurements on the object: (*a*) 1′6″, (*b*) $6\frac{1}{2}$″, (*c*) $3′5\frac{1}{2}$″, (*d*) $11\frac{3}{4}$″, (*e*) 14″.

6. A drawing is to be made to a scale of 1″ = 1′. Compute the dimensions in the drawing that will correspond to the following dimensions on the object: (*a*) 3′9″, (*b*) 2′3″, (*c*) 2′6″, (*d*) 18″, (*e*) 4′0″.

8-13 Finding the Actual Dimension on the Object from the Measured Dimension on the Diagram

If the scale of a diagram is $\frac{1}{8}$, it means that the length of a line on the object is 8 times as long as the corresponding line in the diagram. Thus if a line in a diagram measures $1\frac{3}{4}$ in., the corresponding dimension on the object will be 8 × $1\frac{3}{4}$ = 14 in.; or $1\frac{3}{4}$ in. ÷ scale ratio = $1\frac{3}{4}$ ÷ $\frac{1}{8}$ = $1\frac{3}{4}$ × $\frac{8}{1}$ = 14 in.

To find the measure on an object corresponding to the length of a line on the diagram, divide the diagram measure by the scale ratio.

NOTE. Great care and good judgment must be used when measuring a drawing to determine the actual measure of an object. Temperature, humidity, the reproduction process, the type of paper, and so on all have an effect on the actual lengths of the lines on a drawing. It is best to consider lengths found this way to be a good approximation only.

Problems 8-13

Find the actual dimensions of the following objects drawn to scale by measuring the diagram and dividing the diagram measure by the scale ratio.

1. The diagram in Fig. 8-41 is drawn to a scale of $\frac{3}{4}$. The length of the diagram measures $2\frac{15}{32}''$ and the width measures $1\frac{5}{16}''$. Find the corresponding dimensions of the object.

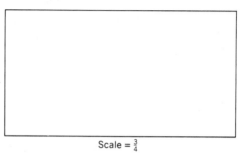

Scale = $\frac{3}{4}$

Fig. 8-41

2. Measure the lines in the diagram in Fig. 8-42 and compute the actual dimensions of the object.
3. Measure the lines in the diagram in Fig. 8-43 and compute the actual dimensions of the object.

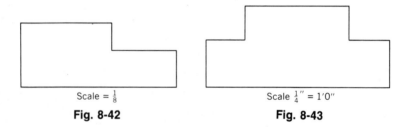

Scale = $\frac{1}{8}$

Fig. 8-42

Scale $\frac{1}{4}'' = 1'0''$

Fig. 8-43

4. Measure the lines in the diagram in Fig. 8-44 and compute the actual dimensions of the object.

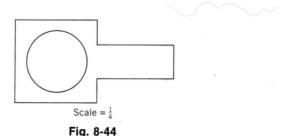

Scale = $\frac{1}{4}$

Fig. 8-44

5. Measure the long and short diameters of the ellipse in Fig. 8-45 and compute the actual diameters of the object represented.

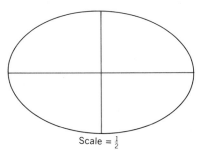

Scale = $\frac{1}{2}$

Fig. 8-45

8-14 Finding the Area of a Figure Drawn to a Certain Scale

Measure each line in the diagram and compute the actual lengths of the corresponding lines in the object. Then proceed to find the area represented according to the rules given for the various figures.

Example
Find the area of the surface represented by the rectangle in Fig. 8-46.

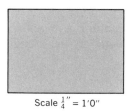

Scale $\frac{1}{4}'' = 1'0''$

Fig. 8-46

SOLUTION Measuring the length and width of the rectangle in the figure, the dimensions are $1\frac{1}{4}$ in. and $\frac{7}{8}$ in., respectively. Since the scale is $\frac{1}{4}$ in. = 1 ft, the scale ratio is $\dfrac{\frac{1}{4} \text{ in.}}{12 \text{ in.}} = \frac{1}{4} \times \frac{1}{12} = \frac{1}{48}$. In other words, the length of a line on the object is 48 times the corresponding length on the diagram. Hence the actual length of the object = $48 \times 1\frac{1}{4}$ in. = $48 \times \frac{5}{4}$ in. = 60 in., and the actual width of the object is $48 \times \frac{7}{8}$ in. = 42 in.

The actual area of the surface represented = $60 \times 42 = 2,520$ sq in. = 17.5 sq ft.

Problems 8-14

Find the areas of the various surfaces represented in the following diagrams.

1. Find the area represented by the rectangle in Fig. 8-47 if $AB = 1\frac{3}{8}''$ and $BC = \frac{3}{4}''$.

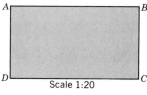
Scale 1:20

Fig. 8-47

2. Find the actual diameter of a sanding disc represented by a circle whose diameter is $1\frac{1}{4}''$ when drawn to a scale of $\frac{1}{8}$.

3. The base of the right triangle in Fig. 8-48 is $1\frac{1}{8}''$ and the altitude is $\frac{5}{8}''$. Find the actual area represented by the diagram.

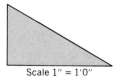
Scale 1" = 1'0"

Fig. 8-48

4. Measure on the diagram in Fig. 8-49 the base and altitude and compute the actual area represented by the triangle.

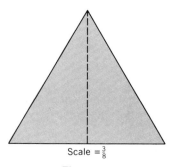
Scale $= \frac{3}{8}$

Fig. 8-49

5. Measure the altitude and bases of the trapezoid in Fig. 8-50 and compute the area represented by the diagram.

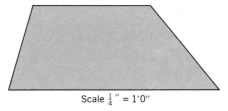

Scale $\frac{1}{4}'' = 1'0''$

Fig. 8-50

8-15   Circles

The perimeter of a circle is called the ***circumference.*** The distance from the center to any point on the circle is called the ***radius.***

The distance across the circle through the center is called the ***diameter.*** The diameter is equal to two radii (see Fig. 8-51).

A 12-in. pulley means a pulley 12 in. in diameter. Similarly, a piece of $\frac{1}{2}$-in. round stock means a piece of stock $\frac{1}{2}$ in. in diameter. A 10-in. circle is a circle whose diameter is 10 in.

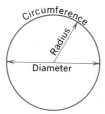

Fig. 8-51

8-16   Circumference

It has been found by measurement and computation that the ratio of the circumference of a circle to its diameter is a constant represented by the Greek letter π (pronounced *pie*). The value of π is nearly equal to 3.1416, and this is the approximation usually used in shop work involving π. If you press the π key on a ten-digit display calculator, you will see that π is given as 3.141592654. The problems in this book will use 3.1416 as a value for π.

The relation between the diameter and the circumference of a circle is expressed by the formula

$$c = \pi d$$

where c is the circumference of the circle and d is its diameter.

Example

Find the circumference of a 16-in. circle.

SOLUTION   Circumference $= \pi d$
$$= 3.1416 \times 16$$
$$= 50.2656$$

Problems 8-16

Find the circumference of the following circles. (Use 3.1416 as an approximation of π.)

1. Find the circumference of a 14″ circle.
2. Find the circumference of a 5-cm circle.
3. Find the circumference of a 2″ circle.
4. Find the circumference of a circle whose radius is 0.5 in.
5. Find the circumference of a $2\frac{1}{8}$″ circle.
6. Find the circumference of a $1\frac{7}{8}$″ circle.
7. Find the circumference of a circle whose radius is 1.625 in.
8. Find the circumference of a circle 4′9″ in diameter.
9. Find the circumference of a circle 135′ in diameter.
10. What is the circumference of a 32″ pulley?
11. What must be the length of a label to fit around a can 3″ in diameter, allowing $\frac{1}{4}$″ for pasting?
12. What must be the length of a bar of steel to make a rim for a carriage wheel 4′2″ in diameter, allowing 1″ for weld?
13. An emery wheel is 8″ in diameter. How many inches will a point on the circumference travel while the wheel makes one revolution?
14. What distance does an automobile travel with each complete turn of the wheels if the LR78-15 tires measure 28 in. in diameter?
15. A circular running track in a gymnasium has an average diameter of 70 ft. Find the circumference.
16. A water tank is 10′6″ in diameter. Find the length of the tie rods around it, allowing 6″ for fastenings.
17. What must be the length of a belt connecting two 16″ pulleys that are 3′8″ center to center?

8-17   Using the Calculator to Find Circumference

The π key on the ten-digit calculator shows π to be 3.141592654. When you use the calculator to solve practical problems, you will have to determine the accuracy needed and where to round off your answers.

Example

Using a calculator, find the circumference of a 16-in. circle. Round off your answer to two decimal places. $C = \pi d$

SOLUTION

The display shows

Turn on the calculator	$0.$
Enter 16	$16.$
Press ×	$16.$
Press π	3.141592654
Press =	50.26548246 (answer)

Round off this answer to 50.27 in.

Problems 8-17

Using a calculator, do all of the problems in Problem Set 8-16 again. Notice the difference in the answers that results from using a more accurate value for π.

8-18 Finding the Diameter of a Circle

When the circumference is given, the diameter is found by dividing the circumference by π. The formula is

$$d = \frac{C}{\pi}$$

Example

Find the diameter of a circle whose circumference is 42.5".

SOLUTION $d = \dfrac{C}{\pi}$

$$= \frac{42.5}{3.1416} = 13.5281 \text{ in.}$$

Problems 8-18

Find the diameters of the following circles. (Use 3.1416 as an approximation of π.)

1. Find the diameter of a circle whose circumference is 12 m.
2. Find the diameter of a circle whose circumference is 280′.
3. Find the diameter of a circle whose circumference is 3.73″.
4. Find the diameter of a circle whose circumference is 3′9″.
5. Find the diameter of a circle whose circumference is 11″.
6. What is the diameter of a shaft if the distance around the shaft is $23\frac{1}{2}$″?
7. What is the pitch diameter of a gear if the circumference of the pitch circle is 20.42″?

8. The circumference of a pulley is 75.4". Find its diameter.
9. What is the diameter of a piece of round stock whose circumference is 8.53"?
10. What is the diameter of a cylindrical tank if the distance around the tank is 62.5'?
11. One lap around a circular running track is 220 yd. What is the diameter of the track?
12. What must be the diameter of an emery wheel if a point on the circumference travels 20.41" for each revolution?
13. The circumference of the trunk of a tree measures 7'4". What is its diameter?

8-19   Finding the Area of a Circle: Method 1

To find the area of a circle, you can multiply the square of the radius by π. The formula is

$$\text{Area} = \pi r^2$$

Example
Find the area of an 11-in. circle.

SOLUTION   Area $= \pi r^2$ $r = \frac{11}{2} = 5.5$ in.
$= 3.1416 \times 5.5^2$
$= 3.1416 \times 30.25 = 95.0334$ sq in.

Problems 8-19

Fill in the following chart. Use 3.14 for π if you are not using a calculator. In either case, estimate the area before solving each problem.

CIRCLES

	Diameter	Radius	Estimated Area	Computed Area
1.	2"	_____	_____	_____
2.	14"	_____	_____	_____
3.	65"	_____	_____	_____
4.	$4\frac{1}{2}$"	_____	_____	_____
5.	$3\frac{3}{8}$"	_____	_____	_____
6.	$1\frac{1}{4}$"	_____	_____	_____
7.	1.85"	_____	_____	_____
8.	3.6"	_____	_____	_____
9.	5'9"	_____	_____	_____
10.	108'	_____	_____	_____

Pg 218

11. A sheet of steel $\frac{1}{8}''$ thick and 1' square weighs 5.08 lb. What is its weight after an 8" hole is cut in it?
12. The top of a gas tank 72'8" in diameter is to be painted at a cost of $1.75 per square yard. Find the cost.
13. Find the area of a piston head $3\frac{3}{4}''$ in diameter.
14. Find the total pressure on the bottom of a cylindrical tank 8'4" in diameter if the pressure is 920 lb per square foot.
15. A splice plate 18" × 6" has 10 rivet holes each $\frac{7}{8}''$ in diameter punched in it. What is the net area of the plate?

8-20 Finding the Area of a Circle: Method 2

To find the area of a circle, multiply the square of the diameter by $\frac{\pi}{4}$. The formula is

$$\text{Area} = \frac{\pi d^2}{4}$$

Since $\frac{\pi}{4} = \frac{3.1416}{4} = 0.7854$, write the formula thus:

$$\text{Area} = 0.7854 d^2$$

Example
Find the area of a 15-cm circle.

SOLUTION   Area = $0.7854 d^2$
                    = 0.7854×15^2
                    = $0.7854 \times 225 = 176.715$ cm^2

Problems 8-20

Find the area of the following circles using Method 2.

1. Find the area of a 5" circle.
2. Find the area of a 16-cm circle.
3. Find the area of a 72' circle.
4. Find the area of a $5\frac{1}{2}''$ circle.
5. Find the area of a 9" circle.
6. Find the area of a $2\frac{1}{2}''$ circle.
7. Find the area of a circle 2.64" in diameter.
8. Find the area of a circle 5.8" in diameter.
9. Find the area of a circle 6'3" in diameter.
10. Find the net area of a manhole cover 26" in diameter with 13 vent holes each 1" in diameter.
11. Find the area of the doorway with the semicircular top represented in Fig. 8-52.

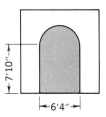

Fig. 8-52

12. Find the area of the bottom of the clothes boiler with semicircular ends illustrated in Fig. 8-53.

Fig. 8-53

8-21   Using the Calculator to Find the Area of a Circle

Example 1

Using a calculator, find the area of an 11-in. circle. Round off your answer to two decimal places.

$$\text{Area} = \pi r^2 \qquad r = \tfrac{11}{2} = 5.5 \text{ in.}$$

SOLUTION

	The display shows
Turn on the calculator	0.
Press π	3.141592654
Press $\times$	3.141592654
Enter 5.5	5.5
Press x^2	30.25
Press $=$	95.03317777 (answer)

Round off to 95.03 sq. in.

Example 2

Using a calculator, find the area of a 15-cm circle. Round off your answer to one decimal place.

$$\text{Area} = \frac{\pi}{4} d^2$$

SOLUTION

The display shows

Turn on the calculator	0.
Press π	3.141592654
Press ÷	3.141592654
Enter 4	4.
Press ×	0.785398163
Enter 15	15.
Press x^2	225.
Press =	176.7145868 (answer)

Round off to 176.7 cm².

Problems 8-21

Use either $A = \pi r^2$ or $A = \dfrac{\pi}{4} d^2$ to find the areas of the circles in Problem Sets 8-19 and 8-20. Compare the two sets of answers and notice the difference when using a more accurate value of π.

8-22 Finding the Diameter of a Circle When the Area Is Given: Method 1

To find the diameter of a circle when the area is known, you can use the formula

$$d = 2\sqrt{\frac{\text{area}}{\pi}}$$

This formula indicates the following operations:

1. Divide the area by 3.1416.
2. Find the square root of the quotient.
3. Multiply the square root of the quotient by 2.

Example
Find the diameter of a circle whose area is 36.8 sq in.

SOLUTION $\quad d = 2\sqrt{\dfrac{\text{area}}{\pi}}$

$$= 2\sqrt{\frac{36.8}{3.1416}}$$

$$= 2\sqrt{11.7138}$$

$$= 2 \times 3.42 = 6.84 \text{ in.}$$

Find the diameter of the following circles whose areas are given.

1. Find the diameter of a circle whose area is 100 m².
2. Find the diameter of a circle whose area is 5,000 sq ft.
3. Find the diameter of a circle whose area is 5.78 sq in.
4. Find the diameter of a circle whose area is 1.19 sq in.
5. Find the diameter of a circle whose area is 30.5 sq in.
6. What is the diameter of a circular disc whose area is $5\frac{1}{4}$ sq in.?
7. A round steel rod can safely bear a pull of 16,000 lb per square inch. What must be the diameter of a rod to carry 125,000 lb?
8. What must be the diameter of a rivet to have a cross-section area of 0.6 sq in.?
9. A copper wire is required to have a cross-section area of $\frac{1}{2}$ sq in. What must be the diameter?
10. What must be the diameter of a pipe to have an area of 4.78 sq in.?
11. What must be the diameter of a cylinder if the base area is required to be 125 sq in.?

8-23   Finding the Diameter of a Circle When the Area Is Given: Method 2

To find the diameter of a circle when the area is known, use the formula

$$d = \sqrt{\frac{\text{area}}{0.7854}}$$

This formula indicates the following operations:

1. Divide the area by 0.7854.
2. Find the square root of the quotient.

Example
Find the diameter of a circle whose area is 25.5 sq in.

SOLUTION $d = \sqrt{\dfrac{\text{area}}{0.7854}} = \sqrt{\dfrac{25.5}{0.7854}}$

$= \sqrt{32.47} = 570$ in.

Find the diameter of the following circles using Method 2.

1. Find the diameter of a circle whose area is 80 sq in.
2. Find the diameter of a circle whose area is 1,000 sq ft.
3. Find the diameter of a circle whose area is 14.62 sq in.

4. Find the diameter of a circle whose area is 3.35 sq in.
5. Find the diameter of a circle whose area is 41.8 sq in.
6. Find the diameter of a circle whose area is 7.5 sq in.
7. What must be the diameter of a steel wire to have a cross-section area of $\frac{1}{8}$ sq in.?
8. A circular wooden post is required to have a cross-section area of 20 sq in. What must the diameter be?
9. What must be the diameter of a steel rod whose cross-section area is $2\frac{1}{2}$ sq in.?
10. The base area of a cylindrical tank is 250 sq ft. What is its diameter?

8-24 Finding the Diameter of a Circle Equal in Area to the Combined Areas of Two or More Circles

This involves the following processes:

1. Find the area of each circle.
2. Find the sum of the areas.
3. Find the diameter of the circle whose area is equal to the sum of the areas of the individual circles.

Example 1
Find the diameter of a circle whose area is equal to the combined areas of a 3-in. circle and a 4-in. circle.

SOLUTION   Area of 3-in. circle = 3.1416 × 1.5 × 1.5 =   7.07 sq in.
Area of 4-in. circle = 3.1416 × 2 × 2   = 12.56 sq in.
Combined areas = 19.63 sq in.

$$d = 2\sqrt{\frac{area}{\pi}} = 2\sqrt{\frac{19.63}{3.1416}} = 2\sqrt{6.2484} = 2 \times 2.50 = 5$$

The diameter of the required circle is 5 in.

Example 2
Find the diameter of a circle whose area is equal to the combined areas of three 5-in. circles and two 4-in circles.

SOLUTION   Area of 5-in. circle = 3.1416 × 2.5 × 2.5
= 19.64 sq in.

Area of three 5-in. circles =   3 × 19.64 =   58.90 sq in.

Area of 4-in. circle = 3.1416 × 2 × 2
= 12.5664 sq in.

Area of two 4-in. circles = 2 × 12.5664 =   25.1328 sq in.
Combined areas =   84.03 sq in.

$$d = 2 \sqrt{\frac{\text{area}}{\pi}} = 2 \sqrt{\frac{84.03}{3.1416}} = 2\sqrt{26.7475} = 2 \times 5.17$$

$$= 10.34$$

The diameter of the required circle is 10.34 in.

Example 3

Find the diameter of a circle whose area is equal to the difference between the area of an 8-in. circle and the area of a 4-in. circle.

SOLUTION   Area of 8-in. circle = 3.1416 × 4 × 4
                                        = 50.26 sq in.

Area of 4-in. circle = 3.1416 × 2 × 2
                                        = 12.57 sq in.

Difference between the two areas = 50.26 − 12.57
                                                        = 37.69 sq in.

$$d = 2 \sqrt{\frac{\text{area}}{\pi}} = 2 \sqrt{\frac{37.69}{\pi}} = 2\sqrt{12} = 6.93 \text{ in.}$$

The diameter of the required circle is 6.93 in.

Problems 8-24

Use your knowledge of finding diameters to solve the following problems.

1. Find the diameter of a circle equal in area to a 6″ circle and a 4″ circle.
2. Find the diameter of a circle equal in area to the combined areas of a 2″ circle, a 3″ circle, and a 5″ circle.
3. Find the diameter of a circle whose area is equal to the area of six circles each 1½″ in diameter.
4. What must be the diameter of a circle whose area is equal to the sum of the areas of three 1″ circles and three ½″ circles?
5. What must be the diameter of a circle whose area is equal to the sum of the areas of ten 2½″ circles?
6. Two 3″ pipes run together into one pipe. What should be the diameter of the large pipe to carry off the flow from the other two pipes?
7. A 2″ pipe and another pipe are used to carry off the flow from a 3″ pipe. Find the size of the other pipe.
8. How many 1″ pipes are required to carry as much water as a 2″ pipe?
9. How many air ducts 10″ in diameter are required to have the same cross-section area as one air duct of rectangular cross section 24″ × 40″?
10. How many copper wires of No. 20 B. & S. gage (0.032″ diameter) are required to have the same cross-section area as a No. 0000 B. & S. gage wire whose diameter is 0.46″?

8-25   Short Method of Comparing Areas of Circles

It is convenient to make use of a shorter method of comparing the areas of circles. The area of a 5-in. square is equal to the combined areas of a 4-in. square and a 3-in. square (see Fig. 8-54).

$$3^2 + 4^2 = 5^2$$
$$9 + 16 = 25$$

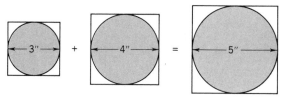

Fig. 8-54

To find the size of a square equal in area to the combined areas of two or more squares, we add those areas and find the square root of the sum:

$$\sqrt{3^2 + 4^2} = \sqrt{25} = 5 \text{ in.} \quad \text{the size of square required}$$

This same rule holds for circles, that is, *the areas of circles bear the same ratio to each other as the squares of their diameters*. The area of a 5-in. circle is equal to the combined areas of a 4-in. circle and a 3-in. circle. Instead of actually computing the areas of the circles, square their diameters and get the square root of the sum of the squares.

Example
Find by the short method the diameter of a circle whose area is equal to the combined areas of a 6-in. circle, an 8-in. circle, and a 9-in. circle.

SOLUTION $\text{Diameter} = \sqrt{6^2 + 8^2 + 9^2}$

$$= \sqrt{36 + 64 + 81}$$

$$= \sqrt{181}$$

$$= 13.45 \text{ in.}$$

A circle 13.45 in. in diameter has an area equal to the sum of the areas of a 6-in. circle, an 8-in. circle, and a 9-in. circle.

Problems 8-25

Find the diameter of a circle equal in area to the sum of the areas of the circles in each problem of Problem Set 8-24 by using the short method of comparing areas of circles. Compare the results with those obtained by the long method.

8-26   Areas of Ring Sections

RULE

To find the area of a ring section, such as Fig. 8-55, subtract the area of the inside circle from the area of the outside circle.

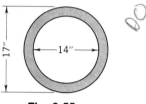

Fig. 8-55

Example
Find the area of the cross section of a pipe whose inside diameter is 14 in. and outside diameter 17 in.

SOLUTION   Area of 17-in. circle = 3.1416 × 8½ × 8½ = 226.98 sq in.
            Area of 14-in. circle = 3.1416 × 7 × 7   = 153.94 sq in.
                                        Net area =   73.04 sq in.

The area of the ring section is 73.04 sq in.

Problems 8-26

Find the areas of the following ring sections.

1. Find the area of a flat ring whose outside diameter is 11″ and inside diameter 7″.
2. Find the area of cross section of a brass pipe 2½″ inside diameter, if the metal is ¼″ thick.
3. How many square feet of pavement are there in a path around a fountain 32′ in diameter if the path is 6′6″ wide?
4. What is the area of cross section of a hollow cast iron column if the outside diameter is 14″ and the metal is 1¼″ thick?
5. At $8.55 per square yard, find the cost of paving a circular walk whose inside diameter is 125′. The walk is 8′6″ wide.
6. Find the area of a washer whose inside diameter is ½″ and whose outside diameter is ⅝″.
7. How many square feet of sod are required for a circular lawn 150′ in diameter if the center of the lawn is occupied by a fountain 35′ in diameter?
8. What is the cross-section area of a pipe whose inside and outside diameters are, respectively, 3½″ and 3¾″?

9. A flat steel ring $12\frac{1}{2}''$ in diameter has a $2\frac{1}{2}''$ hole in the center. What is its area?

10. What is the cross-section area of a steel collar $\frac{3}{16}''$ thick whose inside diameter is $3\frac{3}{8}''$?

8-27 Short Method of Finding the Areas of Ring Sections

There is a short method for finding the area of a ring section.

RULE

Multiply the average circumference of the ring by the thickness of the ring.

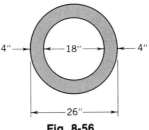

$4''$ → ←$18''$→ ←$4''$

←$26''$→

Fig. 8-56

In Fig. 8-56 the inside diameter is 18 in. and the outside diameter is 26 in. The thickness of the ring is, therefore, $\dfrac{26 - 18}{2} = \dfrac{8}{2} = 4$ in. The average diameter is half the sum of the inside diameter and the outside diameter: $\dfrac{26 + 18}{2} = \dfrac{44}{2} = 22$ in. The circumference of a 22-in. circle is $3.1416 \times 22 = 69.12$ in.

$$\begin{aligned}
\text{Area of ring section} &= \text{average circumference} \times \text{thickness} \\
&= 69.12 \times 4 \\
&= 276.46 \text{ sq in.}
\end{aligned}$$

Example

Find the area of a flat ring 26 in. inside diameter and 33 in. outside diameter.

SOLUTION   Area = average circumference × thickness

$$\text{Average diameter} = \frac{26 + 33}{2} = \frac{59}{2} = 29\tfrac{1}{2} \text{ in.}$$

$$\text{Thickness} = \frac{33 - 26}{2} = \frac{7}{2} = 3\tfrac{1}{2} \text{ in.}$$

$$\text{Area} = 3.1416 \times 29\tfrac{1}{2} \times 3\tfrac{1}{2}$$

$$= 324.4 \text{ sq in.}$$

Find the area of each ring section in Problem Set 8-26 by using the short method. Compare the results with those obtained by the longer method.

8-28   Arcs and Sectors of Circles

A part of a circle such as *AOB* in Fig. 8-57 is called a **sector** of the circle. A part of the circumference such as *AB* is called an **arc**. The sector is bounded by the two radii *OA* and *OB* and the arc *AB*. The angle *AOB* at the center is

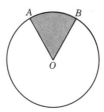

Fig. 8-57

called the **central angle** of the sector. A central angle is measured by the same number of degrees as its intercepted arc. A 40° sector means a sector whose central angle is 40°. Since one degree of arc means $\frac{1}{360}$ of the circumference of the circle, the length of a 40° arc is $\frac{40}{360}$ of the circumference of the circle.

$$\text{Length of arc} = \pi d \left(\frac{x}{360°} \right)$$

where x is the number of degrees of the arc.

$$\text{Area of Sector} = \pi r^2 \left(\frac{x}{360°} \right)$$

where x is the number of degrees of the sector.

Example 1
Find the length of a 35° arc of a 16-in. circle.

SOLUTION   Circumference of 16-in. circle = 3.1416 × 16 = 50.26 in.
          Length of 35° arc = $\frac{35}{360}$ × 50.26 = 4.89 in.

Example 2
Find the area of a 40° sector of a 16-in. circle.

SOLUTION   Area of 16-in. circle = 3.1416 × 8 × 8 = 201.06 sq in.
          Area of 40° sector = $\frac{40}{360}$ × 201.06 = 22.33 sq in.

EXPLANATION   Since a 1° sector contains $\frac{1}{360}$ of the area of the circle, a 40° sector contains $\frac{40}{360}$ of the area of the circle.

Find (*a*) the length of arc and (*b*) the area of sector for each of the following.

1. 60° sector of a 9″ circle
2. 27° sector of a 3½″ circle
3. 94° sector of a 45′ circle
4. 118°30′ of a 4½″ circle
5. 36°30′ sector of a circle 2′8½″ in diameter
6. 90° sector of a 20″ circle
7. 45° sector of a 24″ circle
8. 30° sector of a 3′ circle
9. 15° sector of a circle 18″ in diameter
10. 40°45′ sector of a circle 2′4″ in diameter

8-29 Circles and Regular Figures

A circle is inscribed in another figure when it touches each of the sides of that figure. If a circle is inscribed in a figure, then the sides of that figure are *tangent* to the circle. In Fig. 8-58 circles are inscribed in an equilateral triangle, a square, a regular hexagon, and a regular octagon.

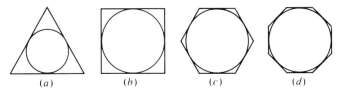

(*a*)　　　(*b*)　　　(*c*)　　　(*d*)

Fig. 8-58

A circle is circumscribed about a figure when all the corners of that figure touch the circle. In Fig. 8-59 the circle is circumscribed about the triangle, square, and so forth. The figures are inscribed in the circle.

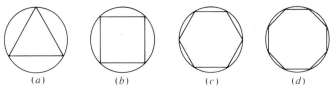

(*a*)　　　(*b*)　　　(*c*)　　　(*d*)

Fig. 8-59

Most of the relations between the circle and the inscribed and circumscribed figures are evident from an inspection of the figure. In Fig. 8-58*b* the diameter of the inscribed circle equals the side of the square. In Figs. 8-58*c* and 8-58*d* the diameters of the inscribed circles are equal to the distance across the flats of the regular hexagon and the regular octagon, respectively. In Fig. 8-59*b* the diameter of the circumscribed circle is equal to the diagonal of the square; in Figs. 8-59*c* and 8-59*d* the diameters of the circles are equal to the distance across the corners of the regular hexagon and the regular octagon, respectively.

Figure 8-60 illustrates the relation between an equilateral triangle and its inscribed and circumscribed circles. The diameter *DB* of the outer circle is divided by the construction into four equal parts: $DC = CO = OA = AB$.

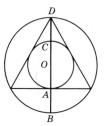

Fig. 8-60

Hence *AC*, the diameter of the inscribed circle, equals $\frac{2}{3}$ of the altitude of the triangle.

Also *AD*, the altitude of the triangle, equals $\frac{3}{4}$ of the diameter of the circumscribed circle.

The expression $2:3:4$ states the relation between these figures.

Example 1

Find the area of a square inscribed in a 9-in. circle (see Fig. 8-61).

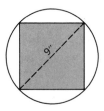

Fig. 8-61

SOLUTION   The diagonal of the square is equal to the diameter of the circle, 9 in.

$$\text{Side of square} = 0.707 \times s = 0.707 \times 9 = 6.363 \text{ in.}$$
$$\text{Area of square} = 6.363 \times 6.363 = 40.49 \text{ sq in.}$$

Example 2

Find the area of a circle inscribed in a regular octagon with $3\frac{1}{2}$-in. sides (see Fig. 8-62).

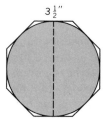

Fig. 8-62

SOLUTION   The diameter of the circle is equal to the distance across the flats of
the octagon.

$$\text{Flats of octagon} = 2.414 \times s = 2.414 \times 3\tfrac{1}{2} = 8.45 \text{ in.}$$
$$\text{Area of circle} = 3.1416 \times 4.23 \times 4.23 = 56.2 \text{ sq in.}$$

Example 3
Find the area of an equilateral triangle inscribed in a $5\tfrac{1}{4}$-in. circle (see Fig.
8-63).

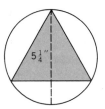

Fig. 8-63

SOLUTION   The altitude of the triangle is equal to $\tfrac{3}{4}$ of the diameter of the
circle.

$$\text{Altitude of triangle} = \tfrac{3}{4} \times 5\tfrac{1}{4} = 3.94 \text{ in.}$$
$$\text{Area of triangle} = 0.577 \times a^2$$
$$= 0.577 \times 3.94 \times 3.94 = 8.88 \text{ sq in.}$$

Problems 8-29

Do the following problems. Draw a neat diagram for each. Label all known
parts. Organize your work as shown in the example problems. Refer to the
Table of Constants on page 172 for the needed ratios.

1. Find the area of a square inscribed in a $6\tfrac{1}{2}''$ circle.
2. Find the area of an equilateral triangle inscribed in a $35'$ circle.
3. Find the area of a regular hexagon inscribed in a $1\tfrac{1}{4}''$ circle.
4. Find the area of a regular octagon inscribed in a $4''$ circle.
5. Find the area of a circle inscribed in a $3\tfrac{1}{2}''$ square.

6. Find the area of a circle inscribed in an equilateral triangle whose sides are 4.72".
7. Find the area of a circle inscribed in a regular octagon with $1\frac{1}{2}$" sides.
8. Find the area of a circle inscribed in a regular hexagon with $5\frac{3}{4}$" sides.
9. Find the flats of a regular hexagon inscribed in a $\frac{3}{4}$" circle.
10. Find the flats of a regular octagon inscribed in a 12" circle.
11. A circle is inscribed in a regular hexagon with $2\frac{1}{2}$" sides. What is the diameter of the circle?
12. Eight holes equally spaced are drilled on the circumference of a $12\frac{1}{2}$" circle. What is the straight-line distance between centers of holes?
13. What is the distance between holes (center to center) if three holes, equally spaced, are drilled on the circumference of a $3\frac{1}{4}$" circle?
14. Find the depth of cut required to mill a square end on a $1\frac{3}{4}$" round shaft.
15. Find the depth of cut required to mill a triangular end on a round shaft $\frac{7}{8}$" in diameter.

8-30   Segments of Circles

A portion of the area of a circle such as the shaded part in Fig. 8-64 is called a *segment* of a circle. It is bounded by an arc and a straight line that joins the ends of the arc. A straight line such as *AB,* that joins the ends of an arc, is called a *chord.* The angle *BOA* formed at the center of the circle by the radii to the ends of the arc is called a *central angle.* A 65-degree segment means a segment whose central angle is 65 degrees.

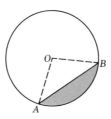

Fig. 8-64

By means of the relations between circles and inscribed figures presented in the previous section, the area of certain segments can be found.

Example
Find the area of a 90° segment of a 24-in. circle (Fig. 8-65).

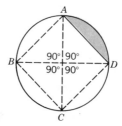

Fig. 8-65

SOLUTION   The diagonal of the square is equal to the diameter of the circle, 24 in.

Side of square = 0.707 × d = 0.707 × 24 in. = 16.97 in.
Area of square = 16.97 × 16.97 = 288.0 sq in.
Area of 24-in. circle = 3.1416 × 12 × 12 = 452.4 sq in.

Subtract area of square from area of circle and divide by 4.

Area of circle = 452.4 sq in.
Area of square = 288.0 sq in.
Area of 4 segments = 164.4 sq in.

Area of one segment = $\dfrac{164.4}{4}$ = 41.10 sq in.

EXPLANATION   There are four 90° angles in a circle. Drawing these four angles and drawing chords between the ends of the arcs, construct the square *ABCD*. Subtracting the area of this square from the area of the circle will leave four segments like the one whose area we are trying to find. Dividing the result by 4 gives the area of one segment.

Problems 8-30

Find the areas of the following segments of circles. Draw net diagrams. Organize your work as shown in the example problem.

1. Find the area of a segment bounded by a chord and a 120° arc of an 8″ circle.
2. Find the area of a segment bounded by a chord and a 90° arc of a 16″ circle.
3. Find the area of a segment bounded by a chord and a 60° arc of a 25″ circle.
4. Find the area of a segment bounded by a chord and a 45° arc of a 36″ circle.
5. In milling a square end on a 1½″ round shaft, how many square inches of cross-section area are removed?

8-31 The Ellipse

The main components of the *ellipse* (see Fig. 8-66) are the *major axis, AB* (also called the long diameter), the *minor axis, CD* (also called the short diameter), the circumference or perimeter, and the area.

The area of the ellipse is found by the formula

$$\text{Area} = \pi ab$$

where $a = \frac{1}{2}$ the long diameter and $b = \frac{1}{2}$ the short diameter.

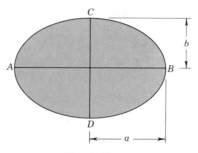

Fig. 8-66

Example 1
Find the area of an ellipse 16 in. long and 12 in. wide.

SOLUTION   Area $= \pi ab$

$$a = \tfrac{16}{2} = 8 \qquad b = \tfrac{12}{2} = 6$$

Area $= 3.1416 \times 8 \times 6 = 150.7$ sq in.

To find the perimeter of an ellipse, use the following approximate formula:

$$\text{Perimeter} = \pi\sqrt{2(a^2 + b^2)}$$

which states the perimeter is equal to π times the square root of twice the sum of a^2 and b^2. The several steps in the process are:

1. Substitute in the formula the values of π, a, and b.
2. Square a and b.
3. Add these squares.
4. Multiply the sum of the squares by 2.
5. Find the square root of the product.
6. Multiply the square root of the product by 3.1416.

Example 2
Find the perimeter of an ellipse with a long diameter of 18 in. and a short diameter of 14 in. (Fig. 8-67).

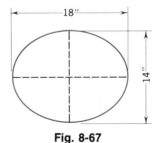

Fig. 8-67

SOLUTION Perimeter $= \pi\sqrt{2(a^2 + b^2)}$

$$= 3.1416\sqrt{2(9^2 + 7^2)}$$

$$= 3.1416\sqrt{2(81 + 49)}$$

$$= 3.1416\sqrt{2 \times 130}$$

$$= 3.1416\sqrt{260}$$

$$= 3.1416 \times 16.12$$

$$= 50.64 \text{ in.}$$

Problems 8-31

Find the areas and perimeters of the ellipses in the following chart. Use 3.14 for π. Round each answer to the nearest hundredth. Try to estimate the answer before solving the problem.

ELLIPSES

	Long Diameter	Short Diameter	a	b	Area	Perimeter
1.	8'	5"				
2.	36"	10"				
3.	$6\frac{1}{2}$"	$4\frac{1}{4}$"				
4.	$7\frac{3}{4}$"	$5\frac{1}{2}$"				
5.	4.2"	2.6"				
6.	10.8"	5.4"				
7.	65'	43'				
8.	5'2"	3'4"				
9.	$16\frac{1}{2}$"	14.8"				
10.	5.62"	3.57"				

8-32 Summary of Formulas for Plane Figures

RIGHT TRIANGLE (Fig. 8-68)

altitude (a) $a = \sqrt{c^2 - b^2}$
base (b) $b = \sqrt{c^2 - a^2}$
hypotenuse (c) $c = \sqrt{a^2 + b^2}$

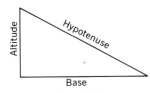

Fig. 8-68

TRIANGLE   (Fig. 8-69)

area (A) $A = \frac{1}{2}ba$
side (s) $A = \sqrt{s(s - a)(s - b)(s - c)}$
altitude (a)
base (b)
hypotenuse (c)

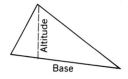

Fig. 8-69

RECTANGLE   (Fig. 8-70)

area (A) $A = lw$
length (l)
width (w)

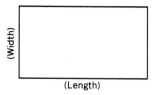

Fig. 8-70

TRAPEZOID   (Fig. 8-71)

area (A)
altitude (a) $\text{average width} = \dfrac{b_1 + b_2}{2}$
upper base (b_1)
lower base (b_2) $A = \left(\dfrac{b_1 + b_2}{2}\right) a$

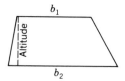

Fig. 8-71

CIRCLE   (Fig. 8-72)
   circumference (C) $C = \pi d$
   area (A) $A = \pi r^2$
   diameter (d)
   radius (r) $A = \dfrac{\pi d^2}{4}$
   pi (π)
   arc length (l_a) $d = \dfrac{C}{\pi}$
   arc degrees (deg_a)
   sector area (A_s)
   sector degrees (deg_s) $d = 2\sqrt{\dfrac{A}{\pi}}$

$$d = \frac{4A}{\pi}$$

$$l_a = \left(\frac{deg_a}{360}\right) C$$

$$A_s = \left(\frac{deg_s}{360}\right) A$$

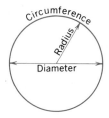

Fig. 8-72

FLAT RING   (Fig. 8-73)
   area (A) average circumference $= \dfrac{d_1 + d_2}{2}$
   inside diameter (d_1)
   outside diameter (d_2)
   thickness (t) $A = \left(\dfrac{d_1 + d_2}{2}\right) t$

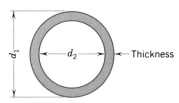

Fig. 8-73

ELLIPSE   (Fig. 8-74)
  area (*A*) $A = \pi ab$
  ½ long diameter (*a*)
  ½ short diameter (*b*) $a = \dfrac{A}{\pi b}$
  perimeter (*P*)
  pi (*π*)

                                $b = \dfrac{A}{\pi a}$

                                $P = \pi\sqrt{2(a^2 + b^2)}$

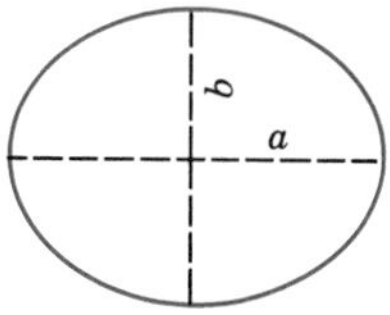

Fig. 8-74

Self-Test

Do the following problems. Check your answers with the answers in the back of the book. Use a copy of the Table of Constants (page 172) and the Summary of Formulas for Plane Figures (pages 219–222) when taking this test. Use 3.1416 for *π* and round your answers to the nearest hundredth. Try to estimate the answer before solving the problem.

1. Find the altitude of an equilateral triangle whose side is 6 in.
2. Find the area of a square whose diagonal is 6 in.
3. Find the distance across the flats of a regular hexagon whose diagonal is 4 in.
4. Find the circumference of a 6-in. circle.
5. Find the area of a 10-in. circle.
6. Find the diameter of a circle which is equal in area to two 10-in. circles.
7. Find the area of a flat ring whose outside diameter is 1 in. and whose inside diameter is ½ in.
8. Find the area of a 60° sector of a 6-in. circle.
9. Find the area of an ellipse with a 12-in. long diameter and a 10-in. short diameter.
10. Find the perimeter of an ellipse with a 6-in. long diameter and a 4-in. short diameter.

Chapter Test

Use a copy of the Table of Constants (page 172) and the Summary of Formulas for Plane Figures (pages 219–222) when taking this test. Use 3.1416 as an approximation for *π*. Make a neat diagram to accompany each problem. Organize your work.

1. Find the altitude of an equilateral triangle whose side is 9".
2. Find the area of an equilateral triangle whose altitude is 12".
3. Find the area of a square whose diagonal is 8".
4. Find the distance across the flats of a regular hexagon whose side is 10".
5. Find the diagonal of a regular octagon whose side is 20".
6. Find the area of a trapezoid whose upper base is 10", lower base 14", and altitude 8".
7. A screwdriver 15" long is drawn to $\frac{1}{4}$ scale. What will be the length of the line in the drawing that represents the length of this screwdriver?
8. Find the circumference of a $10\frac{1}{2}$" circle.
9. Find the diameter of a circle whose perimeter is 30".
10. Find the area of a $5\frac{1}{2}$" circle.
11. Find the diameter of a circle whose area is 1,000 sq in.
12. Find the diameter of a circle whose area is equal to the sum of the areas of two 5" circles.
13. Find the area of a flat ring whose outside diameter is 12" with inside diameter 10".
14. Find the length of the arc of a 60° sector of a 10" circle.
15. Find the area of the sector described in Problem 14.
16. Find the area of a square inscribed in a 15" circle.
17. Find the area of a square circumscribed about a 15" circle.
18. Find the area of a segment bounded by a chord and a 120° arc of a 12" circle.
19. Find the area of an ellipse with a 15" long diameter and a 10" short diameter.
20. Find the perimeter of the ellipse described in Problem 19.

9

Solids

9-1 Definitions

A solid has three dimensions—length, width, and height. This chapter will concern itself with the volume and surface area of some of the common solids that appear frequently in practical work. The vocational and technical student is expected to understand these solids.

9-2 Prisms and Cylinders

Prisms are solid figures with parallel edges and uniform cross section. If the edges are perpendicular to the bases, the prisms are called *right prisms* (Fig. 9-1*a–h*).

When the base of a right prism is a circle, the figure is a *cylinder.*

9-3 Volumes of Prisms

The cubic content or capacity of a solid figure is called the *volume.* The volume of any prism or cylinder is found by multiplying the area of the base

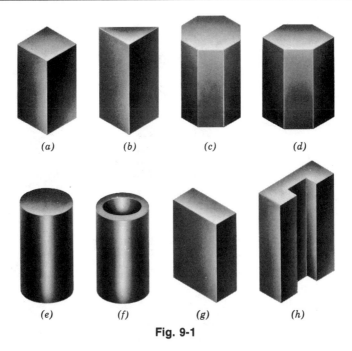

(a) (b) (c) (d)

(e) (f) (g) (h)

Fig. 9-1

by the height. Expressed as a formula,

$$\text{volume} = \text{area of base} \times \text{height}$$

Example 1

Find the volume of a rectangular prism 16 in. high whose base is a $9\frac{1}{2}$-in. square (Fig. 9-2).

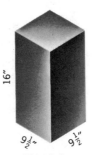

16"

$9\frac{1}{2}$" $9\frac{1}{2}$"

Fig. 9-2

SOLUTION
$$\text{volume} = \text{area of base} \times \text{height}$$
$$\text{area of base} = 9\frac{1}{2} \times 9\frac{1}{2} = 90.25 \text{ sq in.}$$
$$\text{volume} = 90.25 \times 16$$
$$= 1{,}444 \text{ cu in.}$$

Example 2

Find the volume of a cylinder $4\frac{1}{2}$ in. in diameter and 5 in. long (Fig. 9-3).

Fig. 9-3

SOLUTION   volume = area of base × height

area of base = πr^2

= 3.1416 × $2\frac{1}{4}$ × $2\frac{1}{4}$ = 15.90 sq in.

altitude = 6 ft 4 in. = 76 in.

volume = 15.90 × 5

= 79.5 cu in.

Example 3

Find the volume of the solid shell of a hollow cylinder (Fig. 9-4) $16\frac{1}{2}$ in. high; the inside diameter is $8\frac{1}{2}$ in. and the outside diameter 10 in.

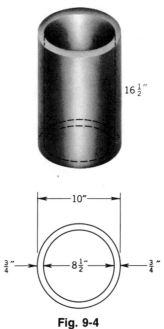

Fig. 9-4

SOLUTION

$$\text{volume} = \text{area of base} \times \text{height}$$
$$\text{area of base (flat ring)} = \text{average circumference} \times \text{thickness}$$
$$= 3.14 \times 9\tfrac{1}{4} \times \tfrac{3}{4}$$
$$= 21.8 \text{ sq in.}$$
$$\text{volume} = 21.8 \times 16\tfrac{1}{2}$$
$$= 359 \text{ cu in.}$$

Problems 9-3

Find the volumes of the following solids.

1. Find the volume of a rectangular prism whose base is 12″ × 12″ and height 24″.
2. Find the volume of a rectangular prism whose base is 12″ × 4″ and height 36″.
3. Find the volume of a rectangular prism whose base is 3½″ × 5″ and height 8″.
4. Find the volume of a rectangular prism whose base is 4½″ × 4½″ and height 6″.
5. Find the volume of a rectangular prism whose base is 5¼″ × 6¾″ and height 8½″.
6. Find the volume of a rectangular prism whose base is 1′2″ × 1′ and height 1′6″.
7. Find the volume of a rectangular prism whose base is 1′3½″ × 1′4½″ and height 1′6″.
8. Find the capacity in gallons of a rectangular tank whose base is 3′ × 3′ and height 5′. (1 gallon = 231 cu in.)
9. Find the capacity in gallons of a rectangular tank whose base is 1′6″ × 2′6″ and height 3′. (1 gallon = 231 cu in.)
10. Find the capacity in gallons of a rectangular tank whose base is 2′5½″ × 2′7″ and height 3′6½″. (1 gallon = 231 cu in.)

Solve the following problems involving the volume of prisms whose bases are not rectangles. (Use the Table of Constants on page 172 to find the area of the base.)

11. Find the volume of a prism 10″ high if the base is a hexagon 5″ across flats.
12. Find the volume of a prism 8½″ high if the base is a hexagon 5½″ across flats.
13. Find the volume of a prism 5¼″ high if the base is a hexagon 8½″ across flats.
14. Find the volume of a prism 6½″ high if the base is a hexagon whose diagonal is 6″.
15. Find the volume of a prism whose base is an equilateral triangle with 5″ sides and whose altitude is 15″.

16. Find the volume of a prism whose base is an equilateral triangle with $4\frac{3}{4}''$ sides and whose height is 20".

17. Find the volume of a prism whose base is an equilateral triangle with $5\frac{3}{4}''$ sides and whose height is $15\frac{1}{2}''$.

18. Find the volume of a prism whose base is an equilateral triangle with 1'2" sides and whose height is 3'3".

19. A prism whose base is an octagon with 15" sides is $4\frac{1}{2}''$ high. Find its volume.

20. A prism whose base is an octagon with $4\frac{1}{2}''$ sides is $10\frac{1}{4}''$ high. Find its volume.

Find the capacity of the following prisms. (1 cu ft = 1,728 cu in.; 1 gallon = 231 cu in.)

pi ex Area X ———➤

21. How many cubic feet are there in a cylinder 16" in diameter and 42'8" long?

22. Find the capacity in gallons of an elliptical tank 31" long if the base is an ellipse 21" × 17".

23. Find the capacity in gallons of a cylindrical tank 14' in diameter and 11'6" high.

24. Find the capacity in gallons of an elliptical tank 12'6" long if the long diameter of the ellipse is 5' and the short diameter is 3'4".

25. Find the capacity in gallons of a cylindrical tank 8'7" in diameter and 10'6" high.

Solve the following problems.

26. How many cubic inches of metal are there in a length of pipe 9'6" long if the inside diameter is 6" and the metal is $\frac{7}{16}''$ thick?

27. Find the weight of a bar of brass 5" in diameter and 6'2" long. (Brass weighs 512 lb per cu ft.)

28. Find the weight of water in the tank described in Problem 23. (Water weighs 62.5 lb per cu ft.)

29. Find the weight of water in the tank described in Problem 24.

30. Find the weight of water in the tank described in Problem 25.

9-4 Finding the Height of a Prism or Cylinder

When the volume of a prism or cylinder and the area of the base are known, the height can be found by using the following formula:

$$\text{height} = \frac{\text{volume}}{\text{area of base}}$$

Example

Find the height of a cylinder 26 in. in diameter to contain 6,500 cu in.

SOLUTION   area of base $= \pi r^2 = 3.14 \times 13 \times 13 = 530.66$ sq in.

$$\text{height} = \frac{\text{volume}}{\text{area of base}} = \frac{6,500}{530.66} = 12.25 \text{ in.}$$

Problems 9-4

Find the height of the following prisms.

1. How high must a cylindrical tank 8′3″ in diameter be in order to have a capacity of 2,000 gallons?
2. A gasoline tank whose cross section is an ellipse 13″ × 10″ is required to contain 16 gallons. What must be its length?
3. A cylindrical container 7″ in diameter is required to have a capacity of 1 gallon. How high must it be?
4. Find the height of a 5-gallon cylindrical container if the diameter is 12″.
5. Find the depth of a tank 4′10″ square to have a capacity of 1,000 gallons.
6. A milk container 3½″ square is required to hold 1 quart. Find the height.
7. How high must a cylindrical can 4½″ in diameter be in order to contain 2 quarts?
8. What should be the depth of the water in an octagonal tank 10′6″ across flats in order to provide for 10,000 gallons?
9. Find the depth of a rectangular vat 2′6″ × 5′10″ to contain 500 gallons.
10. The volume of a steel shaft 6″ in diameter is 2,560 cu in. Find its length.

9-5 Finding the Area of the Base of a Prism or Cylinder

When the volume and the height are known, divide the volume by the height. The formula is

$$\text{area of base} = \frac{\text{volume}}{\text{height}}$$

Example
A cylindrical tank 11 ft high has a volume of 1,050 cu ft. Find the area of the base.

SOLUTION $\text{area of base} = \dfrac{\text{volume}}{\text{height}}$

$$= \frac{1,050}{11} = 95.5 \text{ sq ft}$$

If the diameter of the cylinder is required, use the formula

$$d = 2 \sqrt{\frac{\text{area}}{\pi}}$$

In this example,

$$d = 2 \sqrt{\frac{95.5}{3.14}} = 2 \sqrt{30.4140} = 2 \times 5.51 = 11.02 \text{ ft}$$

Problems 9-5

Find the area of the base of the following prisms.

1. Find the area of the base of a 25-gallon tank 22″ high.
2. If the tank in Problem 1 is a cylinder, find the diameter of the base.
3. If the tank in Problem 1 has a square base, find the length of the sides.
4. A cylindrical tank 6′8″ high is required to contain 200 gallons. What should be its diameter?
5. What should be the diameter of a cylindrical container 10″ high to have a capacity of 1 gallon?
6. Find the diameter of a 1-quart cylindrical container 8″ high.
7. A square water tank 12′8″ high has a capacity of 5,000 gallons. Find the length of the sides.
8. Find the diameter of an oil drum $31\frac{1}{2}″$ high to contain 63 gallons.
9. A cylindrical 10-gallon container is 20″ high. Find the diameter.
10. A circular vat 18″ deep is required to have a capacity of 350 gallons. Find its diameter.

9-6   Lateral Surfaces of Prisms and Cylinders

Lateral surface is the area of the *sides* of the prism. A sheet of paper that can be wrapped around the prism or cylinder to cover the outside will be a rectangle as high as the prism or cylinder and as long as its perimeter (see Fig. 9-5). Thus the formula for the lateral surface of a prism is

lateral surface = perimeter of base × height

Example 1

Find the lateral surface of a cylinder 6 in. in diameter and 12 in. high.

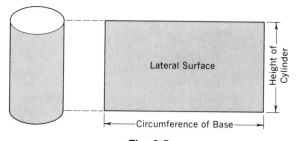

Fig. 9-5

SOLUTION
lateral surface = perimeter of base × height
perimeter of base = 3.1416 × 6 = 18.84 in.
lateral surface = 18.84 × 12
= 226.08 sq in.

Example 2

Find the lateral surface of a prism whose height is $5\frac{1}{2}$ in. and whose base is a hexagon $11\frac{1}{2}$ in. across flats.

SOLUTION
lateral surface = perimeter of base × height
side of hexagon = $0.577 \times f = 0.577 \times 11\frac{1}{2} = 6.64$ in.
perimeter of base = 6 × 6.64 = 39.84 in.
lateral surface = $39.84 \times 5\frac{1}{2}$
= 219.3 sq in.

Problems 9-6

Find the lateral surface area of the following prisms.

1. Find the lateral surface of a cylinder 18″ in diameter and 12″ high.
2. Find the lateral surface of a cylinder $8\frac{1}{2}$″ in diameter and 1′4″ high.
3. Find the lateral surface of a prism with a 16″ square base and 8′4″ high.
4. Find the lateral surface of a prism with a $1'4\frac{1}{2}$″ square base and $3'6\frac{1}{2}$″ high.
5. Find the lateral surface of a triangular prism with 14″ sides and 32″ high.
6. Find the lateral surface of a triangular prism with $8\frac{1}{2}$″ sides and $12\frac{1}{2}$″ high.
7. Find the lateral surface of a prism 7″ high. The base is a hexagon with $1\frac{1}{2}$″ sides.
8. Find the lateral surface of a prism 6′10″ high. The base is a hexagon with 2′6″ sides.
9. Find the lateral surface of an octagonal prism 2′ high. The base is an octagon 1′8″ across flats.
10. Find the lateral surface of an octagonal prism 2′4″ high. The base is an octagon 2′4″ across flats.

Solve the following problems.

11. At $1.55 per square yard find the cost of painting the sides of a cylindrical gas tank 62′ high and 85′ in diameter.
12. A sheet metal duct is 28″ in diameter and 65′6″ long. How many square feet of metal are required for this duct, making no allowances for joints?
13. The sides and bottom of a vat 4′2″ square and 6′8″ high are lined with sheet lead. How many square feet of lead are required?
14. A rectangular tank 4′10″ by 6′8″ and 5′6″ high is lined (sides and bottom) with sheet copper. How many square feet of copper are required?
15. A sheet metal duct of rectangular cross section 22″ by 40″ is 38′6″ long. Find the number of square feet of metal required.

9-7   Pyramids and Cones

Figures like those shown in Figs. 9-5*a*, 9-5*b*, and 9-5*c* are called *pyramids.* When the base is a circle, the figure is called a *cone* (Fig. 9-6).

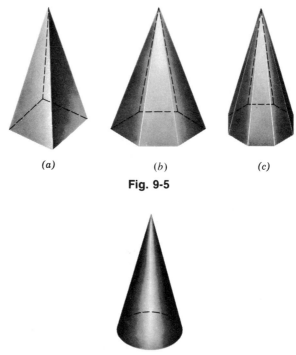

(*a*)   (*b*)   (*c*)

Fig. 9-5

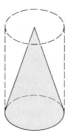

Fig. 9-6

9-8   Volumes of Pyramids and Cones

Figure 9-7 shows a comparison between a cylinder and a cone of the same base and altitude. *The volume of the cone is one-third the volume of the cylinder.* The volume of the cylinder is equal to the product of the area of the base and the altitude; the volume of the cone = $\frac{1}{3}$ × area of base × height. The same rule applies to pyramids.

$$\text{volume of pyramid or cone} = \tfrac{1}{3} \times \text{area of base} \times \text{height}$$

Fig. 9-7

Example 1
Find the volume of a cone 8 in. high; the base is 6 in. in diameter.

SOLUTION $\qquad$ volume $= \frac{1}{3} \times$ area of base $\times$ height
$\qquad$ area of base $= 3.1416 \times 3 \times 3 = 28.26$ sq in.
$\qquad$ volume $= \frac{1}{3} \times 28.26 \times 8$
$\qquad\qquad = 75.4$ cu in.

Example 2
Find the volume of a pyramid $27\frac{1}{2}$ in. high; the base is a rectangle 19 in. $\times$ 16 in.

SOLUTION $\qquad$ volume $= \frac{1}{3} \times$ area of base $\times$ height
$\qquad$ area of base $= 19 \times 16 = 304$ sq in.
$\qquad$ volume $= \frac{1}{3} \times 304 \times 27\frac{1}{2}$
$\qquad\qquad = 2{,}786.7$ cu in.

Problems 9-8

Find the volumes of the following pyramids and cones.

1. Find the volume of a cone 36″ high if the base is 15″ in diameter.
2. Find the volume of a cone 10″ high if the base is $14\frac{1}{2}$″ in diameter.
3. Find the volume of a pyramid whose base is a 28″ square and whose height is 12″.
4. Find the volume of a pyramid whose base is a $15\frac{1}{2}$″ square and whose height is $18\frac{1}{2}$″.
5. Find the volume of a hexagonal pyramid 10″ high; the base is a hexagon with 5″ sides.
6. Find the volume of a hexagonal pyramid $9\frac{1}{2}$″ high; the base is a hexagon with 5″ sides.
7. Find the volume of a pyramid whose base is an equilateral triangle with 30″ sides and whose height is 56″.
8. Find the volume of a pyramid whose base is an equilateral triangle with 1′2″ sides and whose height is 2′5″.
9. Find the volume of a pyramid whose base is an octagon 6″ across flats and whose height is $8\frac{1}{2}$″.
10. Find the volume of a pyramid whose base is a rectangle $16\frac{1}{2}$″ $\times$ $12\frac{1}{2}$″ and whose height is 5′2″.
11. Find the volume of an elliptical cone $18\frac{1}{2}$″ high; the base is an ellipse 39″ $\times$ 56″.
12. A quantity of sand dumped upon the ground assumed a conical shape with a base 32′ in diameter and a height of 11′. How many cubic yards of sand are there in the pile?
13. A quantity of sand dumped upon the ground assumed a conical shape

with a base 40′6″ in diameter and a height of 15′6″. How many cubic yards of sand are there in the pile?
14. An oil container in the shape of an inverted cone is 9½″ high and has a base whose diameter is 11″. Find the capacity in gallons.
15. If cast iron weighs 0.26 lb per cubic inch, what is the weight of a casting in the shape of a square pyramid with a 6½″ square base and 3¼″ high?

9-9   Lateral Surfaces of Pyramids and Cones

In the square pyramid shown in Fig. 9-8, the lateral surface is seen to be made up of four equal isosceles triangles like *ADE*. The base of each triangle is a side of the base of the pyramid. We can obtain the area of one of these triangles and multiply it by 4 to get the lateral surface of the pyramid.

$$\text{area of triangle} = \tfrac{1}{2} \times \text{base} \times \text{altitude}$$

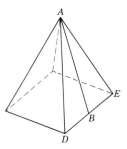

Fig. 9-8

If the pyramid in Fig. 9-9 is 12 in. high and the base is a 10-in. square, then the base of triangle *ADE* is 10 in., but the height of the triangle is the line *AB*, called the **slant height** of the pyramid.

Figure 9-9 shows the pyramid with one-quarter of it removed to display the actual height *AC* and the slant height *AB*. The triangle *ABC* (Fig. 9-10) is a right triangle with the slant height *AB* as the hypotenuse. The height of this right triangle, *AC*, is 12 in., the height of the pyramid. The base *BC* of the triangle is half the distance across the square: ½ of 10 in. = 5 in. = *BC*. The hypotenuse $AB = \sqrt{AC^2 + BC^2} = \sqrt{12^2 + 5^2} = \sqrt{144 + 25} = \sqrt{169} = 13$ in., the slant height.

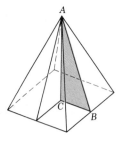

Fig. 9-9

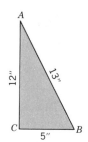

Fig. 9-10

This slant height AB is the altitude of the triangle ADE, which is one of the sides of the pyramid.

$$\text{Area of triangle } ADE = \tfrac{1}{2}ba = \tfrac{1}{2} \times 10 \times 13 = 65 \text{ sq in.}$$

The lateral surface of the pyramid is 4 times the area of one of the sides: $4 \times 65 = 4 \times \tfrac{1}{2} \times 10 \times 13 = 260$ sq in.

But $4 \times 10 = 40$ in., the perimeter of the base. The lateral surface, therefore, equals $\tfrac{1}{2} \times$ the perimeter of the base $\times$ the slant height $= \tfrac{1}{2} \times 40 \times 13 = 260$ sq in.

The formula for finding the lateral surface of a pyramid is lateral surface $= \tfrac{1}{2} \times$ perimeter of base $\times$ slant height.

In each instance the slant height of the pyramid is the hypotenuse of a right triangle whose altitude is the height of the pyramid. The base of this right triangle is seen in Fig. 9-9. In the cone, the base of this triangle is one-half the diameter of the base of the cone. In the hexagonal pyramid and the octagonal pyramid, the base of this triangle is one-half the distance across the flats of the base of the pyramid. In the triangular pyramid the base of this triangle is one-third the height of the triangular base.

Example 1

Find the lateral surface of a pyramid whose base is a hexagon $9\tfrac{1}{2}$ in. across flats and whose height is 16 in.

SOLUTION

$$\begin{aligned}
\text{lateral surface} &= \tfrac{1}{2} \times \text{perimeter of base} \times \text{slant height}\\
\text{side of hexagon} &= 0.577 \times f = 0.577 \times 9\tfrac{1}{2} = 5.48 \text{ in.}\\
\text{perimeter} &= 6 \times 5.48 = 32.89 \text{ in.}\\
\text{slant height} &= \sqrt{4.75^2 + 16^2} = \sqrt{22.56 + 256}\\
&= \sqrt{278.56} = 16.69 \text{ in.}\\
\text{lateral surface} &= \tfrac{1}{2} \times 32.89 \times 16.69 = 274.47 \text{ sq in.}
\end{aligned}$$

Example 2

Find the lateral surface of a cone 1 ft 9 in. high; the base is 6 ft 8 in. in diameter.

SOLUTION

$$\begin{aligned}
\text{lateral surface} &= \tfrac{1}{2} \times \text{perimeter of base} \times \text{slant height}\\
\text{perimeter of base} &= 3.1416 \times 80 = 251.2 \text{ in.}\\
\text{slant height} &= \sqrt{40^2 + 21^2} = \sqrt{1600 + 441}\\
&= \sqrt{2041} = 45.2 \text{ in.}\\
\text{lateral surface} &= \tfrac{1}{2} \times 251.2 \times 45.2\\
&= 5{,}677 \text{ sq in.}
\end{aligned}$$

Problems 9-9

Solve the following problems.

1. Find the lateral surface of a square pyramid with a 4″ square base and a height of 10″.

2. Find the lateral surface of a square pyramid with a $4\frac{1}{2}''$ square base and a height of $10\frac{1}{2}''$.
3. Find the lateral surface of a pyramid whose base is an octagon with $4''$ sides and whose height is $6''$.
4. Find the lateral surface of a cone whose base is $8\frac{1}{2}''$ in diameter and whose height is $18''$.
5. Find the lateral surface of a pyramid whose base is an equilateral triangle with $9''$ sides and whose height is $12''$.
6. Find the lateral surface of a pyramid whose base is a rectangle $14'' \times 20''$ and whose height is $12''$.
7. What is the height of a cone whose base is $36'$ in diameter and whose slant height is $24'$?
8. Find the lateral surface of a pyramid whose base is a hexagon with $10''$ sides and whose height is $30''$.
9. The top of a copper boiler is conical in shape with a base $12'3''$ in diameter and a height of $3'6''$. How many square feet of copper were required to make it, allowing 3% for seams?
10. How many square feet of tin are required to cover a roof in the shape of a square pyramid $9'4''$ high, with the base a $28'$ square?

9-10   Frustums of Pyramids and Cones

When a pyramid or cone is cut at any point below the apex by a plane parallel to the base, the portion below the cutting plane is called a *frustum* of the pyramid or cone. Figure 9-11 shows various frustums.

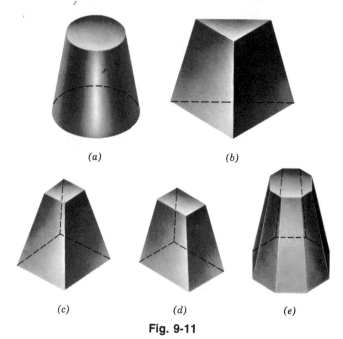

(a)   (b)

(c)   (d)   (e)

Fig. 9-11

The volume of the frustum is obtained by the following formula:

$$V = \tfrac{1}{3}h(B + b + \sqrt{B \times b})$$

where V is the volume, h is the height, B is the area of the large base, and b is the area of the small base. The formula says that the volume $= \tfrac{1}{3} \times$ the height $\times$ the sum of three things: the area of the large base, the area of the small base, and the square root of their product. The process consists of the following steps:

1. Compute the area of B, the large base.
2. Compute the area of b, the small base.
3. Multiply B by b and get the square root of the product.
4. Add the three quantities in the parentheses.
5. Multiply $\tfrac{1}{3}$ of the height by the sum of the three quantities in the parentheses.

Example 1

Find the volume of the frustum of a cone 15 in. high if the upper base is 16 in. in diameter and the lower base 9 in. in diameter.

SOLUTION $V = \tfrac{1}{3} \times h \times (B + b + \sqrt{B \times b})$
$B =$ area of 16-in. circle $= 3.1416 \times 8 \times 8 = 201.0$ sq in.
$b =$ area of 9-in. circle $= 3.1416 \times 4\tfrac{1}{2} \times 4\tfrac{1}{2} = 63.6$ sq in.
$V = \tfrac{1}{3} \times 15 \times (201.0 + 63.6 + \sqrt{201.0 \times 63.6})$
$= \tfrac{1}{3} \times 15 \times (201.0 + 63.6 + \sqrt{12783.6})$
$= \tfrac{1}{3} \times 15 \times (201.0 + 63.6 + 113.1)$
$= \tfrac{1}{3} \times 15 \times 377.7$
$= 1,888.5$ cu in. $= 1.09$ cu ft

This formula is the general formula and applies to all frustrums of cones and pyramids regardless of the shape of the base. For the frustum of the cone, however, the formula can be simplified and much time saved in the computation. The simplified formula for finding the volume of the frustum of a cone follows:

$$V = 0.262h(D^2 + d^2 + Dd)$$

where

$V =$ volume
$D =$ diameter of large base
$d =$ diameter of small base
$h =$ height

Example 2

Using the simplified formula, find the volume of the frustum of a cone with bases 6 in. and 10 in. in diameter and 12 in. high.

SOLUTION $\quad V = 0.262h(D^2 + d^2 + Dd)$
$$= 0.262 \times 12(10^2 + 6^2 + 10 \times 6)$$
$$= 0.262 \times 12(100 + 36 + 60)$$
$$= 0.262 \times 12 \times 196$$
$$= 616.2 \text{ cu in.}$$

Problems 9-10

Find the volume of the following frustums.

1. Find the volume of the frustum of a square pyramid 15″ high; the upper base is a 42″ square and the lower base a 26″ square.
2. Find the volume of the frustum of a hexagonal pyramid 7″ high; the upper base is a hexagon with 3″ sides and the lower base a hexagon with 8″ sides.
3. Find the volume of the frustum of a cone 10″ high; the lower base is 3″ in diameter and the upper base 6″ in diameter.
4. Find the volume of the frustum of an elliptical pyramid 8″ high; the upper base is an ellipse 7″ × 9″ and the lower base an ellipse 14″ × 18″.
5. Find the capacity in gallons of a pail 11″ deep with a top diameter of 10″ and a bottom diameter of 8″.
6. Find the volume of a frustum of a rectangular pyramid $4\frac{1}{2}$″ high; the upper base is a rectangle 12″ × 16″ and the lower base a rectangle 9″ × 12″.
7. What is the capacity in cubic feet of a square coal hopper in the shape of a frustum of a square pyramid 13′0″ high; the upper base is a 22′ square and the lower base a $5\frac{1}{2}$′ square?
8. An oil measure in the shape of a frustum of a cone is $11\frac{1}{2}$″ deep, with a top diameter of $5\frac{1}{4}$″ and a bottom diameter of $7\frac{1}{2}$″. How many quarts of oil can it contain?
9. A concrete base for a heavy machine has the form of a frustum of a rectangular pyramid. The top is 16′4″ × 4′8″ and the bottom is 32′8″ × 9′4″. The depth is 3′8″. How many cubic yards of concrete were used in building this foundation?
10. A basin in the form of a frustum of an elliptical pyramid 8″ deep is 18″ × 15″ at the top and 12″ × 10″ at the bottom. What is its capacity in gallons?

9-11 Finding the Height of the Frustum of a Pyramid or Cone

If, in the designing of a vessel, it is required to find the height for a given capacity, the volume formula is modified to read as follows:

$$h = \frac{3V}{B + b + \sqrt{Bb}}$$

Example 1
Find the height of a frustum of a cone to contain 2,000 cu in. The upper base is 16 in. in diameter and the lower base 12 in. in diameter.

SOLUTION $h = \dfrac{3V}{B + b + \sqrt{Bb}}$

B = area of 16-in. circle = 3.1416 × 8 × 8 = 201.0 sq in.
b = area of 12-in. circle = 3.1416 × 6 × 6 = 113.0 sq in.
$\sqrt{Bb} = \sqrt{201.0 \times 113.0} = 150.7$

$$h = \frac{3 \times 2,000}{201.0 + 113.0 + 150.7}$$

$$= \frac{6,000}{464.7}$$

$$= 12.9 \text{ in.}$$

The simplified formula for finding the height of the frustum of a cone is

$$h = \frac{V}{0.262(D^2 + d^2 + Dd)}$$

Example 2
Find the answer to Example 1 by means of the simplified formula.

SOLUTION $h = \dfrac{V}{0.262(D^2 + d^2 + Dd)}$

$$= \frac{2,000}{0.262(16^2 + 12^2 + 16 \times 12)}$$

$$= \frac{2,000}{0.262 \times 592}$$

$$= \frac{2,000}{155.1} = 12.9 \text{ in.}$$

Problems 9-11

Find the height of the following frustums.

1. Find the height of the frustum of a square pyramid to contain 400 cu in., the upper base to be an 8″ square, and the lower base a 10″ square.
2. Find the height of the frustum of a cone to contain 2,772 cu in., the top to be 18″ in diameter, and the bottom 13½″ in diameter.
3. What must be the height of a frustum of a hexagonal pyramid to contain 200 cu in.; the upper base is a hexagon with 2″ sides and the lower base a hexagon with 5″ sides?

4. Find the height of the frustum of a rectangular pyramid that has a volume of 6,000 cu ft, an upper base 30′ × 15′, and a lower base 40′ × 20′.
5. What must be the height of a gallon measure with a top diameter of $4\frac{1}{4}''$ and a bottom diameter of $7\frac{1}{4}''$?
6. A bin in the form of the frustum of a square pyramid is required to have a capacity of 8,000 bushels. What must be its height if the top is a 50′ square and the bottom a 35′ square? (1 bushel = 2,150.4 cu in.)
7. Find the depth of a 10-gallon pail with a top diameter of 14″ and a bottom diameter of 10″.
8. A hopper whose top is 10′ square and whose bottom is 2′ square is required to have a capacity of 500 cu ft. Find the height.

9-12 Finding the Lateral Surface of the Frustum of a Cone or Pyramid

Figure 9-12 shows the lateral surface of the frustum of a square pyramid to be made up of four trapezoids like *EFHG*. If the upper base is a 9-in. square, the lower base a 14-in. square, and the height 5 in., then each side is a trapezoid with bases of 9 in. and 14 in. and a height *AB* that is the slant height of the frustum. Figure 9-13 shows a cross section through the figure as

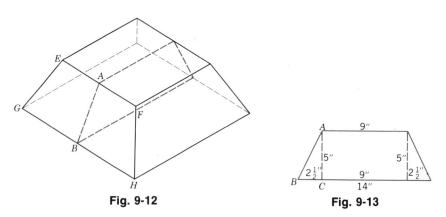

Fig. 9-12   Fig. 9-13

shown by the dotted lines in Fig. 9-12. The slant height *AB* is the hypotenuse of the right triangle *ABC*, in which the height is the height of the frustum, 5 in., and the base, is half the difference between the upper and lower bases of the trapezoid.

$$\text{base } BC = \frac{14 - 9}{2} = \frac{5}{2} = 2\frac{1}{2} \text{ in.}$$

$$\text{slant height } AB = \sqrt{5^2 + 2\frac{1}{2}^2} = \sqrt{25 + 6.25} = \sqrt{31.25} = 5.59 \text{ in.}$$

$$\text{average width of } EFHG = \frac{9 + 14}{2} = 11.5 \text{ in.}$$

area of trapezoid *EFGH* (one side of frustum)
$$= \text{average width} \times \text{slant height}$$
$$= 11.5 \times 5.59 = 64.285 \text{ sq in.}$$

$$\text{lateral surface} = 4 \times \text{area of one side}$$
$$= 4 \times 64.285 = 257.14 \text{ sq in.}$$

Another method of finding the lateral surface is by means of the following formula:

$$\text{lateral surface} = \text{average perimeter of bases} \times \text{slant height}$$

$$\text{perimeter of upper base} = 4 \times 9 \ = 36 \text{ in.}$$
$$\underline{\text{perimeter of lower base} = 4 \times 14 = 56 \text{ in.}}$$
$$\text{sum of perimeters} = 92 \text{ in.}$$

Average perimeter $= \frac{92}{2} = 46$ in., which is the same as 4 times the average width of one side.

$$\text{lateral surface} = 46 \times 5.59 = 257.14 \text{ sq in.}$$

Example
Find the lateral surface of the frustum of a cone 6 in. high with upper and lower bases respectively 11 in. and 7 in. in diameter (Fig. 9-14).

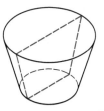 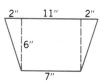

Fig. 9-14

SOLUTION

$$\text{lateral surface} = \text{average perimeter of bases} \times \text{slant height}$$
$$\text{perimeter of upper base} = 3.1416 \times 11 = 34.54 \text{ in.}$$
$$\underline{\text{perimeter of lower base} = 3.1416 \times 7 \ = 21.98 \text{ in.}}$$
$$56.52 \text{ in.}$$

$$\text{average perimeter} = \frac{56.52}{2} = 28.26 \text{ in.}$$

$$\text{altitude of right triangle} = 6 \text{ in.}$$

$$\text{base of right triangle} = \frac{11 - 7}{2} = \frac{4}{2} = 2 \text{ in.}$$

$$\text{slant height} = \sqrt{6^2 + 2^2} = \sqrt{36 + 4} = \sqrt{40} = 6.32 \text{ in.}$$

$$\text{lateral surface} = 28.26 \times 6.32$$

$$= 178.60 \text{ sq in.}$$

Problems 9-12

Find the lateral surface areas of the following frustums.

1. Find the lateral surface of the frustum of a square pyramid 16' high if the upper base is a 28' square and the lower base an 18' square.
2. Find the lateral surface of the frustum of a hexagonal pyramid $9\frac{1}{2}''$ high, the upper base of which is a hexagon with $3\frac{1}{2}''$ sides and the lower base a hexagon with 9'' sides.
3. Find the lateral surface of the frustum of an octagonal pyramid 20'' high if the upper base is an octagon with 14'' sides and the lower base an octagon with 22'' sides.
4. Find the lateral surface of the frustum of a cone 6'10'' high if the upper base is 7'2'' in diameter and the lower base 9'4'' in diameter.
5. Find the lateral surface of the frustum of a triangular pyramid 12'' high whose upper base is an equilateral triangle with 10'' sides and whose lower base is an equilateral triangle with 20'' sides.
6. In the following chart, each set of figures, (a) to (e), represents a frustum of a cone with dimensions as given. Find the lateral surface for each. Give all answers in square feet, to one decimal place.

FRUSTUMS OF CONES

	Large Diameter	Small Diameter	Height	Lateral Surface (sq ft)
(a)	1'10''	1'4''	1'6''	_____
(b)	32''	18''	6''	_____
(c)	14''	6''	12''	_____
(d)	8'4''	2'8''	3'6''	_____
(e)	12''	9''	11''	_____

9-13   Spheres

The volume of a sphere or ball may be found by the formula

$$V = \tfrac{4}{3}\pi r^3$$

in which r is the radius (see Fig. 9-15).

Fig. 9-15

Example

Find the volume of a 6-in. sphere.

SOLUTION $V = \frac{4}{3}\pi r^3$

$= \frac{4}{3} \times 3.1416 \times 3^3$

$= \frac{4}{3} \times 3.1416 \times 27 = 113.1$ cu in.

Problems 9-13

Find the volumes of the following spheres and hemispheres. (A hemisphere equals one-half of a sphere.) Use 3.1416 for π.

1. A 12″ sphere
2. A 1″ sphere
3. A 2″ sphere
4. A 3″ sphere
5. A 15″ sphere
6. A 30′ sphere
7. A sphere $\frac{3}{4}$″ in diameter
8. A sphere $1\frac{1}{2}$″ in diameter
9. A hemisphere 11″ in diameter
10. A hemisphere 3′4″ in diameter
11. A 1′6″ sphere
12. A $2\frac{1}{2}$″ sphere
13. A $3\frac{1}{4}$″ sphere
14. A 1′$2\frac{1}{2}$″ sphere
15. A 16.5″ sphere
16. A 25′ sphere
17. A sphere $\frac{7}{8}$″ in diameter
18. A sphere $2\frac{1}{8}$″ in diameter
19. A hemisphere $10\frac{1}{2}$″ in diameter
20. A hemisphere 2′$6\frac{1}{2}$″ in diameter

9-14   Using the Calculator to Find Volumes of Spheres

Examine your calculator and see if you have a y^x function key. If you do, then you can find volumes of spheres with the following sequence of key-ins.

Example
Find the volume of a sphere 12 in. in diameter (Fig. 9-16). Round off your answer to one decimal place.

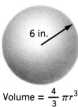

Volume $= \frac{4}{3}\pi r^3$

Fig. 9-16

SOLUTION   Volume $= \frac{4}{3}\pi r^3$

The display shows

Turn on the calculator	0.
Enter 4	4.
Press $\div$	4.
Enter 3	3.
Press $\times$	1.333333333
Press π	3.141592654
Press $\times$	4.188790205
Enter 6	6.
Press y^x	6.
Enter 3	3. $y^x = 6^3$
Press $=$	904.7786842 (answer)

Round off this answer to 904.8 cu in.

Problems 9-14

Using a calculator, do all the problems from Problem Set 9-13 again. Get some experience using the y^x key. Compare your answers to the answers you have from pencil and paper work on these problems.

9-15   Finding the Surface Area of a Sphere

The surface area of a sphere is equal to four times the area of a circle with the same diameter:

$$\text{surface area of sphere} = 4\pi r^2$$

Example

Find the surface area of a 6-in. sphere.

SOLUTION   surface area $= 4 \times \pi \times r^2$
$= 4 \times 3.1416 \times 3^2$
$= 4 \times 3.1416 \times 9 = 113.09$ sq in.

Problems 9-15

Find the surface area of the following spheres:

1. A sphere 1" in diameter
2. A $1\frac{1}{2}$" sphere
3. A sphere 100' in diameter
4. A sphere 1.4" in diameter
5. A 22" sphere

6. A 2'6" sphere
7. A 3'2" sphere
8. A 1.375" sphere
9. A 1'½" sphere
10. A 4'6½" sphere

9-16 Volume of a Ring

The volume of a ring of circular cross section, as shown in Fig. 9-17, is found by multiplying the circumference of the center line circle by the area of cross section.

Example

The ring in Fig. 9-17 has an inside diameter of 3 in. and the cross section of the ring is a ¾-in. circle. Find its volume.

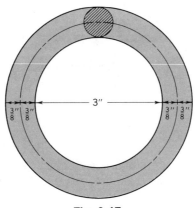

Fig. 9-17

SOLUTION

volume = circumference of center line circle × area of cross section
diameter of center line circle = $3 + \frac{3}{8} + \frac{3}{8} = 3\frac{3}{4}$ in.
circumference of center line circle = $3.1416 \times 3\frac{3}{4} = 11.78$ in.
area of cross section = $3.1416 \times \frac{3}{8} \times \frac{3}{8} = 0.442$ sq in.
volume = 11.78×0.442
= 5.21 cu in.

Problems 9-16

Find the volume of the following rings. Use 3.1416 for π.

1. The rim of a handwheel of circular cross section if the outside diameter is 10" and the inside diameter 9"
2. A ring of circular cross section if the inside diameter is 5" and the ring is $\frac{7}{8}$" thick

3. A ring of circular cross section if the outside diameter is $6\frac{1}{2}''$ and the ring is $\frac{3}{4}''$ thick
4. A ring having a circular cross section with an inside diameter of $7''$ and a thickness of $\frac{1}{2}''$
5. The rim of a handwheel with an outside diameter of $8''$ and $\frac{3}{4}''$ thick

9-17   Volumes of Composite Solid Figures

Many objects are made up of parts, the volumes of which can be found by the rules studied in the preceding articles. In cases of this sort, the volume of each part is found and added to find the total volume of the object.

Example 1

Find the volume of the steeple-head rivet in Fig. 9-18.

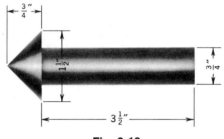

Fig. 9-18

SOLUTION   volume of head (cone) $= \frac{1}{3} \times \frac{3}{4} \times 3.1416 \times \frac{3}{4} \times \frac{3}{4} = 0.442$ cu in.

volume of body (cylinder) $= \frac{7}{2} \times 3.1416 \times \frac{3}{8} \times \frac{3}{8} = \underline{1.547}$ cu in.

Total $= 1.989$ cu in.

Example 2

Find the volume of the taper bushing shown in Fig. 9-19.

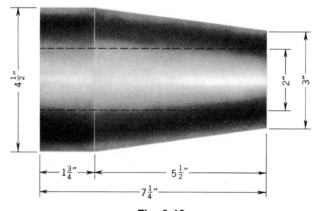

Fig. 9-19

SOLUTION

volume of cylindrical part
$$= 3.1416 \times 2\tfrac{1}{4} \times 2\tfrac{1}{4} \times 1\tfrac{3}{4} \qquad\qquad = 27.8 \text{ cu in.}$$
volume of tapered part
$$= 0.262h(D^2 + d^2 + Dd)$$
$$= 0.262 \times 5\tfrac{1}{2}[(4\tfrac{1}{2})^2 + 3^2 + 4\tfrac{1}{2} \times 3]$$
$$= 0.262 \times 5\tfrac{1}{2}(20.25 + 9 + 13.5)$$
$$= 0.262 \times 5.5 \times 42.75 \qquad\qquad = \underline{61.6} \text{ cu in.}$$
$$\text{Gross volume} = 89.4 \text{ cu in.}$$
volume of cylindrical hole through the bushing
$$= 3.14 \times 1 \times 1 \times 7\tfrac{1}{4} \qquad\qquad = \underline{22.8} \text{ cu in.}$$
$$\text{Net volume or volume of the object} = 66.6 \text{ cu in.}$$

EXPLANATION   First consider the entire object as a solid, then subtract the volume of the cylindrical hole.

Problems 9-17

Find the volume of the following composite figures.

1. Find the volume of the conehead rivet shown in Fig. 9-20.

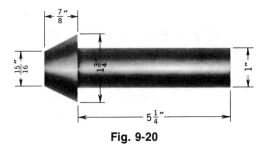

Fig. 9-20

2. Find the volume of the countersunk rivet shown in Fig. 9-21.

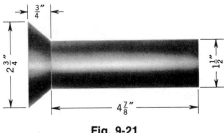

Fig. 9-21

3. Find the volume of the hollow cast iron column shown in Fig. 9-22.

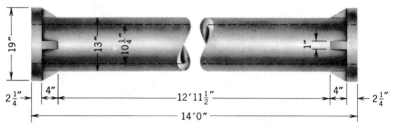

Fig. 9-22

4. Find the capacity of the oil can shown in Fig. 9-23.

Fig. 9-23

5. Find the capacity of a cylindrical pail with a hemispherical bottom if the diameter is 10″ and the over-all depth is 14″.

9-18   Weights of Materials

Table 7 at the back of the book gives the weight per cubic inch and per cubic foot of various materials. The weight of any object may be found by computing its volume in cubic inches or cubic feet and multiplying this volume by the unit weight, that is, the weight per cubic inch or cubic foot of that material.

Example 1
Find the weight of an iron casting whose volume is 7.35 cu ft.

SOLUTION $7.35 \times 450 = 3307.5$ lb

EXPLANATION   Table 7 in the back of this book shows that the weight of cast iron is 450 lb per cu ft. Therefore, the weight of the casting is 7.35×450, or 3,307.5 lb.

Example 2

Find the weight of a sheet of steel 6 ft 8 in. long, 2 ft 4 in. wide, and $\frac{1}{4}$ in. thick (see Fig. 9-24).

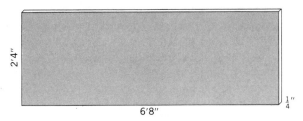

Fig. 9-24

SOLUTION $80 \times 28 \times \frac{1}{4} \times 0.283 \times 158.5$ lb

EXPLANATION The product $80 \times 28 \times \frac{1}{4}$ gives the volume of the sheet in cubic inches. Multiplying by 0.283, the weight per cubic inch of steel, gives the weight of the sheet of steel.

Example 3

Find the weight of the rim of a cast iron flywheel of the following dimensions: inside diameter 48 in., outside diameter $56\frac{1}{2}$ in., face $6\frac{1}{2}$ in. wide (see Fig. 9-25).

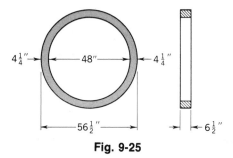

Fig. 9-25

SOLUTION $3.14 \times 52\frac{1}{4} \times 4\frac{1}{4} \times 6\frac{1}{2} \times \frac{450}{1728} = 1,181$ lb

EXPLANATION

volume of a ring section = average circumference × area of cross section

$$\text{diameter of center line circle} = \frac{48 + 56\frac{1}{2}}{2} = \frac{104\frac{1}{2}}{2} = 52\frac{1}{4} \text{ in.}$$

average circumference = $3.14 \times 52\frac{1}{4}$
area of cross section = $4\frac{1}{4} \times 6\frac{1}{2}$

The product of $3.14 \times 52\frac{1}{4} \times 4\frac{1}{4} \times 6\frac{1}{2}$ gives the volume in cubic inches. Dividing by 1,728 changes this to cubic feet. Since each cubic foot of cast iron weighs 450 lb, we multiply by 450, giving 1,181 lb.

Find the weight of the following. Use Table 7 in the back of this text.

1. Find the weight of a rectangular steel plate $3'3'' \times 2'1''$ and $\frac{3}{4}''$ thick.
2. Find the weight of a 20' length of wrought iron pipe whose inside diameter is $3.548''$ and outside diameter $4''$.
3. Find the weight of 1,000 round U.S. Standard wrought iron washers whose outside diameter is $3\frac{3}{4}''$, with a $1\frac{3}{4}''$ hole and a thickness of $\frac{11}{64}''$.
4. Find the weight of 500 wrought iron washers $5''$ square and $\frac{3}{8}''$ thick. The hole is $1\frac{1}{2}''$ in diameter.
5. Find the weight of a cast iron manhole cover with semicircular ends if the overall length is $4'2''$, the width $2'1''$, and the thickness $1\frac{1}{2}''$.
6. Find the weight of a triangular steel plate $8\frac{1}{2}''$ on a side and $1''$ thick.
7. Find the weight of a steel ball $3''$ in diameter.
8. Find the weight of 500 steel bearings $1''$ in diameter.
9. Find the weight of 1,000 lead discs $\frac{1}{2}''$ in diameter and $\frac{1}{16}''$ thick.
10. Find the weight of a 4' length of cast iron pipe whose inside diameter is $3\frac{1}{2}''$ and whose outside diameter is $4\frac{1}{4}''$.

9-19 Weights of Castings from Patterns

It is possible to estimate very closely the weight of a casting from the weight of the pattern.

Example
Find the weight of a cast iron casting made from a white pine pattern weighing $9\frac{1}{4}$ lb.

SOLUTION   volume of pattern $= \dfrac{9\frac{1}{4}}{25} = 0.37$ cu ft

weight of casting $= 0.37 \times 450 = 166.5$ lb

EXPLANATION   Since a cubic foot of white pine weighs approximately 25 lb, the volume of the pattern will be $\dfrac{9\frac{1}{4}}{25}$ or 0.37 of a cubic foot. A cubic foot of cast iron weighs 450 lb. Hence 0.37 of a cubic foot will weigh 0.37×450 or 166.5 lb.

Using Table 7, find the weight of the following castings from the weight of the patterns.

1. Find the weight of a brass casting made from a cherry pattern that weighs $7\frac{3}{4}$ lb.

2. A pattern made of white oak weighs 5 lb $7\frac{1}{2}$ oz. What will a copper casting made from it weigh?

3. The ratio of the weight of white pine to cast iron is 1 to 16. If the pattern weighs 4 lb 14 oz, what will an iron casting weigh?

4. Bell metal weighs 14.2 times as much as mahogany. If a pattern of mahogany weighs 8 lbs 6 oz, what will the bell metal casting weigh?

5. Compute the following weight ratios: (a) pine to brass, (b) maple to cast iron, (c) mahogany to copper, (d) pine to bell metal, and (e) oak to aluminum.

6. Find the weight of a cast iron casting made from a maple pattern that weighs $6\frac{1}{2}$ lb.

7. Find the weight of a brass casting made from a white pine pattern that weighs 4 lb 6 oz.

8. Find the ratio of the weight of silver to gold.

9. Find the ratio of the weight of gold to silver.

10. Find the weight of a maple pattern if the brass casting weighs 85 lb.

9-20 Board Measure

Lumber is measured in terms of board feet. A **board foot** of lumber is a piece of wood having an area of 1 sq ft and a thickness of 1 in. Feet board measure may be abbreviated **fbm.**

From the definition of a board foot we obtain the following rule.

RULE

To find the number of board feet in a piece of lumber, multiply the length in feet by the width in feet by the thickness in inches, counting a thickness less than 1 in. as 1 in.

Lumber is either rough or dressed. **Rough stock** is lumber that is not dressed or planed. **Dressed stock** is lumber that is planed on one or more sides.

In measuring lumber, we compute the full size, that is, the rough stock required to make the desired piece.

Allowance for Dressing

When lumber is dressed or planed, $\frac{1}{16}''$ is taken off each side if the lumber is less than $1\frac{1}{2}''$ thick. If the lumber is $1\frac{1}{2}''$ or more in thickness, $\frac{1}{8}''$ is taken off each side. Sometimes a little more is planed off in dressing lumber. Lumber for framing houses measures $\frac{1}{2}$ in. less than the name of the piece. Thus standard 2 in. × 4 in. studs actually measure about $1\frac{1}{2}$ in. × $3\frac{1}{2}$ in.

Standard Lengths and Widths

Building lumber is cut in even lengths, such as 8 ft, 10 ft, 12 ft, 14 ft, and 16 ft, up to 22 ft. Lumber longer than 22 ft is more expensive and more difficult to find than the shorter lengths. Building lumber is also cut in even widths, such as 2 in., 4 in., 6 in., and 8 in., up to 12 in. Lumber wider than 12 in. is also more expensive and more difficult to find. The thickness of building lumber is usually 1 in., 2 in., 3 in., or 4 in.

Hardwoods, such as ash, maple, Honduras mahogany, and walnut, are cut in the widths and lengths that get the most lumber from the tree from which it is cut. This lumber is usually dressed on only two sides. The widths and lengths are not standard. The thickness of this lumber is counted in $\frac{1}{4}$-in. graduations, such as $\frac{1}{2}$ in., $\frac{3}{4}$ in., $\frac{4}{4}$ in., $\frac{5}{4}$ in., and $\frac{6}{4}$ in., up to $\frac{8}{4}$ in. or 2 in. In measuring the width of lumber a fraction less than $\frac{1}{2}$ in. is disregarded, whereas $\frac{1}{2}$ in. or more is regarded as 1 in. Thus a board $5\frac{1}{4}$ in. wide is considered 5 in. wide, whereas a board $7\frac{1}{2}$ in. in width is figured as 8 in.

Example 1
Find the number of board feet in a piece of lumber 2 in. × 6 in. × 20 ft.

SOLUTION $20 \times \frac{6}{12} \times 2 = 20$ board feet

EXPLANATION   Convert the width into feet by dividing 6 in. by 12, that is, 6 in. $= \frac{6}{12}$ of a foot. Multiply the length 20 ft by the width $\frac{6}{12}$ ft, by the thickness 2 in., and obtain 20 board feet.

Example 2
Find the number of board feet in a piece of lumber $\frac{7}{8}$ in. × 14 in. × 22 ft.

SOLUTION $22 \times \frac{14}{12} \times 1 = 25\frac{2}{3}$ board feet

EXPLANATION   Dividing 14 by 12 changes the width to feet. Since the thickness is less than 1 in., it is figured as 1 in. Multiplying the length in feet by the width in feet by the thickness 1 in., we get the board feet in the piece.

Example 3
Find the number of board feet in a board $1\frac{1}{2}$ in. × $9\frac{3}{4}$ in. × 16 ft.

SOLUTION $16 \times \frac{10}{12} \times \frac{3}{2} = 20$ board feet

EXPLANATION   Since the width is $9\frac{3}{4}$ in., we regard it as 10 in. Dividing 10 in. by 12 reduces it to feet. We then multiply the length in feet by the width in feet by the thickness in inches and obtain the number of board feet in the piece.

Problems 9-20

Find the number of board feet in the following pieces.

1. 2″ × 4″ × 12′
2. 2″ × 6″ × 10′
3. 1″ × 8″ × 14′
4. 2″ × 4″ × 8′

   5. $1'' \times 10'' \times 12'$ 6. $2'' \times 10'' \times 16'$
   7. $2'' \times 3'' \times 12'$ 8. $2'' \times 12'' \times 16'$
   9. $2'' \times 8'' \times 20'$ 10. $1'' \times 6'' \times 12'$
11. $\frac{3}{4}'' \times 10'' \times 6'$ 12. $\frac{7}{8}'' \times 10'' \times 10'$
13. $\frac{1}{2}'' \times 8'' \times 16'$ 14. $.1'' \times 8'' \times 8'$
15. $2'' \times 3'' \times 16'$ 16. $2'' \times 2'' \times 12'$
17. $4'' \times 4'' \times 10'$ 18. $4'' \times 6'' \times 12'$
19. $3'' \times 4'' \times 10'$ 20. $4'' \times 4'' \times 18'$

Find the number of board feet of lumber in the following practical problems.

21. A truckload of lumber is $16'$ long, $5'4''$ wide, and $4'8''$ high. How many board feet does it contain?
22. How many board feet are there in 72 boards $2\frac{1}{4}'' \times 16'' \times 28'$?
23. A building lot $120'$ long and $30'$ wide is to be enclosed by a fence $10'$ high. The boards are $6''$ wide and $1''$ thick. Bracing pieces $3'' \times 4'' \times 12'$ are spaced $10'$ apart. Three longitudinal nailing pieces $2'' \times 4''$ are used. How many board feet of lumber are required?
24. How many board feet of lumber are required for a boardwalk $200'$ long and $10'$ wide if $2''$-thick planks are used? The planks are to be nailed to three pieces $3'' \times 4''$ running the full length of the walk.
25. Find the number of board feet required to build a boardwalk a mile long and $12'$ wide. The boards are $6''$ wide and $1\frac{3}{4}''$ thick, dressed on one side, and spaced $\frac{1}{2}''$ apart. Four planks $3'' \times 4''$ run the full length of the walk.
26. How many board feet of lumber are required to floor a loft with $1\frac{1}{2}''$ lumber dressed on one side if the floor is $60'6''$ long and $22'3''$ wide, allowing 2% for squaring?
27. Find the number of board feet required for flooring a circular bandstand $24'$ in diameter if $1\frac{3}{4}''$ stock dressed on one side is used. Allow 3% for waste.
28. A porch $50'6''$ long and $8'3''$ wide is to be floored with $1''$ stock planed on one side. If the allowance for squaring is 3%, how many board feet are required?
29. A loading platform is laid with $2''$ rough stock. The platform is $42'9''$ long and $8'6''$ wide. How many board feet are required if 2% is allowed for waste?
30. A factory floor is laid with $1\frac{5}{8}''$ stock dressed on one side. If the floor is $36'6'' \times 61'4''$, how much lumber is required, allowing 3% for waste?

9-21 Flooring

Some hardwood flooring is tongued and grooved. Hence in figuring the amount of lumber required for covering surfaces such as the floor or ceiling of a room, allowance must be made for matching as well as for waste. The allowance for matching varies from 24 to 50% of the surface to be covered, depending on the face width of the lumber used. For waste, the usual allowance is 3 to 5%.

The following table shows the percentages customarily allowed for matching.

Face Width of Boards	Percentage Allowed for Matching
$1\frac{1}{2}$ in.	50%
2 in.	$37\frac{1}{2}$%
$2\frac{1}{4}$ in.	$33\frac{1}{3}$%
$3\frac{1}{4}$ in.	24%

The standard thicknesses of hardwood flooring are: $\frac{3}{8}$ in., $\frac{1}{2}$ in., $\frac{5}{8}$ in., and $\frac{25}{32}$ in.

Example

Find the number of board feet of flooring required for laying a floor 12'6″ × 21'3″ with $2\frac{1}{4}$″ matched flooring $\frac{5}{8}$″ thick. Allow 5% for waste.

SOLUTION
$$12.5 \times 21.25 \times 1 = 265.62$$
$$\text{(Allowance for matching) } 33\tfrac{1}{3}\% \text{ of } 265.62 = 88.54$$
$$\text{(Allowance for waste) } 5\% \text{ of } 265.62 = \underline{13.28}$$
$$367.44$$

Total flooring required is 368 board feet.

EXPLANATION Multiplying the width of the floor by the length gives the area in square feet. Since the thickness of the flooring is less than 1 in., it is figured as 1 in. Hence multiplying the area of the floor by 1 in. gives 265.62, the number of board feet that would be required if no allowances were made for matching or waste. Since $2\frac{1}{4}$-in. matched flooring is used, the allowance for matching, according to the table, is $33\frac{1}{3}$% of 265.62, or 88.54 fbm. Five percent for waste = 0.05 × 265.62, or 13.28 fbm. Adding these three quantities gives the total of 367.44, or 368 board feet.

Problems 9-21

Find the amount of flooring required in each of the following problems. Use the foregoing table for tongue and groove allowance.

1. Find the number of board feet of flooring required for a room 14'8″ × 24'0″ if 2″ matched flooring $\frac{1}{2}$″ thick is used. Allow 3% for waste.
2. How many board feet of flooring are needed for a room 13'6″ × 22'9″ if $2\frac{1}{4}$″ matched boards $\frac{1}{2}$″ thick are used? Allow 5% for waste.

3. An apartment consists of three rooms having the following dimensions: the living room is 11′6″ × 20′6″, the bedroom is 10′9″ × 12′3″, the dinette is 7′10″ × 11′8″, the foyer is 6′3″ × 15′6″. Find the number of board feet required for the floors of the apartment, using 2″ boards $\frac{5}{8}$″ thick. Allow 3% for waste.

4. A room 14′3″ × 24′0″ is to be floored with $2\frac{1}{4}$″ matched lumber $\frac{5}{8}$″ thick. Allowing 3% for waste, how many board feet are required?

5. A room 24′6″ × 32′6″ is to be floored with $3\frac{1}{4}$″ matched boards $\frac{25}{32}$″ thick. How many board feet are required if 5% is allowed for waste?

6. At $1,850.00 per M board feet, that is, 1,000 board feet, find the cost of covering the floor of a room 18′3″ × 28′6″. Matched lumber $2\frac{1}{4}$″ wide and $\frac{5}{8}$″ thick is used. Allow 3% for waste.

7. The walls of a dining room 21′3″ × 12′6″ are to be wainscoted for a height of 4′ above the floor. There are two doors 3′0″ wide. How many board feet of lumber are required if 2″ boards $\frac{5}{8}$″ thick are used, allowing 2% for waste?

8. A semicircular platform 48′ in diameter is to be floored with $2\frac{1}{4}$″ stock $\frac{25}{32}$″ thick. Allowing 8% for waste, how many board feet are needed?

9. Compute the amount of lumber required to cover the floor of a dance hall 70′ × 42′ if $3\frac{1}{4}$″ boards $\frac{5}{8}$″ thick are used. Allow 2% for waste.

10. How many board feet of hardwood flooring are required for a classroom 32′6″ × 28′9″ if $2\frac{1}{4}$″ boards $\frac{5}{8}$″ thick are used? Allow 3% for waste.

9-22 Summary of Formulas for Solids

PRISM AND CYLINDER

$$\text{Volume} = \text{area of base} \times \text{height}$$

$$\text{Area of base} = \frac{\text{volume}}{\text{height}}$$

$$\text{Altitude} = \frac{\text{volume}}{\text{area of base}}$$

$$\text{Lateral surface} = \text{perimeter of base} \times \text{height}$$

PYRAMID AND CONE

$$\text{Volume} = \tfrac{1}{3} \times \text{area of base} \times \text{height}$$

$$\text{Area of base} = \frac{3 \times \text{volume}}{\text{height}}$$

$$\text{Height} = \frac{3 \times \text{volume}}{\text{area of base}}$$

$$\text{Lateral surface} = \tfrac{1}{2} \times \text{perimeter of base} \times \text{slant height}$$

FRUSTUMS OF PYRAMIDS AND CONES

$$\text{Volume} = \tfrac{1}{3}h(B + b + \sqrt{B \times b})$$

$$h = \frac{3 \times \text{volume}}{B + b + \sqrt{B \times b}}$$

Lateral surface = average perimeter of bases × slant height

SPHERE

$$\text{Volume} = \tfrac{4}{3}\pi r^3$$

$$\text{Diameter} = \sqrt[3]{\frac{6 \text{ volume}}{\pi}}$$

$$\text{Surface} = 4\pi r^2$$

SOLID RING Volume = circumference of the center line circle × area of cross section.

EQUILATERAL TRIANGLE, SQUARE, HEXAGON, AND OCTAGON

For constants in reference to these figures, see the Table of Constants on page 172.

Self-Test

Solve the following problems. Use the tables in the back of the book, the Table of Constants on page 172, and the Summary of Formulas for Solids on pages 255–256. Check your answers with the answers in the back of the book. Try to estimate the answer to the problem before solving it.

1. Find the volume of a rectangular prism whose base is a 4 in. square and whose height is 10 in.
2. Find the volume of a cylinder 9 in. high and 3 in. in diameter.
3. Find the volume of a cone whose base is 4 in. in diameter and whose height is 8 in.
4. Find the lateral surface of a pyramid whose base is a 3 in. square and whose height is 4 in.
5. Find the volume of a 6 in. sphere.
6. Find the volume of a ring whose outside diameter is 4 in. and whose cross section is a 1-in. circle.
7. Find the weight of a steel bar 4 in. square and 20 ft. long. Steel weighs 490 lb per cu ft.
8. Find the weight of a cast iron casting made from a white pine pattern weighing 10 lb. Cast iron weighs 450 lb per cu ft and white pine weighs 25 lb per cu ft.
9. Find the number of board feet in 48 pieces of 2 × 6 ceiling joists 14 ft long.

10. Find the number of board feet of flooring required for laying a floor 10 ft by 12 ft with matched flooring $2\frac{1}{4}$ in. wide and $\frac{5}{8}$ in. thick. Allow 33% for matching and 5% for waste.

Chapter Test

Solve the following problems. Use the tables in the back of this book, the Table of Constants on page 172, and the Summary of Formulas for Solids on pages 255–256. Organize your work carefully and avoid making arithmetical errors. Estimate the answer before solving each problem.

1. Find the volume of a rectangular prism whose base is $10'' \times 15\frac{1}{2}''$ and height $22''$. (Answer in cu in.)
2. Find the volume of a prism whose base is an equilateral triangle with $10\frac{1}{2}''$ sides and whose height is $16''$. (Answer in cu in.)
3. Find the volume of a cylinder $14''$ in diameter and $10'6''$ long. (Answer in cu ft; 1 cu ft = 1,728 cu in.)
4. Find the weight of a brass bar $6''$ in diameter and $5'2''$ long. (Brass weighs 512 lb per cubic foot.)
5. What should the height of a rectangular prism be if the base is $10'' \times 10''$ and the volume is 1,500 cu in.?
6. A cylindrical 5-gallon container is $12''$ high. Find the diameter. (1 gallon = 231 cu in.)
7. Find the lateral surface of a cylinder $15''$ in diameter and $20''$ high.
8. Find the volume of a cone $18''$ high if the base is $15''$ in diameter.
9. Find the volume of a pyramid whose base is a square $12''$ on a side and whose height is $16''$.
10. Find the lateral surface of the cone in Problem 8.
11. Find the lateral surface of the pyramid in Problem 9.
12. Find the volume of a sphere $8\frac{1}{2}''$ in diameter.
13. Find the volume of a sphere whose radius is $10''$.
14. Find the surface area of a sphere whose diameter is $12''$.
15. Find the volume of a ring of circular cross section if the outside diameter is $6''$ and the ring is $1''$ thick.
16. The inside dimensions of a rectangular tank are $12'' \times 12''$ and $20''$ long and the tank is filled with water. What would be the weight of the water? (Water weighs 62.5 lb per cu ft.)
17. How many gallons of water would the tank in Problem 16 hold? (231 cu in. = 1 gallon.)
18. Find the weight of an iron casting whose volume is 7.82 cu ft. (Iron weighs 450 lb per cubic foot.)
19. Find the weight of 500 brass washers $2''$ in diameter and $\frac{1}{16}''$ thick. The hole is $1''$ in diameter.
20. Find the weight of 500 brass discs $1''$ in diameter and $\frac{1}{16}''$ thick.
21. Find the number of board feet in a floor joist $2'' \times 10'' \times 16'$.

22. A truckload of lumber is 16′ long, 7′6″ wide, and 4′6″ high. How many board feet does it contain?

23. How many board feet are in a 4″ × 4″ × 12′-long fir post?

24. Find the number of board feet required for a room 14′6″ × 18′0″ if $2\frac{1}{4}$″ matched flooring $\frac{5}{8}$″ thick is used. Allow $33\frac{1}{3}$% for matching and 3% for waste.

25. At $1,850.00 per thousand board feet, what will be the cost of the flooring in Problem 24?

10

Metric Measure

10-1 Introduction

Today almost all the industrialized world uses the *metric system* of measure-ment. The metric system is easier to use (once it is learned) than the *English system* of measure that the United States currently uses. The metric system has 10 as a base, and you can change from one unit to another by shifting the decimal point.

In the United States, Congress has decreed that the changeover to the metric system is voluntary. It is a matter of economics, because goods sold to other countries should be based on their system of measurement. The automobile is a good example. Suppose you were living in France and you purchased an American-manufactured car. All the nuts and bolts would be based on sixteenths of an inch, while all your wrenches and sockets would be based on millimeters. Hardly any of your tools would fit the parts of your new car. You would need two sets of tools. You would also not find it easy to get replacement parts, because the thread sizes in France are based on millimeters, and metric bolts are not compatible with American thread stan-dards. You can see that it would be a lot of trouble for you to maintain the car and you would probably be reluctant to purchase American-made goods

SOME reasons to go metric

SACRE-BLEU! These American cars! My metric tools won't fit. Those Americans will have to walk for a while!

unless their parts were compatible with the parts and tools available to you. The same analogy carries over to other American-made trade goods. Other countries are not going to convert to the measurement system based on inches, feet, yards, miles, and so on, because they find the metric system is easier to use and is more logical. The metric system is their official system of measurement and incidentally is also our *official* system of measurement. The French developed the metric system and in 1799 adopted it as their official system of measurement. Its use became mandatory in France in 1837.

The unit of length in the metric system is the meter. The French scientists decided that the length should be one-ten millionth of the distance from the north pole to the equator and used this distance as the standard of length of the meter. Their calculations were slightly in error and a better way was needed to define the length of a meter. Now the meter is defined as the length of 1,650,763.73 wavelengths in vacuum of the orange-red spectrum of krypton-86. Scientists use an interferometer to measure this length using light waves. In this way the length of a meter can be duplicated anywhere in the world. The meter is about 39.37 in. long (Fig. 10-1).

1 meter = 39.37 inches = 3.28084 feet = 1.09361 yards

The metric unit of area is the square meter. A square meter is the area of a square whose sides are each one meter long.

1 meter = 39.37 inches

1 yard = 36 inches

Fig. 10-1

The unit of volume is the cubic meter. A cubic meter is the volume of a cube whose length, width, and height are each one meter long. The liter is 0.001 part of a cubic meter, that is, a cube whose length, width, and height are each 0.1 meter long. Although the liter is not an official unit of the International System of Units, it is commonly used to measure the volume of fluids.

The standard unit of weight (technically mass) is the kilogram, a cylinder of platinum-iridium alloy kept at the International Bureau of Weights and Measures in Paris. There is a duplicate of this kilogram mass standard at the National Bureau of Standards in Washington, D.C.

Measurement of small items is usually done in millimeters (mm). For example, the width of film in 35 mm cameras is 35 millimeters—that is the reason for the name of this type of camera. The measure of nuts and bolts in foreign-made cars is also in millimeters. You can see from these two examples that smaller items are usually measured in millimeters. The standard for measurements on shop drawings is millimeters. All measurements are given in millimeters on shop drawings, no matter how large—this is an industry standard.

If the number of millimeters becomes larger, then it becomes more common to use centimeters (cm). A centimeter is about the width of the fingernail on your small finger. The width and length of this page could be given in centimeters.

Larger items, such as the room size in houses and rugs, and so on, use meters (m) as a measure. The length of a competitive swimming pool is 25 meters. Track events use 50 meters, 100 meters, 400 meters, and so forth for the sprints and distance events. Scientists use 300,000,000 meters per second as the velocity of light.

Distances between two cities use kilometers (km). The speedometer on some cars is calibrated in kilometers per hour, and the odometer registers the number of kilometers the car has been driven.

Time measure is still in our present system, although there is a suggestion to base time units on tens rather than use sixty as a base. Time will probably not be changed to a metric base in the near future.

Example 1
Find the total length of the part shown in Fig. 10-2. (All measurements are given in millimeters.)

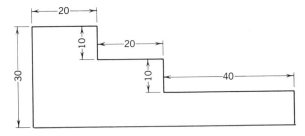

Fig. 10-2

SOLUTION   20 mm + 20 mm + 40 mm = 80 mm

EXPLANATION   The total length is the sum of all the dimensions in the horizontal direction for this piece. There are three dimensions to be added: 20 mm + 20 mm + 40 mm = 80 mm.

Example 2

The distance from Philadelphia to New York is about 150 kilometers. If a salesman has to commute daily from Philadelphia to New York and back, how many kilometers does he travel in 20 days?

SOLUTION   2 × 150 km/day × 20 days = 6,000 km

EXPLANATION   Since he must travel to New York and back to Philadelphia each day, he travels 300 kilometers each day. In 20 days he travels 20 × 300 or 6,000 km.

Problems 10-1

Use your knowledge of the metric system to answer the following questions.

1. The thickness of a dime is about: (a) 1 mm, (b) 1 cm, (c) 1 m, (d) 1 km.
2. The length of a baseball bat is about: (a) 1 mm, (b) 1 cm, (c) 1 m, (d) 1 km.
3. If a high school senior boy was 150 cm tall, would the basketball coach think he was tall?
4. The height of a classroom is about: (a) 3 mm, (b) 3 cm, (c) 3 m, (d) 3 km.
5. The distance from home plate to first base on an official baseball field is about: (a) 27 mm, (b) 27 cm, (c) 27 m, (d) 27 km.
6. The distance from Philadelphia to Pittsburgh is about: (a) 500 mm, (b) 500 cm, (c) 500 m, (d) 500 km.
7. Our inch is a little greater than: (a) 25 mm, (b) 25 cm, (c) 25 m, (d) 25 km.
8. Find the total length of the piece shown in Fig. 10-3. (All dimensions are in millimeters.)

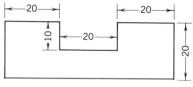

Fig. 10-3

9. The distance from the earth to the sun is about 150,000,000 km. Light travels about 300,000,000 meters per second. How many minutes does it take the light from the sun to reach the earth using these measurements? (Note: 1,000 meters equals 1 kilometer.)
10. Film is manufactured in wide rolls and then slit to make the narrower

widths to fit modern cameras. If the film is manufactured in 7-meter width rolls, how many slits would be needed to make 35-millimeter roll film from the entire 7-meter width? (This question is a little bit tricky.)

The ruler in Fig. 10-4 is divided into millimeters. It is a metric ruler. Each division is a millimeter and each number represents a centimeter (10 millimeters = 1 centimeter).

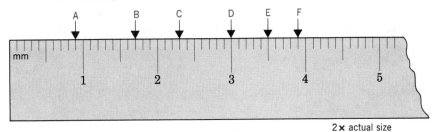

2× actual size

Fig. 10-4

11. What is the measure at A in millimeters?
12. Give the measure at B.
13. What is the measure at C?
14. Give the measure at D.
15. How many millimeters does the mark at E represent?
16. What is the measure at F?
17. What is the measure of A added to the measure at C?
18. Is 10 times the measure at C greater than or less than a meter?
19. How would you express 100 times the measure at E?
20. Subtract the measure at B from the measure at F.

10-2 Units of Length

The metric unit of length is the meter. The following table shows how the meter is subdivided. The units are related to one another by factors of 10.

METRIC LINEAR MEASURE
1,000 meters = 1 kilometer (km)
100 meters = 1 hectometer (hm)
10 meters = 1 dekameter (dam)
1 meter = 1 meter (m)
0.1 meter = 1 decimeter (dm)
0.01 meter = 1 centimeter (cm)
0.001 meter = 1 millimeter (mm)

To change a unit of length to the next larger unit of length, divide by 10; this is done by moving the decimal point one place to the left. To change a unit of length to the next smaller unit of length, multiply by 10; this is done by moving the decimal point one place to the right.

The commonly used units of length are the millimeter (mm), the centimeter (cm), the meter (m), and the kilometer (km).

Example 1

Change 80 millimeters (mm) to centimeters.

SOLUTION   80 mm = 8 cm

EXPLANATION   Since centimeters is the next larger unit to millimeters, divide by 10 by moving the decimal point one place to the left, thus converting millimeters to centimeters.

Example 2

Change 15.4 centimeters (cm) to millimeters (mm).

SOLUTION   15.4 cm = 154 mm

EXPLANATION   Since the millimeter is the next smaller unit to centimeters, multiply by 10 by moving the decimal point one place to the right; thus converting the centimeters to millimeters.

Example 3

Change 8,465 meters (m) to kilometers (km).

SOLUTION   8,465 m = 846.5 dam = 84.65 hm = 8.465 km

EXPLANATION   For each step up divide by 10 by moving the decimal point one place to the left. Since kilometers are three steps greater than meters, the decimal point is moved three places to the left, effectively dividing by 1,000.

Problems 10-2

Write the following lengths in meters.

1. 5 millimeters (mm)	2. 4.1 kilometers (km)
3. 6.3 hectometers (hm)	4. 2 decimeters (dm)
5. 276 millimeters (mm)	6. 25.4 centimeters (cm)
7. 47,635 millimeters (mm)	8. 15.3 millimeters (mm)
9. 0.05 kilometers (km)	10. 0.8 decimeters (dm)

Do the following problems. Be sure to include the units of measure in your answers.

11. What is the total length of eight pieces of steel each 21.6 cm long?
12. What is the total length of 25 pieces of drill rod each 12.95 cm long?
13. To find the circumference of a circle, multiply the diameter by 3.14. Find the circumference of a circle whose diameter is 25 cm.
14. A planer takes a 2-mm cut on a piece of steel 100 mm thick. What is the remaining thickness?
15. A steel casting 22.2 mm thick is finished by taking a 0.5-mm cut. What is the final thickness?
16. From a bar of brass 42 cm long the following three pieces are cut: 31.75 mm, 88.9 mm, and 85.5 mm. What is the final length of the bar allowing 1.55 mm for each cut?
17. The diagonal of a square is found by dividing a side by 0.707. What is the diagonal of a square whose sides are each 15 cm?
18. A space 12 m long is to be divided into 16 equal parts. How long is each part?
19. The diameter of a circle can be found by dividing the circumference by 3.14. What is the diameter of a circle whose circumference is 79.75 mm?
20. How many holes spaced 36.5 mm center to center can be drilled in an angle iron 57.4 cm long, allowing 31.5 mm end distances?

10-3 Units of Area

The unit of area is the square meter (Fig. 10-5).

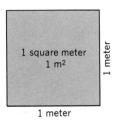

Fig. 10-5

The following table shows how the units of area are subdivided. Since area is the product of the length and width, the units are related to one another by a factor of 100. There are two common measures of area called *are* and *hectare*. These measures are names of a measure of area so that the term "square" is not used with them. The *hectare* (ha) is equal to a square hectometer (hm²), and the *are* is equal to a square dekameter (dam²).

To change a unit of square measure to the next larger unit of square measure, divide by 100; this is done by moving the decimal point two places to the left. To change a unit of square measure to the next smaller unit of square measure, multiply by 100; this is done by moving the decimal point two places to the right.

The commonly used units are square millimeter, square centimeter, square meter, and square kilometer. The hectare is used in land measure and is equivalent to 10,000 square meters.

METRIC SQUARE MEASURE
1,000,000 square meters = 1 square kilometer (km²)
10,000 square meters = 1 square hectometer (hm²)
100 square meters = 1 square dekameter (dam²)
1 square meter = 1 square meter (m²)
0.01 square meter = 1 square decimeter (dm²)
0.0001 square meter = 1 square centimeter (cm²)
0.000001 square meter = 1 square millimeter (mm²)

Example 1

Change 50,000 square centimeters (cm²) to square meters (m²).

SOLUTION   50,000 cm² = 5 m²

EXPLANATION   Since square meters is two units larger than square centimeters, divide by 100 twice, that is, divide by 10,000. This is done by moving the decimal point four places to the left, thus converting square centimeters to square meters.

Example 2

Change 7.75 hectares (ha) to square meters.

SOLUTION   7.75 ha = 77,500 m²

EXPLANATION   Since square meters (m²) is two units smaller than square hectometers (hm²), which in land measure is called hectares (ha), multiply by 100 twice, that is, multiply by 10,000. This is done by moving the decimal point four places to the right and adding zeros that are necessary, thus converting square hectometers to square meters.

Problems 10-3

Change the following to the units indicated.

1. 176.5 cm² = _____ dm²   2. 47 cm² = _____ dm²
3. 86.2 mm² = _____ cm²   4. 4.67 dm² = _____ cm²
5. 0.752 m² = _____ cm²   6. 0.75 km² = _____ m²
7. 0.005 m² = _____ mm²   8. 976.3 mm² = _____ m²
9. 0.56 km² = _____ ha   10. 46,255 m² = _____ ha

Do the following problems. Be sure to include the units of measure in your answers.

11. Find the floor area of a room 5 m long and 3.25 m wide.
12. Find the area of the walls of a room 2.5 m high if the room is 6 m long and 3.75 m wide. Make no allowances for door or window openings.
13. What is the total surface area of a cubical box if the box is 5.5 cm on an edge?
14. Find the cost of covering a floor 4.2 m by 6.5 m with carpet costing $12.25 per square meter.
15. A building lot has an area of 500 m². If its width is 15 m, how long is the lot?
16. Find the area of a right triangle if the altitude is 5 cm and the base is 4 cm.
17. Find the area of an isosceles triangle if one of the equal sides is 7.8 cm and the base is 6 cm.
18. The area of a square is 156.25 mm². What is the length of its side?
19. Find the area of a right triangle if one side is 20 mm and the hypotenuse is 25 mm.
20. The area of a circle is equal to 3.14 times the radius squared. Find the area of a circle whose radius is 7.5 mm.

Find the area of the following composite figures. Make a sketch of the figure and show how you have divided it into smaller figures whose area can be found. Label all parts of your computation.

21.

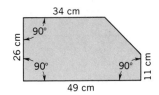

22.

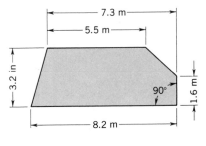

23.

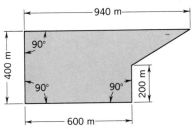

24.

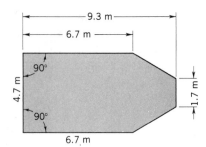

25.

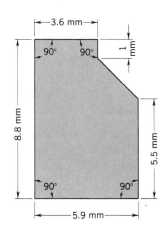

26.

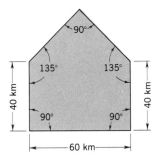

27.

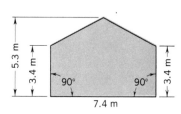

28. Find the area of the ring.

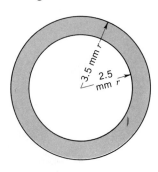

29. Find the area of the shaded portion.

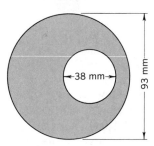

30. Find the area of the 60° sector of the circle whose radius is 37 mm as shown.

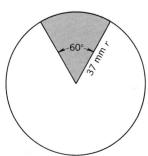

10-4   Units of Volume

The unit of volume measure is the cubic meter (Fig. 10-6).

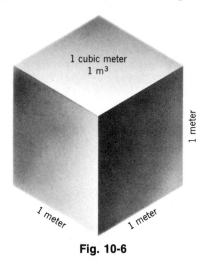

1 cubic meter
1 m³

1 meter

1 meter

1 meter

Fig. 10-6

The following table shows how the units of volume are subdivided. Since volume is the product of the length, width, and height, the units are related to one another by a factor of 1,000. The liter is a common term for a cubic decimeter and is frequently used when measuring volumes of liquids. The cubic meter is the largest volume measure used in this book.

METRIC VOLUME MEASURE
1 cubic meter = 1 cubic meter (m^3)
0.001 cubic meter = 1 cubic decimeter (dm^3)
0.000001 cubic meter = 1 cubic centimeter (cm^3)
0.000000001 cubic meter = 1 cubic millimeter (mm^3)

To change a unit of volume measure to the next larger unit of volume measure, divide by 1,000; this is done by moving the decimal point three places to the left. To change a unit of volume measure to the next smaller unit of volume measure, multiply by 1,000; this is done by moving the decimal point three places to the right.

The commonly used units of measure are the cubic millimeter (mm^3), the cubic centimeter (cm^3), the liter (l), which is a cubic decimeter (dm^3), and the cubic meter (m^3).

Example 1

Change 587.3 cubic millimeters (mm³) to cubic centimeters (cm³).

SOLUTION   587.3 m³ = 0.5873 cm³

EXPLANATION   Since cubic centimeters is the next larger unit to cubic millimeters, divide by 1,000 by moving the decimal place three places to the left and renaming cubic millimeters to cubic centimeters.

Example 2

Change 1.5 cubic meters (m³) to liters (l), that is, cubic decimeters (dm³).

SOLUTION   1.5 m³ = 1,500 l = 1,500 dm³

EXPLANATION   Since cubic decimeters is the next smaller unit to cubic meters, multiply by 1,000 by moving the decimal point three places to the right and renaming the cubic meters to liters.

Example 3

Find the volume of a rectangular prism whose base is a rectangle 5.5 cm by 5.25 cm and whose height is 6.5 cm.

SOLUTION $V = l \times w \times h$
$= 5.5 \times 5.25 \times 6.5$
$= 187.6875$ cm³ ≈ 187.7 cm³

EXPLANATION   Volume is the product of the length, width, and height. Since the measures are all in the same units, multiply the three together and round off the answer to 187.7 cm³.

Problems 10-4

Change the following to the units indicated.

1. 3.5 dm³ = _____ m³
2. 4.8 dm³ = _____ cm³
3. 487 mm³ = _____ cm³
4. 0.6 cm³ = _____ mm³
5. 0.56 m³ = _____ l
6. 763.2 l = _____ m³
7. 547,345 mm³ = _____ m³
8. 0.005 l = _____ mm³
9. 0.0067 cm³ = _____ mm³
10. 1.35 m³ = _____ mm³

Find the volumes of the following solids.

11. Find the volume of a rectangular prism 42 cm high whose base is a square 24 cm on each edge.
12. Find the volume of a cylinder 12 cm in diameter and 18 cm long.
13. Find the volume of the solid shell of the hollow cylinder in Fig. 10-7.

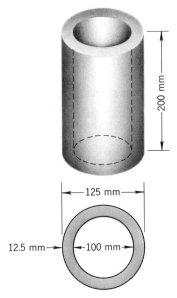

Fig. 10-7

14. Find the volume of a rectangular prism whose base is 30.5 cm × 37.5 cm and whose height is 45.7 cm.
15. Find the volume of a cylindrical tank in cubic meters if the diameter of the base is 2.45 m and the height is 7.6 m.
16. Find the volume of a prism 30.2 cm high whose base is a regular hexagon 2.5 cm across the flats. (The area of a hexagon is $0.866\,f^2$.)
17. Find the volume of a tank 7.5 m high if the base is a regular octagon whose side is 2.75 m long. (The area of a regular octagon is $4.828s^2$.)
18. Find the volume of a prism whose base is an equilateral triangle with 127 mm sides and whose height is 35 cm.
19. Find the volume (in liters) of a tank whose height is 122 cm and whose base is a rectangle 12.7 cm × 75 cm (1 = 1 dm³).
20. Find the volume of a cylindrical tank (in liters) if the height is 2.5 m and the diameter of the base is 1.25 m.

Do the following problems.

21. Find the capacity in liters of an elliptical tank 4 m long if the base is ellipse 55 cm × 50 cm.
22. Find the weight of a brass bar 75 mm in diameter and 250 mm long (brass weighs 8.2 g per cm³).
23. A rectangular prism with a square base 70 mm on edge is to hold 1 liter of fluid. How tall must the container be?
24. Find the depth of a rectangular vat 2.8 m × 4.7 m to contain 100 m³ of fluid.
25. The volume of a steel shaft 75 mm in diameter is 93.75 cm³. What is its length?

26. Find the lateral surface of a cylinder 36 cm in diameter and 24 cm high.
27. The base of a cylindrical tank is a circle whose area is 188.4 dm² and the tank has a capacity of 2.826 m³. What is the lateral surface area?
28. Find the volume of a right circular cone 1 m high whose base is 50 cm in diameter.
29. A quantity of sand dumped upon the ground assumed a conical shape with a circumference about the base of 31.4 m and a height of 3.3 m. How many cubic meters of sand are in the pile?
30. If the sand weighs 1,600 kg per m³, what is the weight of the pile in Problem 29?

10-5 Units of Weight

The standard unit of weight is the kilogram, the cylindrical platinum-iridium bar at Paris. This unit is the only metric unit defined by an artifact. The following table shows how the units of weight are subdivided. The units are related to one another by a factor of 10.

METRIC WEIGHT (MASS) MEASURE
1,000 kilograms = 1 metric ton (T)
1,000 grams = 1 kilogram (kg)
100 grams = 1 hectogram (hg)
10 grams = 1 dekagram (dag)
1 gram = 1 gram (g)
0.1 gram = 1 decigram (dg)
0.01 gram = 1 centigram (cg)
0.001 gram = 1 milligram (mg)

To change a unit of weight to the next larger unit of weight, divide by 10; this is done by moving the decimal point one place to the left. To change a unit of weight to the next smaller unit of weight, multiply by 10; this is done by moving the decimal point one place to the right. The metric ton measure in our table is three steps up from the kilogram.

The commonly used units of weight are the milligram (mg), the centigram (cg), the gram (g), the kilogram (kg), and the metric ton.

Example 1
Change 5,648 milligrams to grams.

SOLUTION   5,648 mg = 5.648 g

EXPLANATION   Since grams are three units larger than milligrams, divide by 10
three times, that is, divide by 1,000. This is done by moving the decimal
point three places to the left, thus converting milligrams to grams.

Example 2
Change 3.22 centigrams (cg) to milligrams (mg).

SOLUTION   3.22 cg = 32.2 mg

EXPLANATION   Since milligrams is the next smaller unit to centigrams, multiply
by 10 by moving the decimal point one place to the right and rename the
centigrams to milligrams.

Problems 10-5

Write the following weights as grams.

1. 4.2 dekagrams (dag)   2. 2.3 hectograms (hg)
3. 45.8 decigrams (dg)   4. 39.2 milligrams (mg)
5. 0.78 decigrams (dg)   6. 5.4 decigrams (dg)
7. 8.7 centigrams (cg)   8. 1.3 kilograms (kg)
9. 0.09 kilograms (kg)   10. 1 metric ton (T)

Do the following problems. Be sure to include the units of measure in your
answers.

11. What is the sum of the following weights: 3.4 g, 4.9 g, 18.7 g, and 3.05 g?
12. What is the sum of the following weights: 7.6 g, 5.9 dg, 4.7 dg, and 3.5 g?
13. Distribute 56 g into eight equal weights.
14. What is the weight of 47 bolts if each bolt weighs 97.3 mg?
15. A piece of steel plate weighs 1 kg. If 25 holes are drilled through the steel
    plate where the amount of material removed for each hole is 2.67 g, what
    is the final weight of the steel plate?
16. What is the weight of 687 castings if each casting weighs 3.45 kg?
17. What is the total weight of 45 pieces of steel where each piece weighs
    6.85 kg and 35 pieces of brass where each piece weighs 10.5 kg?
18. What is the sum of the following weights: 2.4 kg, 4.8 hg, 8.7 dag, 9.2 hg,
    and 8.8 hg?
19. If a brick measured 5.7 cm × 9.5 cm × 20.3 cm, what would it weigh if
    bricks weighed 2.4 grams per cm³?
20. How many bricks in Problem 19 would it take to weigh a metric ton?

10-6 English Length to Metric Length

One meter is equal to 39.37 in. One meter is equal to 100 cm. Hence, 1 cm is equal to 0.3937 in. Using ratio and proportion (page 96), we can find the number of centimeters in 1 in. (Fig. 10-8).

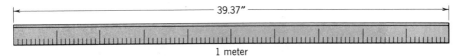

1 meter

Fig. 10-8

$$0.3937 \text{ in.}: 1 \text{ cm} = 1 \text{ in.}: x \text{ cm}$$

The extreme (x) can be found by multiplying the means (1 cm and 1 in.) and dividing by the other extreme (0.3937):

$$\frac{1 \times 1}{0.3937} = x$$

$$2.54 = x$$

Therefore, 2.54 cm is equal to 1 in.

Example 1
How many millimeters are in 1 in?

SOLUTION   2.54 cm = 1 in.
            25.4 mm = 1 in.

EXPLANATION   Since millimeters is the next smaller unit to centimeters, multiply by 10 by moving the decimal point one place to the right and rename centimeters to millimeters.

Example 2
A basketball player is 6′7″ tall. How tall is he in meters and parts of meters?

SOLUTION   6′7″ = (6 × 12)″ + 7″ = 79″
            2.54 × 79″ = 200.66 cm
        200.66 cm = 2.0066 m ≈ 2.01 m

EXPLANATION   Change the 6′7″ to inches. The 79″ is then multiplied by the number of centimeters in each inch, that is, 2.54 × 79″ = 200.66 cm. Change the centimeters to meters and round off the result to the nearest centimeter.

Example 3
How many centimeters are in one foot?

SOLUTION $1'' = 2.54$ cm
                     $1' = 12''$
        $2.54 \times 12 = 30.48$ cm

EXPLANATION   In 1 in. there are 2.54 cm, hence there are $12 \times 2.54 = 30.48$ cm
    in 1 ft.

ENGLISH MEASURE TO METRIC MEASURE SUMMARY

1 inch = 2.54 cm	1 yard = 36 inches
= 25.4 mm	= 91.44 cm
1 foot = 12 inches	= 0.9144 m
= 30.48 cm	1 mile = 5,280 feet
	= 1,609.344 m

Problems 10-6

Change the English measure in the following table to millimeters, centimeters, and meters as indicated.

ENGLISH MEASURE TO METRIC MEASURE

	English Measure	Millimeters (mm)	Centimeters (cm)	Meters (m)
1.	$4''$	_____	_____	_____
2.	$1'6''$	_____	_____	_____
3.	$8''$	_____	_____	_____
4.	$6'8''$	_____	_____	_____
5.	$4\frac{1}{2}''$	_____	_____	_____
6.	$2\frac{3}{4}''$	_____	_____	_____
7.	$\frac{1}{32}''$	_____	_____	_____
8.	$1'7\frac{1}{2}''$	_____	_____	_____
9.	$2'3\frac{1}{4}''$	_____	_____	_____
10.	Your height	_____	_____	_____

Do the following problems.

11. What is the length in centimeters of the piece of stock required for eight
    taper pins each $6\frac{1}{2}''$ long, allowing $\frac{1}{8}''$ waste for each cut?
12. In Fig. 10-9, what is the total length of the bolt in centimeters?
13. Find the difference in millimeters between the diameters at the ends of
    the tapered piece shown in Fig. 10-10.

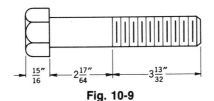

Fig. 10-9

Fig. 10-10

14. A planer takes a $\frac{3}{32}''$ cut on a piece of steel $\frac{3}{8}''$ thick. What is the remaining thickness in millimeters?
15. The diameter of a circle is $1''$. What is its circumference in centimeters? (Use 3.14 for π.)
16. The diameter of a circle is $6\frac{1}{2}''$. What is its circumference in centimeters?
17. The circumference of a circle is $26.69''$. What is its radius in centimeters?
18. A common brick measures $2\frac{1}{4}'' \times 3\frac{3}{4}'' \times 8''$. What is its measure in metric units?
19. The length of a compact car is $164''$. What is its length in meters?
20. An engine has a stroke of $3.5''$. What is its stroke in metric measure?

10-7 English Area to Metric Area

Area measure is the product of length and width. If the length and width are in inches, then the area is in square inches. If the length and width are in centimeters, then the area is in square centimeters. To change from the English area measure to the metric area measure, change the length and width measures to the same metric units and find their product. There are metric conversion tables in the Appendix of this book.

Example 1
Find the area of a rectangle in square centimeters if its length is $4\frac{1}{2}$ in. and its width is $3\frac{1}{2}$ in.

SOLUTION $2.54 \times 4\frac{1}{2} = 2.54 \times 4.5 = 11.430$ cm
$2.54 \times 3\frac{1}{2} = 2.54 \times 3.5 = 8.890$ cm
$8.890 \times 11.430 = 101.6127$ cm^2
$= 101.6$ cm^2

EXPLANATION   Change both of the inch measures to centimeters by multiplying them by 2.54 (the number of centimeters in 1 in.), then find the product of the two, that is, $8.890 \times 11.430 = 101.6127$. The answer is in square centimeters and is rounded off to one decimal place.

Example 2
Find the area of a circle in square centimeters if the radius of the circle is 1.875".

SOLUTION $A = \pi r^2$
$= 3.14 \times (2.54 \times 1.875)^2$
$= 71.22 \text{ cm}^2$

EXPLANATION   Use 3.14 as the approximation for π and change the 1.875" to centimeters by multiplying it by 2.54. This product is the radius in centimeters. Square the radius and multiply by 3.14. The answer is rounded off to two decimal places.

Problems 10-7

Find the area (in square centimeters) of the rectangles in the following chart. Change the length and width to centimeters, then find the area to one decimal place accuracy.

RECTANGLES

	Length	Width	Length (cm)	Width (cm)	Area
1.	3"	2"	_____	_____	_____
2.	6"	4"	_____	_____	_____
3.	3.5"	3"	_____	_____	_____
4.	6.7"	4.5"	_____	_____	_____
5.	$3\frac{1}{4}$"	2"	_____	_____	_____
6.	$2\frac{3}{4}$"	$1\frac{1}{2}$"	_____	_____	_____
7.	1'2"	9"	_____	_____	_____
8.	$1'4\frac{1}{2}$"	$8\frac{1}{2}$"	_____	_____	_____
9.	$1'4\frac{1}{4}$"	1'3"	_____	_____	_____
10.	$2'6\frac{5}{8}$"	$1'3\frac{1}{2}$"	_____	_____	_____

Find the following areas. Use metric units in your answer. Select the units that would be reasonable for the particular problem.

11. Find the area of a page of a book $7\frac{1}{2}$" long and 5" wide.
12. Find the area of a room 16'4" long and 10'8" wide.
13. How many square meters are there in a sidewalk 198' long and 4' wide?
14. Find the area of the walls of a room if the walls are 8' high and the room is 18' long and 12' wide. Make no allowance for doors or windows.
15. Allowing 2.5 m² per child, how many children may safely play in a playground 200' long and 140' wide?

16. How many square centimeters are there in the six sides of a box $8\frac{1}{2}''$ × $6''$ × $3\frac{1}{2}''$?
17. How many square meters are needed for the walls of a garage $15'6''$ × $8'9''$ and $8'3''$ high? (Subtract 6 m² as door and window allowance.)
18. Find the cost of covering a floor $13'9''$ × $21'6''$ with carpet costing $15.25 per square meter.
19. Find the area of a circle whose radius is $12'$.
20. Find the area of a 60° sector of a circle whose radius is $8\frac{1}{2}''$.

10-8   English Area to Metric Area Using Constants

If we have an area in square yards and we wish to change the measure to square meters, we can use the following procedure. A square meter in English units is a square whose sides are each 39.37″ long. A square meter is equal to 1,549.9969 sq in. A square yard is a square whose sides are each 36″. A square yard is equal to 1,296 sq in. A square yard is smaller than a square meter. The ratio of the square yard to the square meter is

$$\frac{1296}{1549.9969} = 0.8361307$$

To change square yards to square meters, multiply the square yards by 0.83613 (rounded off to the nearest hundred thousandth).

To find the ratio of the square foot to the square meter, compare the number of square inches in a square foot with the number of square inches in a square meter. The ratio is

$$\frac{144}{1549.9969} = 0.0929034$$

To change square feet to square meters, multiply the square feet by 0.09290 (round off the ratio to the nearest hundred thousandth).

To change square inches to square centimeters, compare the square inch to the number of square inches in a square centimeter, that is, 1 cm² = 0.3937″ × 0.3937″. The ratio is

$$\frac{1}{0.15499969} = 6.4516$$

To change square inches to square centimeters, multiply by 6.4516.

ENGLISH AREA TO METRIC AREA SUMMARY
1 sq in. = 6.4516 cm²
1 sq ft = 0.09290 m²
1 sq yd = 0.83613 m²
1 acre = 0.4047 hectares

Example 1

Change 950 sq yd to square meters.

SOLUTION $950 \times 0.83613 = 794.3235$
$= 794.3 \text{ m}^2$

EXPLANATION   Multiply the square yards by the constant 0.83613 to change square yards to square meters and round the answer to one decimal place.

Example 2

Change 2.5 sq ft to square meters.

SOLUTION $2.5 \times 0.09290 = 0.23225$
$= 0.23 \text{ m}^2$

EXPLANATION   Multiply the square feet, 2.5, by the constant 0.09290 to change square feet to square meters. Round the answer to two decimal places.

Example 3

Change 4.75 sq in. to square centimeters.

SOLUTION $4.75 \times 6.4516 = 30.6451$
$= 30.65 \text{ cm}^2$

EXPLANATION   Multiply the square inches by 6.4516 to change square inches to square centimeters. Round the answer to two decimal places.

Problems 10-8

Use the constants in this section to change the following square measures.

1. Change 1,985 sq yd to square meters.
2. Change 456.87 sq yd to square meters.
3. Change 45.8 sq ft to square meters.
4. Change 8.75 sq ft to square meters.
5. Change 40 sq in. to square centimeters.
6. Change 1.5 sq in. to square centimeters.
7. Change 34,450 sq yd to hectares. (1 hectare equals 10,000 sq m.)
8. Change 4,840 sq yd (the area of an acre) to ares. (An are equals 100 sq m.)
9. Change 20 acres to ares.
10. Change 5,000 acres to hectares.

10-9   English Volume to Metric Volume

Volume measure is the product of the length, width, and height. If these measures are in inches, then the volume is in cubic inches; if these measures are in centimeters then the volume is in cubic centimeters. To change from the English volume measure to the metric volume measure, change the length, width, and height measures to the same metric units and find their product.

Example 1

Find the volume in cubic centimeters of a rectangular prism whose height is 3.5″ and whose base is a rectangle 1.6″ by 2.2″.

SOLUTION $V = l \times w \times h$

$= (2.2 \times 2.54) \times (1.6 \times 2.54) \times (3.5 \times 2.54)$

$= 5.588 \times 4.064 \times 8.890$

$= 201.88862 \text{ cm}^3 \approx 201.9 \text{ cm}^3$

EXPLANATION   Volume is the product of the length, width, and height. Since the volume is wanted in cubic centimeters, the inch measures were changed to centimeter measures by multiplying each by 2.54, the number of centimeters in one inch. The answer was obtained by multiplying and rounding the result to the nearest tenth, that is, 201.9 cm^3.

If the volume of a solid is known in English units, the metric equivalent of the volume can be found by using constants that change English volume units to metric volume units.

ENGLISH VOLUME TO METRIC VOLUME SUMMARY
1 cu in. = 16.387 cm^3
1 cu ft = 28.317 dm^3
= 0.0283 m^3
1 cu yd = 0.7646 m^3

Example 2

Change 500 cu in. to cubic centimeters.

SOLUTION $500 \times 16.387 = 8{,}193.5 \text{ cm}^3$

EXPLANATION   To change cubic inches to cubic centimeters, multiply by the constant 16.387, since the constant table shows that there are 16.387 cm^3 in every cubic inch.

Problems 10-9

Find the volume in cubic centimeters of the rectangular prisms in the following chart. Use 3.14 as an approximation for π when finding the volume of the cylinders.

RECTANGULAR PRISMS

	English Measures			Metric Measures			
	Length	Width	Height	Length	Width	Height	Volume
1.	4″	3″	5″	_____	_____	_____	_____
2.	3.7″	2″	4″	_____	_____	_____	_____
3.	$2\frac{1}{2}″$	$2\frac{1}{2}″$	3″	_____	_____	_____	_____
4.	7.5″	4.2″	4.2″	_____	_____	_____	_____
5.	1′	8″	9″	_____	_____	_____	_____
6.	1′3″	1′1″	1′6″	_____	_____	_____	_____

Find the volume in cubic centimeters of the cylinders in the following chart. Use 3.14 as an approximation for π.

CYLINDERS

	English Measures		Metric Measures		
	Diameter	Height	Diameter	Height	Volume
7.	4″	3″	_____	_____	_____
8.	2.8″	4″	_____	_____	_____
9.	$3\frac{1}{2}″$	$2\frac{1}{2}″$	_____	_____	_____
10.	7.6″	3.5″	_____	_____	_____
11.	1′	8″	_____	_____	_____
12.	1′6″	1′2″	_____	_____	_____

Use the constants shown in the English volume to metric volume summary to find the following volumes in metric measure.

13. The volume of a rectangular prism is 1,444 cu in. What is its volume in cubic centimeters?
14. The volume of a right circular cylinder is 80 cu in. What is its volume in cubic centimeters?
15. The volume of a tank is 5 cu ft. What is its volume in liters? (One cubic decimeter equals one liter.)
16. What is the capacity of a gas tank in liters if it holds 20 gallons? (One gallon equals 231 cu in.)
17. A concrete truck holds 13 cu yd of concrete. How many cubic meters is this?
18. A circular vat has a capacity of 575 gallons. What is its capacity in cubic meters?
19. A concrete walkway is 4 ft wide, 80 ft long, and 6 in. deep. How many cubic meters of concrete was needed for this walk?

20. A swimming pool is 40 feet wide, 75 feet long, and 4 feet deep at the shallow end, and 12 feet deep at the deep end. How many cubic meters of water does it hold?

10-10   English Weight to Metric Weight

One kilogram equals 2.20462 lb, rounded off to 2.2 lb for calculations that do not demand great accuracy. One pound equals 0.453592 kg, rounded off to 0.454 kg for calculations that do not demand great accuracy.

Example 1
Change 3.5 lb to kilograms.

SOLUTION $3.5 \times 0.454 = 1.5890$ kg
$$= 1.6 \text{ kg}$$

EXPLANATION   To change pounds to kilograms, multiply by the number of kilograms in 1 lb, that is, 0.454. Round off the product to 1.6 kg.

Problems 10-10

Find the following weights in kilograms.

1. Change 8.75 lb to kilograms.
2. Change $5\frac{1}{2}$ lb to kilograms.
3. Find the sum of the following weights in kilograms; $2\frac{1}{2}$ lb, 4.375 lb, 1.875 lb, 3.25 lb.
4. Distribute 64 lb into 10 equal weights in metric measure.
5. What is the weight in grams of 20 bolts if each bolt weighs 1.1 oz?
6. What is the weight in kilograms of 875 castings if each casting weighs 21.5 lb?
7. Twelve hundred and fifty (1,250) castings weigh 5,625 lb. What is the average weight of each casting in kilograms?
8. One cubic foot of water weighs 62.5 lb. What is its weight in kilograms?
9. A metric ton is 1,000 kg. How many metric tons are there in 275 short tons? (One short ton equals 2,000 lb.)
10. A long ton weighs 2,240 lb. How many metric tons are there in 8,753 long tons?
11. What is your weight in kilograms?
12. A baby weighs between 6 and 9 lb at birth. What is this range in kilograms?

10-11   English and Metric Temperature

Water boils at 212°F and freezes at 32°F in the Fahrenheit temperature scale. The °F is called ''degrees Fahrenheit.'' In the metric system, water boils at

100°C and freezes at 0°C. This is called the Celsius temperature scale and the °C is called "degrees Celsius." At 4°C, water is more dense than water at 0°C (ice), and therefore the ice floats on the more dense water and the fish and other aquatic life are able to live through the cold weather.

In comparing the two scales, we see there are 108 divisions in the Fahrenheit scale that cover the same range as 100 divisions in the Celsius scale (Fig. 10-11). If 32 is subtracted from the degrees in the Fahrenheit scale, then both scales start at 0°, the temperature at which water freezes. Using these two ideas the conversions °F to °C and °C to °F can be made. To change °F to °C, subtract 32 from the °F and multiply the difference by the ratio $\frac{100}{180}$. This ratio is equal to $\frac{5}{9}$. To change a °C to °F, multiply the °C by $\frac{9}{5}$ and then add 32.

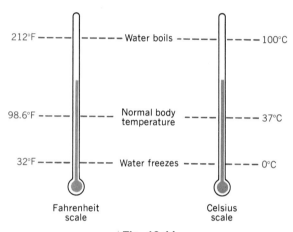

Fig. 10-11

Example 1
  Change 68°F to °C.

SOLUTION $(68 - 32) \times \frac{5}{9} = 36 \times \frac{5}{9} = 20°C$

EXPLANATION   Subtract 32 from the °F and then multiply the difference by $\frac{5}{9}$.

Example 2
  Change 50°C to °F.

SOLUTION $(50 \times \frac{9}{5}) + 32 = 122°F$

EXPLANATION   Multiply the °C by $\frac{9}{5}$ and add 32 to that product.

Example 3
  Change 14°F to °C.

SOLUTION $(14 - 32) \times \frac{5}{9} = -18 \times \frac{5}{9} = -10°C$

EXPLANATION   The answer is negative and indicates that the temperature is below the freezing point of water (0°C). The procedure is to subtract 32 from the °F and then multiply the difference by $\frac{5}{9}$. The difference is always negative in temperatures below 32°F.

Problems 10-11

Change to the temperature scale as indicated.

1. Change 115°F to °C.
2. Change 35°F to °C.
3. Change 210°F to °C.
4. Change 20°F to °C.
5. Change   0°F to °C.
6. Change 20°C to °F.
7. Change 45°C to °F.
8. Change 10°C to °F.
9. Change 90°C to °F.
10. Change 30°C to °F.

Self-Test

Do the following problems. Check your answers with the answers in the back of the book. Estimate the answer to each problem before solving it.

1. Change 4.8 km to meters.
2. Find the area of a right triangle whose altitude is 12 cm and whose base is 5 cm.
3. Find the volume of a rectangular prism whose base is 18.5 × 16.5 cm and whose height is 12 cm.
4. Distribute 81 cg into 90 equal weights.
5. The radius of a circle is 4 in. What is its circumference in centimeters?
6. What is the total surface area in cubic centimeters of a cube 3 in. on an edge?
7. Change 2.5 sq in. to square centimeters.
8. What is the volume of a cube 6 in. on an edge in metric measure?
9. Change 50 lb to kilograms.
10. Change 100°F to °C.

Chapter Test

Do the following problems. Use the constants developed in this chapter. Make sure that you use units that are reasonable for the problem. Try to estimate the answer before solving each problem.

1. What is the total length of 12 pieces of brass each 13.4 cm long?
2. A shaper takes a 1.2 mm cut on a piece of steel 1.27 cm thick. What is the remaining thickness?

3. The following pieces were cut from a 6 m length of copper tubing: five pieces 30 cm long and four pieces 42.5 cm long. What is the remaining length of the copper tubing?

4. A space 10 m long is to be divided into 14 equal parts. How long is each part?

5. Find the floor area of a house 7.5 m by 15.5 m.

6. Find the cost of covering a floor 5.2 m by 6 m with a rug at $15.75 per square meter.

7. A building lot has an area of 665 m². What is its length if its width is 17.5 m?

8. Find the area of a triangle if its base is 27 cm and its altitude is 13.5 cm.

9. Find the area of the splice plate shown below.

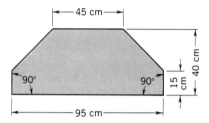

10. Find the area of the plate shown below.

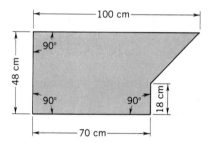

11. Find the area of the ring shown below.

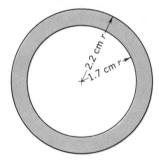

12. Find the volume of a rectangular prism whose base is 27 cm × 35.6 cm and whose height is 40 cm.

13. Find the volume of a cylindrical tank if the radius of the base is 1.5 m and the height is 3 m.
14. Find the volume in liters of a tank whose base is a square 60 cm on a side and whose height is 40 cm.
15. Distribute 72 kg into 100 equal weights.
16. What is the weight of 150 bolts if each weighs 35 g?
17. What is the length in centimeters of 12 dowels each 4.5″ long?
18. The circumference of a circle is 30 in. What is its radius in centimeters?
19. How many square meters are there in a sidewalk 70′ long and 3.5′ wide?
20. Change 15 sq in. to square centimeters.
21. Find the area in square centimeters of a circle whose radius is 4.6 in.
22. Find the volume in cubic centimeters of a right circular cylinder whose height is 5″ and the radius of the base is 2″.
23. 456 similar castings weigh a total of 2,052 lb. What is the average weight of each casting in kilograms?
24. Change 40°F to °C.
25. Change 70°C to °F.

11

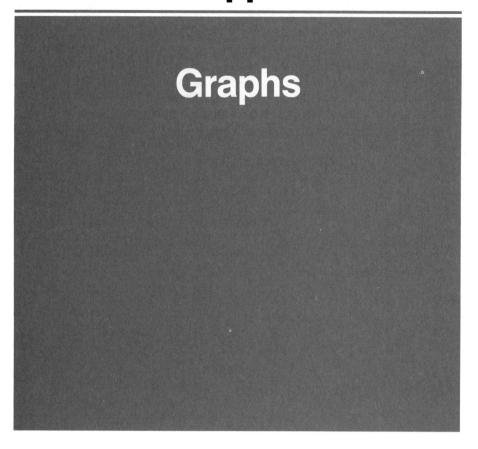

Graphs

11-1 Definitions

When there are two sets of related facts, the relations between them can often be shown clearly by means of diagrams, or **graphs.** Such graphs are frequently used in all types of business and technical work. Squared paper, sometimes called **cross-section paper,** is generally used for the purpose, and convenient values are assigned to the horizontal and vertical divisions. Two lines at right angles to each other are selected as **axes** or reference lines; the horizontal line is called the **X-axis** and the vertical line the **Y-axis.** The intersection of these two lines is called the **origin** (see Fig. 11-1). Points are plotted in relation to the two axes. These points are then connected by straight lines or curves, forming a broken line or a continuous curve, which shows graphically the relations that exist and the changes that take place.

11-2 Types of Graphs

There are many kinds of graphs, each adapted to a particular purpose. In general, however, they may be divided into two types, as follows.

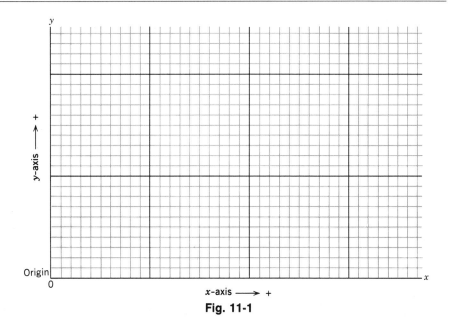

Fig. 11-1

Where no causal relation exists, that is, where the graph is merely a picture of a changing condition, is the first type of graph. Graphs showing changes of temperature or of population within a given period are examples of this type. The method of construction and the use of such graphs are illustrated by the following example.

Example 1

The population of a certain town, taken at intervals of ten years, varied as shown in the following table. Draw a graph showing the growth of this town.

Year	Population
1910	12,000
1920	16,000
1930	22,400
1940	26,000
1950	27,000
1960	27,200
1970	29,500
1980	29,200

CONSTRUCTION   Draw the horizontal and vertical reference lines as shown in Fig. 11-2. On the horizontal line, or *X*-axis, mark off the years 1910, 1920, and so on; on the vertical line, or *Y*-axis, mark off the populations by thousands. Distances on the vertical lines above the *X*-axis now represent

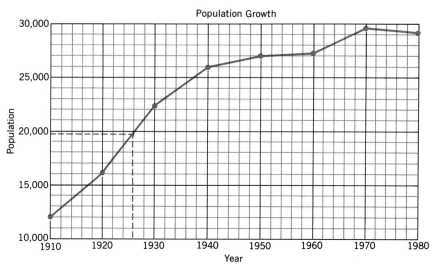

Fig. 11-2

populations, and distances on the horizontal lines to the right of the Y-axis represent years. At the intersection of the 1910 year line with the 12,000 population line, locate the first point. Locate other points at the intersection of the 1920 year line with the 16,000 population line, the 1930 line with the 22,400 line, and so forth. Connecting successive points with straight lines produces a broken line, which is the required graph of this information.

Note that the 10,000 line is used as the base line of this graph. This is done merely to save space, since nothing is gained by starting at zero, which is the real base line.

EXPLANATION   In this instance there is no necessary relation between the two sets of figures used in plotting the graph. The values do not depend on any law or rule, and the graph is merely a picture of the increase in the population. Instead of making a study of the tabulated figures, we can see at a glance how the population has varied from 1910 to 1980.

USE OF GRAPH   In connecting successive points by straight lines, we assume that the change from one point to the next has been uniform. To find the probable population for any year other than those given, we locate a point on the graph corresponding to that year. For example, to find the probable population in the year 1926, we note that the 1926 year line intersects the graph at a point whose height represents a population of approximately 19,700. We therefore say that the population of the town in the year 1926 was about 19,700.

Where causal relation does exist, that is, where the figures are related to each other by some definite rule or formula, is the second type of graph. Graphs showing the electrical resistance of a wire in relation to its diameter and graphs showing the variation in the surface speed of an emery wheel due to variation in the number of revolutions per minute are examples of this type. The following examples will illustrate the construction and use of such graphs.

Example 2
The surface speed in feet per minute of a 6-in. emery wheel, at various speeds, is given in the accompanying table. Draw a graph showing the relation between the surface speed and the revolutions per minute.

Revolutions per Minute	Surface Speed in Feet per Minute
0	0
500	785.4
1,000	1,570.8
1,500	2,356.2
2,000	3,141.6
2,500	3.927.0
3,000	4,712.4

CONSTRUCTION   Mark off the rpm on the *X*-axis and the surface speed in feet per minute on the *Y*-axis. The first point is at the origin, zero; the next point is at the intersection of the 500 rpm line with the 785.4 ft per minute line; the third point is at the intersection of the 1,000 rpm line with the 1,570.8 ft per minute line, and so on. Drawing a line through these points gives the required graph (see Fig. 11-3).

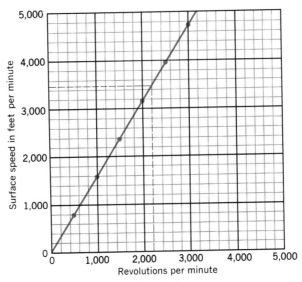

Fig. 11-3

EXPLANATION   This graph is a straight line beginning at zero and passing through all the plotted points. This is always true when the two sets of figures used in plotting the graph are related to each other by a simple proportion. The two figures used in plotting any point form the same ratio as those used for any other point; thus 500 : 785.4 :: 1,000 : 1,570.8 :: 1,500 : 2,356.2, and so forth. When we know that such a proportion exists, we can plot the graph by locating only two points and drawing a straight line through them, continuing the line as far as required. All of the points will lie on this line. It is advisable, however, in cases of this sort, to select the two points as far apart as the limits of the graph will permit. If possible, the divisions on the X-axis should equal the divisions on the Y-axis. The graph then shows the rate of change between the two.

USE OF THE GRAPH   This graph may be used for finding the surface speed produced by a given number of revolutions per minute or for finding the revolutions per minute required to produce a given surface speed. To find the surface speed resulting from 2,200 rpm, for example, we note the intersection of the 2,200 rpm line with the graph and read the corresponding value of the surface speed, 3,440 ft per minute, on the Y-axis. To find the number of revolutions per minute required to give a surface speed of 3,000 ft per minute, we note the intersection of the 3,000 ft per minute line with the graph and read the corresponding value of the revolutions per minute, 1,910 on the X-axis.

Example 3

The following table shows the electrical resistance in ohms per thousand feet of copper wires of different diameters. Plot a graph showing how the resistance of the wire varies with the diameter.

Diameter of Wire in Mils	Resistance in Ohms per 1000 ft
36	8.04
51	4.01
72	2.00
102	1.00
144	0.50
204	0.25

CONSTRUCTION   Mark off the diameters in mils on the X-axis and the resistances in ohms per 1,000 ft on the Y-axis. The intersection of the 36-mil line with the 8.04-ohm line is the first point; the second point is the intersection of the 51-mil line with the 4.01-ohm line, and so on. The graph is obtained by drawing a *smooth curve* through all the points (Fig. 11-4).

EXPLANATION   Since the several resistances are related to their respective di-

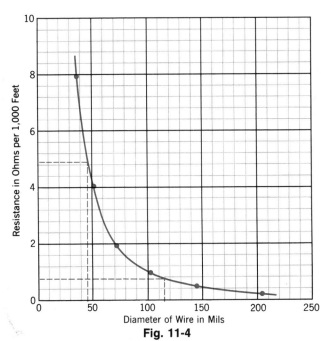

Fig. 11-4

ameters by a certain rule or formula (in this instance the law of inverse squares), there will be no sudden breaks or changes in the direction of the next point, as we can see by plotting a number of intermediate points. It is, however, unnecessary to do this since a smooth curve through the plotted points will also pass through all the intermediate points. The application of this rule enables us to reduce considerably the labor involved in plotting a graph representing a lengthy tabulation. For plotting we can select every third or fifth point and by drawing a smooth curve through them obtain the required graph.

USE OF THE GRAPH   Values on this graph are found in the same way as on the graph of Example 2. To find the diameter of a wire whose resistance is 5 ohms per 1,000 ft, we note the intersection of the 5-ohm line with the graph and read the corresponding value of the diameter, 46 mils, on the X-axis. To find the resistance per 1,000 ft of a wire whose diameter is 115 mils, we note the intersection of the 115-mil line with the graph and read the corresponding value of the resistance, 0.8 ohm, on the Y-axis.

11-3   Use of Graphs in Experimental Work

In making tests of materials and in experimental work of various kinds, we frequently find that we obtain points on the graph that do not seem to lie on the straight line or curve we know to represent correctly the conditions involved. Such points are accidental and may be due to inaccuracies in the measuring instruments or in the measurements, or to small imperfections in the material. When this happens, we disregard the *off* points and draw a smooth curve approximating the position of the majority of the points. Let us take, for example, a tension test of a steel wire where we know that, below the elastic limit, the load applied and the elongation are directly proportional to each other. This condition is represented by a straight line. We therefore draw a straight line approximating as nearly as possible the position of the majority of the points, disregarding the off points as inaccurate and not representing the actual conditions.

11-4   Two or More Graphs Combined

Such a combination is often of great value in showing the relations existing between two or more series of figures. A number of distinct sets of points may be plotted in reference to the same pair of axes, each set of points determining a separate graph.

Example

The average daily register and attendance of a certain school are given by the following table. Plot a graph showing these statistics.

Year	Register	Attendance
1960	711	683
1963	818	783
1966	901	827
1969	915	895
1972	961	933
1975	972	950
1978	1,003	978
1981	1,115	1,052
1984	1,008	982
1987	920	902
1990	864	833

CONSTRUCTION   Taking 600 pupils as a convenient base line, in Fig. 11-5 we plot two points on each vertical year line for which we have data, one point for the register and one for the attendance. Two distinct sets of points are thus obtained and lines drawn through them. To distinguish the two lines, we use a full line for the register and a dotted line for the attendance with an explanatory note, as in Fig. 11-5. Sometimes different colors are used to distinguish one graph from another.

Problems 11-1 to 11-4

Construct graphs from the given information. Choose your divisions on the X-axis and Y-axis and connect the points with a smooth curve. Be neat as well as accurate.

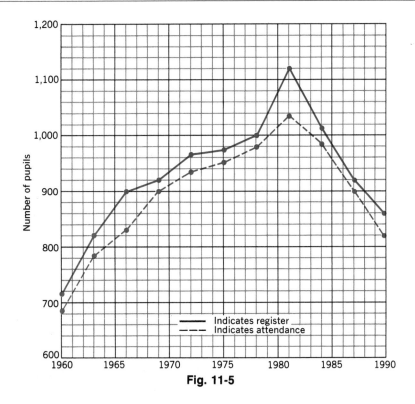

Fig. 11-5

1. The average temperature in New York City at different times of the year is as follows. Plot a graph showing the variation of temperature during the year.

Month	Temperature (°F)
January	30
February	31
March	38
April	48
May	59
June	68
July	74
August	72
September	66
October	56
November	44
December	34

2. Illustrate by means of a graph the Income Tax collections for the years indicated in the following data. (Source: U.S. Internal Revenue Service.)

Year	Collections in Dollars
1890	142,594,697
1900	295,316,108
1910	289,957,220
1920	4,256,456,929
1930	3,040,145,733
1940	11,487,934,290
1950	38,957,131,768
1960	91,774,802,823
1970	195,722,096,497
1980	359,927,162,449
1986	412,162,490,003

3. The weather report for a certain town shows the temperature, taken at 3-hour intervals, as follows.

Time	Temperature (°F)
12 A.M.	27
3 A.M.	24
6 A.M.	22
9 A.M.	25
12 P.M.	30
3 P.M.	38
6 P.M.	38
9 P.M.	37
12 A.M.	34

    Draw a graph showing the variation in temperature during the 24-hour period. Find from the graph the probable temperature at 7 A.M., at 2 P.M., and at 8 P.M.

4. In a tension test of a bar of wrought iron, the elongation of the bar for applied loads was as indicated in the following table. Plot a curve showing the relation between the load applied and the elongation of the bar.

Load (Pounds)	Elongation (Inches)	Load (Pounds)	Elongation (Inches)
200	0.012	2,600	0.230
500	0.035	2,700	0.272
1,000	0.065	2,800	0.357
1,500	0.140	2,900	0.418
2,000	0.172	3,000	0.473
2,500	0.215		

5. Illustrate by means of a graph the number of people immigrating to the United States as given in the following table. (Source: U.S. Immigration and Naturalization Service.)

Year	Number of People
1979	460,348
1980	530,639
1981	596,600
1982	594,131
1983	559,763
1984	543,903
1985	570,009
1986	601,708

6. Plot a curve showing the relationship between the whole numbers from 0 to 30 and their squares.
7. From the information in the following table of wages, construct three graphs showing the rise in average union hourly scale from 1960 to 1985 of the building trade; of the printing trade; of the local trucking trade. (Source: U.S. Department of Labor-Bureau of Labor Statistics.)

Year	Building		Printing		Local Trucking		
	Jour-ney-men	Helpers and Laborers	Book and Job	News-papers	Drivers	Helpers	Local Transit Workers
1960	$ 3.86	$ 2.88	$3.08	$ 3.48	$2.68	$2.38	$2.37
1965	4.64	3.54	3.58	3.94	3.26	2.90	2.88
1970	6.54	4.86	4.65	5.13	4.41	3.91	4.03
1975	9.32	7.06	6.86	7.57	6.87	6.27	6.25
1980	9.94	7.60	7.37	9.14	7.99	7.15	7.02
1984	12.13	10.44	9.53	10.88	9.36	8.48	8.35
1985	12.31	10.53	9.77	11.02	9.50	8.63	8.41

8. The following results were obtained when ten samples of a certain cement were prepared and tested for tensile strength at different time intervals.

Age (Weeks)	Tensile Strength (Pounds per Square Inch)
1	120
2	150
3	175
4	195
5	210
6	223
7	233
8	242
9	250
10	255

Draw a graph showing the relation between the age of the cement and its tensile strength.

9. The following table shows the number of tons of freight carried in the United States during certain years.

Year	Tons of Freight
1890	631,740,000
1895	686,615,000
1900	1,081,983,000
1905	1,427,732,000
1910	1,849,900,000
1915	1,802,018,000
1920	2,234,548,000

Draw a graph showing the growth of the freight traffic in the United States from 1890 to 1920.

10. The following table gives the normal weight of children from birth until they are 16 years old. Using the same pair of axes, draw a graph showing the weight increase for boys and for girls. From the graph find the normal weight of boys, and the normal weight of girls, 7 years old and $13\frac{1}{2}$ years old.

| Age | Normal Weight (Kilograms) | |
---	Boys	Girls
Birth	3.43	3.25
6 months	7.27	7.04
1 year	9.32	9.00
2 years	12.0	11.6
3 years	14.2	13.6
4 years	15.9	15.5
5 years	18.7	18.1
6 years	20.9	19.9
8 years	24.8	24.0
10 years	30.3	29.1
12 years	36.3	37.0
14 years	45.1	45.6
16 years	56.2	51.4

11. The following table gives the number of people in the United States who are 17 years old and the number of graduates from high school for that particular year. Using the same set of axes, draw a graph showing the relationship between the 17-year-old population and the number of high school graduates. (Source: U.S. Office of Education.)

School Year	Population 17 Years Old	High School Graduates
1974/1975	4,210,000	3,140,000
1975/1976	4,216,000	3,153,000
1976/1977	4,208,000	3,150,000
1977/1978	4,195,000	3,134,000
1978/1979	4,152,000	3,134,200
1979/1980	4,143,000	3,058,000
1980/1981	4,027,000	3,020,000
1981/1982	3,975,000	3,001,000
1982/1983	3,864,000	2,888,000
1983/1984	3,698,000	2,773,000
1984/1985	3,660,000	2,683,000

12. A water wheel was tested for efficiency at different speeds and the following results were obtained.

Speed	Efficiency
50 rpm	18%
100 rpm	37%
150 rpm	51%
200 rpm	63%
250 rpm	68%
300 rpm	68½%
350 rpm	65%
400 rpm	57%

Plot the efficiency curve and find the greatest efficiency and the speed at which it occurs.

11-5   Circle Graphs

An effective way of showing the percentage distribution of a given quantity is by means of a *circle graph*. The following example illustrates the method.

Example

The cost of governing a certain city amounted to $24,652,000, apportioned as follows: education, $4,659,228; police and fire departments, $2,317,288; health, hospital, and sanitation, $2,859,632; pensions, $2,243,322; debt service, $4,511,316; others, $8,061,204. Plot a circle graph showing the percentage allotments.

SOLUTION

Education =

$$\frac{4,659,228}{24,652,000} = 18.9\% \text{ of total} = 0.189 \times 360° = 68.04°$$

Police and fire =

$$\frac{2,317,288}{24,652,000} = 9.4\% \text{ of total} = 0.094 \times 360° = 33.84°$$

Health, etc. =

$$\frac{2,859,362}{24,652,000} = 11.6\% \text{ of total} = 0.116 \times 360° = 41.76°$$

Pensions =

$$\frac{2,243,322}{24,652,000} = 9.1\% \text{ of total} = 0.091 \times 360° = 32.76°$$

Debt service =

$$\frac{4{,}511{,}316}{24{,}652{,}000} = 18.3\% \text{ of total} = 0.183 \times 360° = 65.88°$$

Others =

$$\frac{8{,}061{,}204}{24{,}652{,}000} = 32.7\% \text{ of total} = 0.327 \times 360° = 117.72°$$

Total = 100%   360°

CONSTRUCTION   Compute the percentage of the total that is spent for each item. Since there are 360° in the circumference of a circle, the amount spent on education, which is 18.9% of the total, is represented by a sector whose arc is 18.9% of 360° or 68.04°. Since 9.4% of the total is allotted to the police and fire departments, that item is represented by a sector whose arc is 9.4% of 360° or 33.84°, and so forth. With a protractor lay off the respective arcs on the circumference of a circle and draw the radii, forming the sectors as shown in Fig. 11-6.

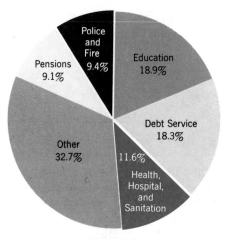

Fig. 11-6

Problems 11-5

Draw a circle graph illustrating the following data. (Use a compass and a protractor for this work.)

1. Show by means of a circle graph the percentage of the total subscribed to the International Bank by each of the countries listed in the following table.

Country	Percentage of Total
United States	41.4
United Kingdom	17.0
China	7.8
France	5.9
India	5.2
Canada	4.2
Netherlands	3.6
Other countries	14.9
Total =	100.0%

2. The U.S. Department of Energy estimates that by 1990 the sources of energy will be as follows.

Source	Percentage of Total
Hydropower	4.1
Coal	31.9
Nuclear power	9.2
Oil	34.9
Gas	19.9
Total =	100.0%

Draw a circle graph using this information.

3. In 1971 the personal expenditures of the population of the United States was $103.5 billion for durable goods such as automobiles, furniture, and the like; $278.1 billion for nondurable goods such as food, clothing, and gas; and $283.3 billion for services such as housing, transportation, and household operation. Draw a circle graph showing the relationships of personal expenditures. (Source: U.S. Dept. of Commerce.)

4. The following table gives the production of agricultural commodities in 1986 in bushels. Construct a circle graph showing what percentage of the total each item constituted. (Source: U.S. Dept. of Commerce.)

Commodity	Bushels
Corn	8,252,834,000
Wheat	2,086,780,000
Soybeans	1,764,112,000
Oats	1,185,936,000

5. The outlay for education in a city in 1965 was 29.5% of the entire budget; for police and fire, 13.6%; health, hospital, and sanitation, 9.7%; pensions, 5.2%; debt service, 9%; others, 33%. Illustrate the given percentages by means of a circle graph.

11-6   Bar Graphs

Another method of comparing quantities is by means of a *bar graph*.

Example

The United States budget in millions of dollars for the years indicated is as follows. (Source: U.S. Bureau of Budget.)

Year	Budget
1940	6,879
1950	40,940
1960	92,470
1970	198,686
1980	579,011

Draw a bar graph showing the budget increase in these ten-year periods.

CONSTRUCTION   On the horizontal axis, mark off the millions of dollars and, on the vertical axis, mark off the years as shown in the diagram (Fig. 11-7). The length of each bar represents the millions of dollars budgeted in that year. The graph shows at a glance how the U.S. budget has increased.

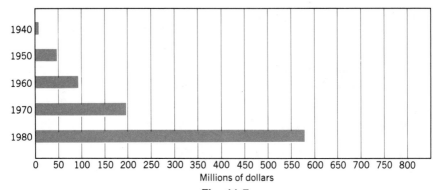

Millions of dollars

Fig. 11-7

The graph can also be constructed with the bars drawn vertically as shown in Fig. 11-8. In this instance, the years are marked off on the horizontal axis and the millions of dollars on the vertical axis.

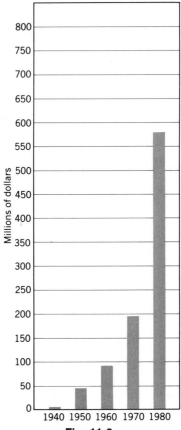

Fig. 11-8

Problems 11-6

Draw bar graphs showing the following data.

1. Draw a bar graph showing years of expected life at birth. (Source: U.S. Dept. of Health, Education and Welfare.)

Year of Birth	Life Expectancy
1920	54.1
1930	59.7
1940	62.9
1950	68.2
1960	69.7
1970	70.8
1980	71.0

2. Show by means of a bar graph the amount of Pennsylvania anthracite coal produced as given in the following table. (Source: Bureau of Mines.)

Year	Thousands of Tons
1945	54,934
1950	44,077
1955	26,205
1960	18,817
1965	14,866
1970	9,729
1975	10,144
1980	10,367
1985	7,222

3. The following table shows the wool produced in millions of pounds for the years 1960 to 1970. Draw a vertical bar graph showing these data. (Source: U.S. Dept. of Agriculture.)

Year	Wool Produced (Millions of Pounds)
1960	298.9
1962	261.2
1964	237.4
1966	219.2
1968	198.1
1970	176.9

4. Illustrate the following data by a bar graph. (Source: Interstate Commerce Commission.)

Year	Railroad Domestic Freight Traffic (Millions of Ton-Miles)
1965	708,700
1970	771,168
1975	759,000
1980	932,000
1985	895,000
1990 (est.)	850,000

12

Measuring Instruments

12-1 The Micrometer

For making fine measurements, an instrument called the *micrometer* is used. This consists essentially of a 40-pitch screw and a thimble whose circumference is divided into 25 equal parts (see Fig. 12-1).

Since the pitch of the screw is $\frac{1}{40}$ of an inch, each complete turn of the thimble advances the spindle $\frac{1}{40}$ in. or 0.025 in. The barrel or body of the instrument is therefore calibrated by dividing the inch into 40 equal parts, each 0.025 in. For convenience in reading, every fourth division, representing 0.100 in., is numbered, so that number 1 represents 0.100 in., number 2, 0.200 in., and so on. The reading is taken at the edge of the thimble. The first mark after the zero represents 0.025 in., the second mark after the zero, 0.050 in., the first mark after the 4, 0.425 in., the third mark after the 8, 0.875 in., and so forth. To read between these figures, the beveled edge is divided into 25 equal parts, each representing $\frac{1}{25}$ of 0.025 in., which is 0.001 in. When the zero on the thimble coincides with the gage line on the barrel, the reading is simply so many complete turns as shown by the barrel reading. If the zero on the thimble does not coincide with the gage line on the barrel, the line

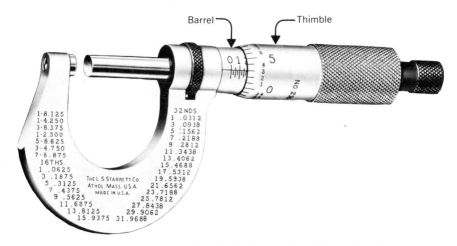

Fig. 12-1 (Courtesy of the L. S. Starrett Co., Athol, Mass.)

which does coincide must be found and that many thousandths must be added to the barrel reading. For example, if the line marked 15 coincides with the gage line, add 0.015 in.

Example 1

What is the reading when the edge of the thimble is between the 0.125 and the 0.150 lines on the barrel, and the 18 line on the thimble is the coinciding line?

SOLUTION   Micrometer reading = barrel + thimble
$$= 0.125 \text{ in.} + 0.018 \text{ in.} = 0.143 \text{ in.}$$

Example 2

Find the reading if the thimble edge is between 0.850 and 0.875 on the barrel and the 3 line coincides.

SOLUTION   Micrometer reading = barrel + thimble
$$= 0.850 \text{ in.} + 0.003 \text{ in.} = 0.853 \text{ in.}$$

Example 3

Find the reading if the thimble edge is between 0.050 and 0.075 and the gage line is between the 13 and 14 thimble lines, but nearer the 14.

SOLUTION   Micrometer reading = barrel + thimble
$$= 0.050 \text{ in.} + 0.014 \text{ in.} = 0.064 \text{ in.}$$

EXPLANATION   When the gage line falls between two of the thimble lines without exactly coinciding with either of them, select the nearest one.

Problems 12-1

Supply the micrometer readings for each of the micrometer settings in the following chart.

	Thimble is between	Coinciding Line on Thimble	Micrometer Reading
1.	0 and 0.025	8	_____
2.	0.125 and 0.150	18	_____
3.	0.275 and 0.300	17	_____
4.	0.850 and 0.875	16	_____
5.	0.125 and 0.150	5	_____
6.	0.975 and 1.000	1	_____
7.	0.025 and 0.050	22	_____
8.	0.375 and 0.400	4	_____
9.	0.150 and 0.175	2	_____
10.	0.650 and 0.675	16	_____

Set the following micrometer readings:

11. 0.109″ 12. 0.073″ 13. 0.218″ 14. 0.813″ 15. 0.401″
16. $\frac{3}{8}''$ 17. $\frac{1}{4}''$ 18. $\frac{53}{64}''$ 19. $\frac{21}{32}''$ 20. $\frac{9}{16}''$

12-2 The Ten-thousandths Micrometer

When more accurate measurements are required, we employ a micrometer that has an extra scale added to the barrel, enabling us to read to ten-thousandths of an inch. This scale consists of a series of lines on the barrel, parallel to the axis of the barrel (see Fig. 12-2). A part of the circumference of the barrel equal to 9 of the thimble divisions is divided into ten equal parts. Each part therefore represents $\frac{1}{10}$ of 0.009 in. or 0.0009 in. The difference between one of these divisions and a thimble division is 0.001 in. − 0.0009 in. = 0.0001 in. We find the number of ten-thousandths of an inch to add to the barrel and thimble readings by looking for the line of this auxiliary scale that exactly coincides with a thimble line. A scale of this sort which depends on the *difference between dimensions* for its principle is called a *vernier scale.*

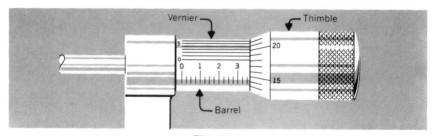

Fig. 12-2

The reading on a ten-thousandths micrometer consists of three parts:

1. The barrel, on which each division represents 0.025 in.
2. The thimble, on which each division represents 0.001 in.
3. The vernier, on which each division represents 0.0001 in.

Example

The thimble is between 0.075 and 0.100, the gage line on the barrel is between 19 and 20 on the thimble, and line 8 on the vernier coincides with a line on the thimble. What is the reading?

SOLUTION   Micrometer reading = barrel + thimble + vernier
$$= 0.075 \text{ in.} + 0.019 \text{ in.} + 0.0008 \text{ in.}$$
$$= 0.0948 \text{ in.}$$

Problems 12-2

Supply the micrometer reading for each of the micrometer settings in the following chart.

Thimble is between	Gage Line between Thimble Lines	Coinciding Vernier Line	Micrometer Reading
1. 0.125 and 0.150	16 and 17	3	———
2. 0.925 and 0.950	1 and 2	1	———
3. 0 and 0.025	9 and 10	2	———
4. 0 and 0.025	24 and 25	9	———
5. 0.025 and 0.050	14 and 15	7	———
6. 0.800 and 0.825	11 and 12	8	———
7. 0.625 and 0.650	9 and 10	2	———
8. 0.975 and 1.000	18 and 19	3	———
9. 0.850 and 0.875	7 and 8	6	———
10. 0.225 and 0.250	3 and 4	5	———

Set the following micrometer readings:

11. 0.9815″   12. 0.2713″   13. 0.1043″   14. 0.0008″   15. 0.2108″
16. 0.0315″   17. 0.0071″   18. 0.6835″   19. $\frac{3}{16}$″   20. $\frac{31}{64}$″

12-3   The Vernier Caliper

Another instrument that depends on the vernier principle is the vernier caliper (see Fig. 12-3). This consists essentially of a main scale with a fixed jaw and an auxiliary or vernier scale attached to the sliding jaw. On the main scale each inch is divided into 40 parts, so that each part is $\frac{1}{40}$ in. or 0.025 in. For convenience in reading, every fourth division is numbered 1, 2, 3, and so on, indicating 0.100 in., 0.200 in., 0.300 in., and so forth.

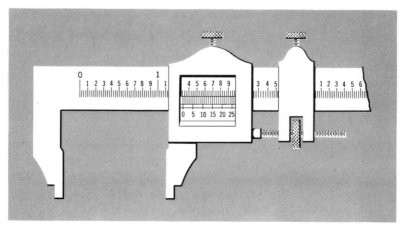

Fig. 12-3

The zero of the sliding scale is the index or gage line, and its location in reference to the main scale determines part of the reading. The remainder of the reading is determined by the vernier scale, which is constructed as follows: A length equal to 24 main scale divisions or 0.600 in. is divided into 25 equal parts on the vernier scale; every fifth division is numbered to facilitate reading. Each space on the vernier is therefore equal to $\frac{1}{25}$ of 0.600 in. or 0.024 in. The difference between a scale division and a vernier division is 0.025 in. − 0.024 in. = 0.001 in. Figure 12-4 shows an enlarged view of the scale and vernier showing a zero reading, the zero of the vernier coinciding with the zero of the scale. If the sliding jaw is now moved until the first division of the vernier coincides with the first scale division, we have moved 0.001 in. because that is the difference between a vernier division and a scale division. If we move the sliding jaw until the fifth line coincides, we have moved 0.005 in. and set the caliper for 0.005 in. reading. The vernier number

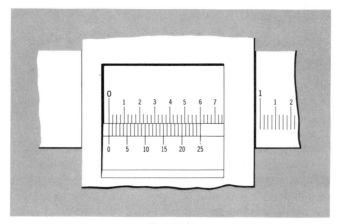

Fig. 12-4

that coincides with a scale division shows the number of thousandths of an inch to add to the scale reading.

Example 1

The index is between 0.275 in. and 0.300 in. on the scale, and line 23 on the vernier coincides with a scale line. What is the reading?

SOLUTION   Reading = scale + vernier
$$= 0.275 \text{ in.} + 0.023 \text{ in.} = 0.298 \text{ in.}$$

Example 2

Set the vernier caliper for 0.131 in.

SOLUTION   0.131 = 0.125 + 0.006. Bring the index over 0.125 in., then move the sliding jaw until line 6 on the vernier coincides with a scale line.

Problems 12-3

Supply the reading for each of the settings of the vernier caliper in the following chart.

Index between Lines	Coinciding Vernier Line	Reading
1. 0.125 and 0.150	16	_____
2. 0.175 and 0.200	9	_____
3. 0.250 and 0.275	18	_____
4. 0.900 and 0.925	12	_____
5. 0.975 and 1.000	6	_____
6. 0.350 and 0.375	1	_____
7. 0.100 and 0.125	13	_____
8. 0.000 and 0.025	5	_____
9. 0.075 and 0.100	24	_____
10. 0.050 and 0.075	11	_____

Set the vernier caliper for the following measurements.

11. 0.020″	12. 0.218″	13. 0.173″	14. 0.101″	15. 0.908″
16. 1.003″	17. 0.125″	18. 0.099″	19. $\frac{3}{4}$″	20. $\frac{3}{16}$″

12-4   The Protractor

Figure 12-5 shows an ordinary protractor used for measuring and laying out angles. In use, the base line 0° to 180° is laid on one of the lines with the center of the protractor at the point that is to be the vertex of the required angle. A mark is made at the point on the circumference where the required number of degrees is indicated. A line is then drawn between these two points. The illustration shows a typical plastic protractor.

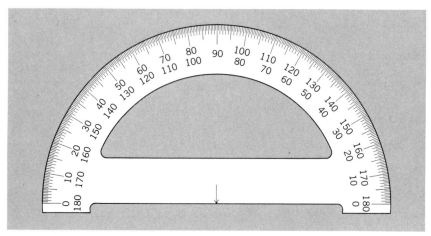

Fig. 12-5

12-5   The Vernier Protractor

For more accurate work, a steel protractor is used, carrying a vernier scale in addition to the regular scale (see Fig. 12-6). In this particular instrument, the smallest division on the main scale is $\frac{1}{2}°$, or $30'$. By means of the vernier scale, we can divide one of these divisions into ten parts, enabling us to read to $3'$. Figure 12-7 shows an enlarged view of the vernier scale in relation to the main scale. A length of arc equal to 9 of the smallest scale divisions, or $9 \times 30' = 270'$, is divided into ten equal parts on the vernier scale, making each division equal to $\frac{1}{10}$ of 270, or $27'$. The difference between a scale division and a vernier division is $30' - 27' = 3'$.

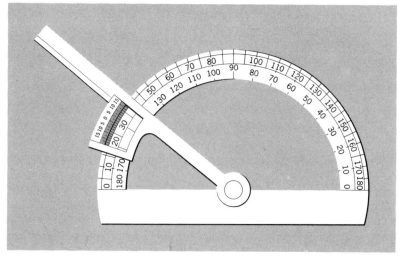

Fig. 12-6

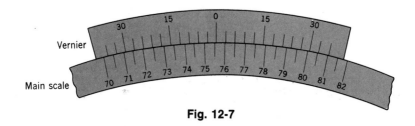

Fig. 12-7

When the vernier zero, which is the index, exactly coincides with a scale division, the scale division shows the reading. But when the index does not exactly coincide with any scale division, we take the lower scale division and add the vernier reading as indicated by the coinciding vernier line.

Example 1
The index is between 27°30′ and 28°, and the 21 on the vernier is the coinciding line. What is the reading?

SOLUTION   Angle = scale + vernier
             = 27°30′ + 21′ = 27°51′

Example 2
Set the vernier protractor for 72°42′.

SOLUTION   Set the index between 72°30′ and 73° and move the vernier until the 12 line on the vernier coincides with a scale division.

Problems 12-5

Supply the reading for each of the vernier protractor settings in the following chart.

	Index between	Coinciding Vernier Line	Reading
1.	12°00′ and 12°30′	15	_____
2.	91°30′ and 92°00′	21	_____
3.	11°30′ and 12°00′	27	_____
4.	60°00′ and 60°30′	3	_____
5.	27°00′ and 27°30′	0	_____
6.	41°00′ and 41°30′	15	_____
7.	125°30′ and 126°00′	27	_____
8.	16°30′ and 17°00′	6	_____
9.	3°30′ and 4°00′	21	_____
10.	75°00′ and 75°30′	18	_____

Set the vernier protractor for the following angles.

11. 27°54′	12. 2°15′	13. 45°45′	14. 15°15′	15. 52°03′
16. 12°12′	17. 9°27′	18. 55°57′	19. 21°33′	20. 91°12′

12-6   The Planimeter

The *planimeter* is an instrument used for measuring irregular areas such as indicator card diagrams. Figure 12-8 shows the Amsler Polar Planimeter, one of the simplest forms of this instrument.

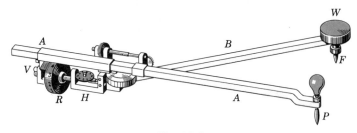

Fig. 12-8

The following are the essential parts and their functions:

The long beam *AA* carrying at one end the tracing point *P*.

The short beam *B*, pivoted at one end so as to allow free movement at the joint and carrying at the other end a needle point *F*, that acts as a fixed center around which the entire instrument turns.

A small weight *W* to hold this needle point in place.

The registering apparatus.

The rolling disk *R* of the registering apparatus rotates as the entire instrument is dragged around after the tracing point *P*, which is tracing the outline of the figure whose area is being measured. One complete turn of the rolling disk indicates an area of 10 sq in. The circumference of the disk, therefore, is divided into ten parts, each representing 1 sq in. Each space is further divided into ten parts, each division representing $\frac{1}{10}$ of a square inch. Hundredths of a square inch are read by means of the vernier *V*, which is placed next to the roller. The zero of this vernier is the index or gage line of the instrument.

Complete turns of the roller, each representing 10 sq in., are counted by means of the horizontal disk *H*, which is connected to the roller by means of a worm and gear. Ten complete turns of the roller are required to produce one complete turn of the counter. Its surface is therefore divided into ten parts, each representing 10 sq in.

12-7   Use of the Planimeter

The planimeter is used as follows: The diagram in Fig. 12-9 shows a view of the planimeter in reference to the figure whose area is being measured. The

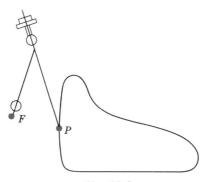

Fig. 12-9

fixed point F is located conveniently near the figure. The tracing point P is placed at a marked point on the perimeter of the figure and carefully drawn over the entire perimeter until it returns to the starting point. This motion of the tracing point causes the roller to rotate and register the area. Readings are taken before and after the operation, and the difference between the readings is the area of the figure. To minimize errors it is customary to repeat the operation several times and take the average of the results.

The initial reading is taken with the tracer P at the starting point, and the final reading is taken when the tracer has returned to the starting point.

The following example will illustrate the method of reading the planimeter. If the gage line of the counter is between 6 and 7, the reading is between 60 and 70 sq in., that is, it is 60 plus, each division on the counter representing 10 sq in. If the vernier index is between 71 and 72 on the roller, the roller reading is between 7.1 and 7.2 sq in., that is, 7.1 plus. If line 6 on the vernier coincides with a roller line, the vernier reading is 0.06 sq in.

$$\text{Total reading of instrument} = \text{counter} + \text{roller} + \text{vernier}$$
$$= 60 + 7.1 + 0.06 = 67.16 \text{ sq in.}$$

If 67.16 sq in. is the initial reading and 73.81 is the final reading, the area of the figure measured is $73.81 - 67.16 = 6.65$ sq in.

Problems 12-7

Supply the planimeter reading for each set of conditions in the following chart.

Counter between	Roller between	Coinciding Vernier Line	Reading
1. 2 and 3	56 and 57	2	_____
2. 9 and 10	31 and 32	3	_____
3. 5 and 6	00 and 01	1	_____
4. 6 and 7	66 and 67	5	_____

Counter between	Roller between	Coinciding Vernier Line	Reading
5. 5 and 6	45 and 46	8	_____
6. 0 and 1	98 and 99	9	_____
7. 3 and 4	11 and 12	6	_____
8. 2 and 3	03 and 04	1	_____
9. 7 and 8	33 and 34	4	_____
10. 0 and 1	15 and 16	5	_____

13

Geometrical Constructions

13-1 Applications of Geometry

Many of the fundamental facts of geometry as related to practical work have already been studied in the chapters on areas and volumes, and the space available in this book will not permit a more complete discussion. There are, however, certain geometrical constructions that are common and of great value in the solution of problems. Some of these will be discussed in the following pages.

The work in this chapter requires the use of the following instruments: a compass, a pair of dividers, a rule or straightedge, and a scale.

13-2 Bisecting a Line Segment

To *bisect* simply means to cut or divide exactly in half.

Example

Bisect the line segment *AB* in Fig. 13-1.

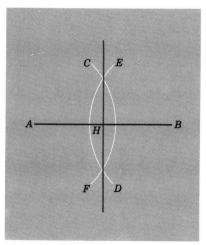

Fig. 13-1

SOLUTION   With a radius more than half the length of *AB*, use *A* as a center and draw the arc *CF*; with the same radius use *B* as center and draw the arc *ED*. Draw a line through the points where these two arcs intersect. This line, crossing *AB* at *H*, cuts *AB* into two equal parts, making *H* the *midpoint*.

13-3   Bisecting an Angle

Example

Bisect the angle *ABC* in Fig. 13-2.

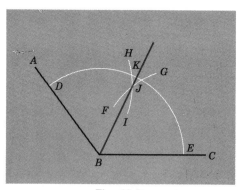

Fig. 13-2

SOLUTION   With any radius, such as *BD*, use *B* as a center and draw the arc *DE*. With the same radius, use *E* as a center and draw the arc *FG*. With the same radius again, use *D* as a center and draw the arc *HI*. Arcs *FG* and *HI* intersect at *J*. The line you draw through *J* and *B* bisects angle *ABC*, which means that angle *ABJ* = angle *CBJ*.

13-4   Bisecting an Arc

The method used for bisecting an arc is exactly the same as that used for bisecting an angle. In Fig. 13-2, the line *JB* cuts the arc *DE* into two equal arcs, *DK* and *EK*.

13-5   Constructing a Perpendicular to a Line at a Given Point on the Line

A *perpendicular* is a line or line segment that creates right angles to another line or line segment.

Example
From point *P* construct a perpendicular to the line *AB* in Fig. 13-3.

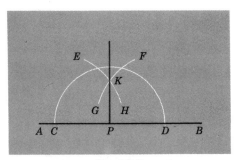

Fig. 13-3

SOLUTION   Using *P* as a center and any convenient radius, draw the semicircle *CD*. With a radius somewhat larger than *PC* and with *C* as a center, draw the arc *EH*. With the same radius and *D* as a center, draw the arc *GF*. These arcs intersect at *K*. The line through *K* and *P* is the required perpendicular. Therefore angles *APK* and *BPK* each equal 90°.

13-6   Constructing a Perpendicular to an Endpoint of a Line Segment

Example
Construct a perpendicular to the line *AB* at *A* in Fig. 13-4.

SOLUTION 1   Take any point as *C*, not on the line *AB*, as a center and *CA* as a radius, and draw the arc *EAD*. From *E* draw the line *EC* and prolong it until it cuts the arc at *D*. Draw a line through *A* and *D*. This is the required perpendicular. Therefore angle *DAE* = 90°.

SOLUTION 2   Lay off on *AB* (Fig. 13-5), a distance $AC = \frac{4}{4}$ in. = 1 in. With a radius of $\frac{5}{4}$ in. = $1\frac{1}{4}$ in. and using *C* as a center, draw the arc *DE*. With a radius of $\frac{3}{4}$ in. and *A* as a center, draw the arc *FG*. These two arcs intersect

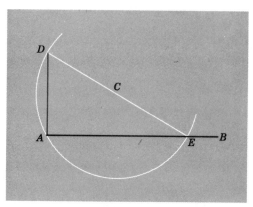

Fig. 13-4

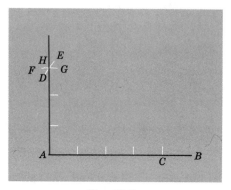

Fig. 13-5

at H. The line through H and A is the perpendicular required. Therefore angle $HAC = 90°$.

13-7   Constructing a Perpendicular to a Line through a Point Not on the Line

Example

Construct the perpendicular to the line BC (Fig. 13-6) that passes through point A.

SOLUTION   Using A as a center and any convenient radius, draw an arc cutting line BC at D and E. Using the same radius and D as a center, draw the arc GH. Using E as a center and the same radius, draw the arc IJ. Arcs GH and IJ intersect at F. The line through A and F is the required perpendicular.

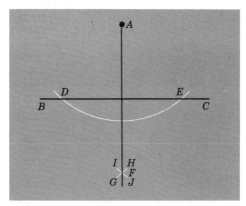

Fig. 13-6

13-8 Constructing a Line Parallel to Another Line

Example
Through the point A construct a line parallel to BC (Fig. 13-7).

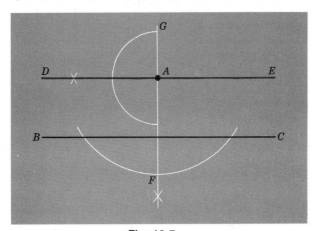

Fig. 13-7

SOLUTION By the method of Section 13-7 draw the line FG perpendicular to BC. Then by the method of Section 13-5 draw the line DE through A and perpendicular to FG. Then DE is parallel to BC and is the required line.

RULE _____

If two lines are perpendicular to the same line and are coplanar, then they are parallel.

13-9   Dividing a Line Segment into a Number of Equal Parts

Example

Divide line segment AB into seven equal parts (Fig. 13-8).

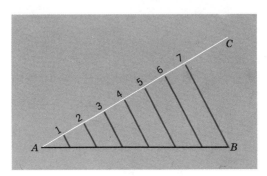

Fig. 13-8

SOLUTION   Draw the line AC at an angle to AB, not exceeding 45°. Lay off, with the dividers or scale, seven equal parts on the line AC. Connect the last point, 7, with B. Draw lines through the other points parallel to $7B$. The intersections of these lines with line segment AB divide AB into seven equal parts.

13-10   Constructing an Angle Equal to a Given Angle

Example

Construct an angle equal to the angle ABC in Figs. 13-9 and 13-10.

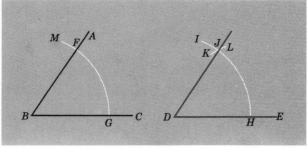

Fig. 13-9

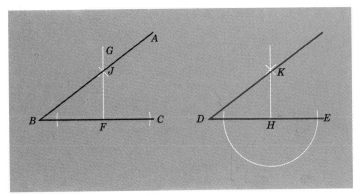

Fig. 13-10

SOLUTION 1   (Fig. 13-9.) Draw the line *DE*. Then, with *B* as a center and any radius, draw the arc *MG* cutting the line *AB* at *F*. Using the same radius and *D* as a center, draw the arc *HI*. Using *FG* as a radius and *H* as a center, draw the arc *KL* intersecting *HI* at *J*. Draw a line through *J* and *D*. The angle *JDE* is equal to *ABC* and is the required angle.

SOLUTION 2   (Fig. 13-10.) At any point on the line *BC*, as *F*, erect a perpendicular *FG* intersecting the line *BA* at *J*. Draw the line *DE* and lay off *DH* equal to *BF*. Erect a perpendicular to *DE* at *H*. On this perpendicular, lay off a distance *HK* equal to *FJ*. Draw a line through *K* and *D*, giving the required angle *KDE*.

13-11 Constructing an Equilateral Triangle of Given Size

Example
Construct an equilateral triangle with sides equal to *AB* in Fig. 13-11.

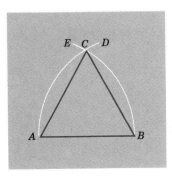

Fig. 13-11

SOLUTION With *A* as a center and *AB* as a radius, draw the arc *BE*. With *B* as a center and the same radius, draw the arc *AD* intersecting arc *EB* at *C*. Join *A* to *C* and *B* to *C*, giving the required equilateral triangle *ABC*.

13-12 Constructing a Circle through Three Given Points

Example

Draw a circle through the points *A*, *B*, and *C* (Fig. 13-12).

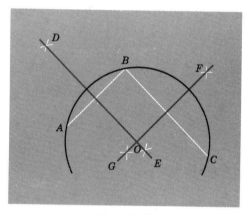

Fig. 13-12

SOLUTION Connect the points *A* and *B*, and *B* and *C*. Draw *DE* bisecting *AB*; draw *FG* bisecting *BC*. The intersection of these two lines at *O* is the center of the circle. With *O* as center and radius *OA*, draw the required circle through the three given points.

13-13 Finding the Center of a Circle or Arc

Example

Find the center of the arc *ABC* in Fig. 13-13.

SOLUTION Draw any two chords of the arc as *AB* and *BC*. Bisect each chord. The intersection *O* of the bisecting lines *DE* and *FG* is the center of the circle.

NOTE. This construction is similar to that in Section 13-12.

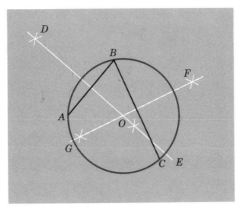

Fig. 13-13

13-14 Inscribing a Square in a Circle

Example

Inscribe a square in the circle in Fig. 13-14.

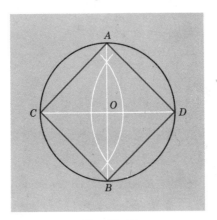

Fig. 13-14

SOLUTION Draw the diameters *AB* and *CD* at right angles to each other. Connecting the four points *A*, *C*, *B*, and *D* gives the required square *ACBD*.

13-15 Constructing a Square of a Given Size

Example

Construct a square with sides equal to length *AB* in Fig. 13-15.

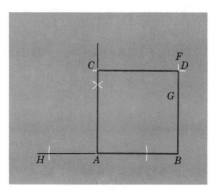

Fig. 13-15

SOLUTION   Extend line *BA* to *H*. Erect a perpendicular to line *AB* at *A*. With *A* as center and *AB* as radius, draw an arc cutting this perpendicular at *C*. With the same radius and *C* as center, draw arc *FG*. With the same radius and *B* as center, draw an arc cutting *FG* at *D*. Connect points *ACDB* to produce the required square.

13-16   Constructing a Square Equal in Area to the Sum or Difference of Two Given Squares

Example 1

Construct a square equal in area to the *sum* of the two squares *ABCD* and *EFGH* (Fig. 13-16).

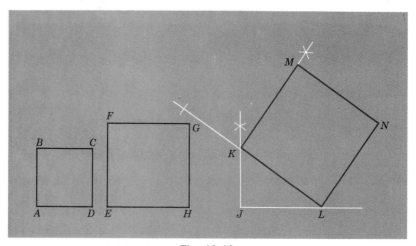

Fig. 13-16

SOLUTION   Construct a right triangle whose base *JL* is equal to *EH* and whose altitude *JK* is equal to *AB*. On the hypotenuse *KL* construct a square *KMNL*, which is the required square.

Example 2

Construct a square equal in area to the *difference* between the two squares *ABCD* and *EFGH* from Fig. 13-16.

SOLUTION   Construct the right triangle *RST* (Fig. 13-17) with *ST* equal to *EH*, the side of the larger square, as hypotenuse, and *RT* equal to *AD*, the side of the smaller square, as base. Lay off length *RT* equal to *AD* on straight line *XY*. Draw a perpendicular at *R*. With radius *EH* and *T* as center, draw an arc cutting the perpendicular at *S*. On the side *RS* construct the required square *QPSR*.

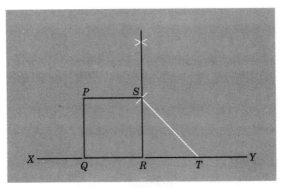

Fig. 13-17

13-17   Constructing a Circle Equal in Area to the Sum or Difference of the Areas of Two Given Circles

Example

Construct a circle equal in area to the combined areas of a $\frac{1}{2}$-in. circle and a $\frac{3}{4}$-in. circle (Fig. 13-18).

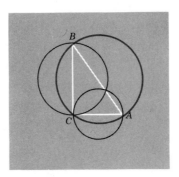

Fig. 13-18

SOLUTION   Construct the right triangle ACB, making $CA = \frac{1}{2}$ in. and $CB = \frac{3}{4}$ in. The circle drawn with the hypotenuse AB as diameter is equal in area to the sum of the areas of the two circles drawn with the sides CA and CB as diameters. Also, either of the side circles is equal to the difference between the hypotenuse circle and the other side circle.

13-18   Inscribing a Hexagon in a Circle

Example
Inscribe a hexagon in the circle in Fig. 13-19.

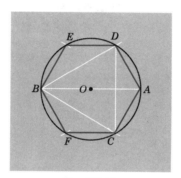

Fig. 13-19

SOLUTION   Draw the diameter AB. With a radius OA equal to the radius of the circle and with A as a center, draw arcs cutting the circle at C and D; with the same radius and B as a center, draw arcs cutting the circle at E and F. Connect the successive points, producing the hexagon $BEDACF$.

13-19   Inscribing an Equilateral Triangle in a Circle

Example
Inscribe an equilateral triangle in the circle in Fig. 13-19.

SOLUTION   Using the same construction as in Section 13-18, draw lines connecting the alternate points B, D, and C, producing the equilateral triangle BDC.

13-20   Constructing a Hexagon Whose Sides Will Be a Given Length

Example
Construct a hexagon with sides equal to the given length AB in Fig. 13-20.

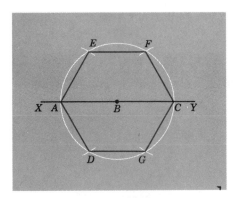

Fig. 13-20

SOLUTION   On line *XY* in Fig. 13-20 lay off the required length *AB*. With *B* as center and *AB* as radius, draw a circle through *A* cutting *XY* at *C*. With the same radius and *A* as center, draw arcs cutting the circle at *D* and *E*. With the same radius and *C* as center, draw arcs cutting the circle at *F* and *G*. Connecting the successive points gives the required hexagon *AEFCGD*.

13-21   Constructing a Hexagon with One of the Sides on a Given Line

Example
Construct a hexagon on a given line with sides equal to the given length *AB* in Fig. 13-21.

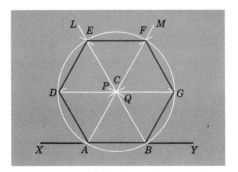

Fig. 13-21

SOLUTION   On the line *XY* in Fig. 13-21 lay off the length *AB*. With *A* as center and *AB* as radius, draw the arc *PQ*. With the same radius and *B* as center, draw an arc cutting the first arc at *C*. With the same radius and *C* as center, draw a circle through *A* and *B*. Draw the line *AM* through *A* and *C*, cutting the circle at *F*; draw the line *BL* through *B* and *C*, cutting the circle at *E*. With the same radius and *A* as center, drawn an arc cutting the circle at D; with the same radius and B as center, draw an arc cutting the circle at G. Connecting the successive points gives the required hexagon $ADEFGB$.

13-22   Inscribing an Octagon in a Circle

Example

Inscribe an octagon in the circle in Fig. 13-22.

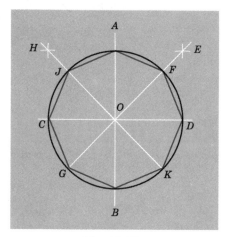

Fig. 13-22

SOLUTION   Draw the diameters AB and CD at right angles to each other. Bisect angle AOC with line HK, cutting the circle at J and K. Bisect angle AOD with line EG, cutting the circle at F and G. Connecting the successive points gives the octagon $AFDKBGCJ$.

13-23   Inscribing an Octagon in a Square

Example

Inscribe an octagon in the square in Fig. 13-23.

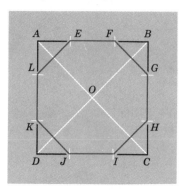

Fig. 13-23

SOLUTION Draw the diagonals *AC* and *DB*, intersecting at *O*. With radius *OA* equal to half the diagonal distance and *A* as center, draw arcs cutting the sides of the square at *F* and *K*. Using the same radius, repeat this operation at the other corners of the square, *B*, *C*, and *D*. Connecting successive points produces the octagon *EFGHIJKL*.

13-24 Constructing an Octagon of a Given Size

Example
Construct an octagon with sides equal to *AB* in Fig. 13-24.

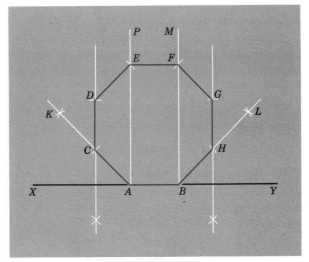

Fig. 13-24

SOLUTION Erect perpendiculars to line *XY* at *A* and *B*. Bisect the angles *PAX* and *MBY*. With *AB* as radius and *A* and *B* as centers, draw arcs cutting the bisectors at *C* and *H*. Draw perpendiculars to *XY* through *C* and *H*. With the same radius and *C* and *H* as centers, draw arcs cutting these perpendiculars at *D* and *G*. With the same radius and *D* and *G* as centers, draw arcs cutting *AP* and *BM* at *E* and *F*. Connecting the successive points gives the required octagon *ACDEFGHB*.

13-25 Constructing a Pentagon

Example
Inscribe a pentagon in the circle in Fig. 13-25.

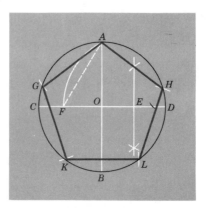

Fig. 13-25

SOLUTION   Draw the diameters *AB* and *CD* perpendicular to each other. Bisect *OD* at *E*. With *E* as a center and a radius equal to *EA*, describe the arc *AF*. The line *AF* is the side of the inscribed pentagon. Starting at *A* with a radius equal to the line *AF*, strike arcs around the circumference, giving the points *A, G, K, L,* and *H*. Connecting these points produces the pentagon *AGKLH*.

13-26 Constructing a Tangent to a Circle

A *tangent* is defined as a line that intersects with a circle at one, and only one, point.

Example

Draw a tangent to the circle at the point *A* in Fig. 13-26.

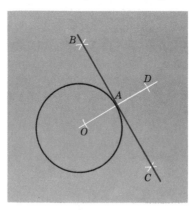

Fig. 13-26

SOLUTION   Draw the line *OD* through the center of the circle and point *A*. Draw *BC* perpendicular to *OD* at *A* by the method previously studied. Then *BC* is the required tangent.

13-27   Constructing a Tangent to a Circle through a Point outside the Circle

Example

Construct a tangent to the circle from the point A in Fig. 13-27.

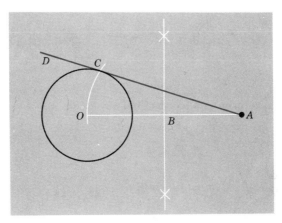

Fig. 13-27

SOLUTION   Draw the line OA connecting the center of the circle with the point A. Bisect the line OA, obtaining the midpoint B. Using B as a center and BO as a radius, draw the arc OC cutting the given circle at C. The line AD drawn through A and C is the required tangent.

13-28   Constructing a Tangent to Two Circles of Equal Size

Example

Construct a tangent to the two equal circles shown in Fig. 13-28.

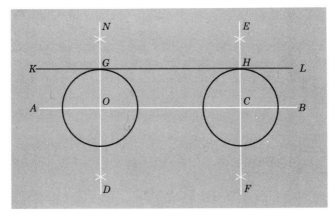

Fig. 13-28

SOLUTION   Draw the line *AB* through the centers *O* and *C* of the circles. At *O* and at *C* draw the lines *ND* and *EF* perpendicular to *AB*. Draw the line *KL* through the points *G* and *H*, where these perpendiculars intersect the circles. Then *KL* is the required tangent.

13-29 Constructing an Internal Tangent to Two Equal Circles

Example
Construct an internal tangent to the two equal circles shown in Fig. 13-29.

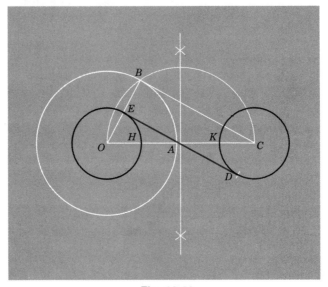

Fig. 13-29

SOLUTION   Draw the line *OC* connecting the centers of the two circles. On this line lay off *HA* = *OH*. With *O* as a center and *OA* as a radius, draw the large circle shown in the figure. By means of the method of Section 13-27, draw the tangent *CB* to the large circle through *C*. Draw *OB* intersecting the small circle at *E*. With *E* as center and *BC* as radius, draw an arc cutting circle *C* at *D*. Draw *ED*, the required tangent.

13-30 Constructing an External Tangent to Two Circles of Unequal Size

Example
Construct an external tangent to the two circles in Fig. 13-30.

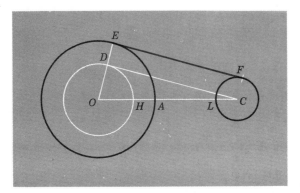

Fig. 13-30

SOLUTION   Connect the centers of the two circles by the line OC. On OA, the radius of the large circle, lay off $AH = LC$, the radius of the small circle. Using O as a center and OH as a radius, draw the construction circle shown. By the method of Section 13-27, draw the tangent CD to the construction circle and through C. Prolong the radius OD until it cuts the outer circle at E. With E as center and DC as radius, draw an arc cutting circle C at F. Connecting E and F gives the required tangent.

13-31   Constructing an Internal Tangent to Two Unequal Circles

Example
Construct an internal tangent to the two circles shown in Fig. 13-31.

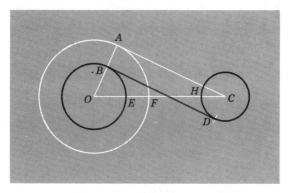

Fig. 13-31

SOLUTION   Connect the centers by the line OC. Lay off on the line of centers the distance EF equal to the small radius, making OF equal to the sum of the two radii OE and CH. With O as center and radius OF, draw the large circle shown in the figure. Draw tangent CA to the large circle through point C. Connect O and A, cutting the given circle at B. With B as center and AC as radius, draw an arc cutting circle C at D. Connecting points B and D gives the required tangent.

13-32   Constructing an Ellipse

Example

Construct an ellipse with a long axis of 2 in. and a short axis of $1\frac{1}{4}$ in.

SOLUTION 1   (Fig. 13-32.) Draw two concentric circles with diameters equal respectively to the long and short axes of the ellipse. Draw two axes XY and LM at right angles to each other. Draw any radius as OA. Through

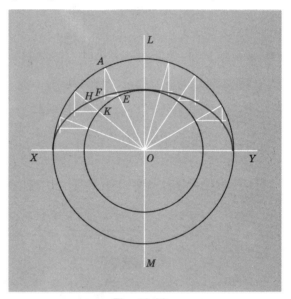

Fig. 13-32

point A, where this radius meets the large circle, draw a line AK parallel to LM. Through point E, where radius OA crosses the small circle, draw a line EH parallel to axis XY. The intersection of these two lines AK and EH at F is a point on the perimeter of the ellipse. Other points on the perimeter may be found in a similar manner and, by obtaining a sufficiently large number of points, the ellipse may be drawn through these points.

SOLUTION 2   (Fig. 13-33.) Draw the 2 in. and the $1\frac{1}{4}$ in. axes at right angles to each other and bisecting each other. Using B as center and AC, half the major axis, as radius, draw an arc cutting the major axis at two points F and G. Insert thumb tacks into the paper at the points F, G, and B; stretch a piece of thread so that it is fairly taut around these three points. Now remove the tack from point B, and keeping the string uniformly taut with the point of a pencil, draw the ellipse.

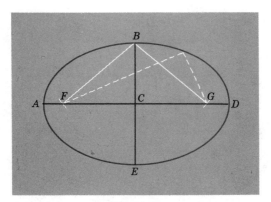

Fig. 13-33

14

Logarithms

14-1 Definitions

Common *logarithms* are particularly useful in problems involving multiplication, division, raising a number to a power, or extracting roots. The fundamental laws that are valid for common logarithms are also valid for logarithms to other bases that occur in higher mathematics.

The logarithm of a number is the exponent indicating the power to which it is necessary to raise a given number (called the base) to produce the number. Zero and one are not used as bases.

In the *common* or *Briggs system* of logarithms the base is 10. The common logarithm of a number is the power to which 10 must be raised to produce the given number. Thus since $10^2 = 100$, that is, since 10 must be raised to the second power to equal 100, we say the logarithm of 100 is 2, written in abbreviated form, log 100 = 2. Similarly, log 10,000 = 4, since 10 must be raised to the fourth power to produce 10,000.

The logarithm of a number that lies between 1,000 and 10,000 lies between 3 and 4; that is, it is 3 plus a fraction. The logarithm of a number that lies between 10 and 100 lies between 1 and 2; that is, it is 1 plus a fraction. The logarithm of a number that lies between 0.1 and 1 lies between $^-1$ and 0; that is, it is $^-1$ plus a fraction. The logarithm of a number that lies between 0.001 and 0.01 lies between $^-3$ and $^-2$; that is, it is $^-3$ plus a fraction. For example, log 0.008 = $\bar{3}$.90309. The negative sign is placed above the 3 instead of in front of it to show that the 3 only is negative, whereas the decimal part of the logarithm is plus, or positive.

$10^4 = 10{,}000$	log 10,000 = 4	$10^{-1} = 0.1$	log 0.1 = $^-1$
$10^3 = 1{,}000$	log 1,000 = 3	$10^{-2} = 0.01$	log 0.01 = $^-2$
$10^2 = 100$	log 100 = 2	$10^{-3} = 0.001$	log 0.001 = $^-3$
$10^1 = 10$	log 10 = 1	$10^{-4} = 0.0001$	log 0.0001 = $^-4$
$10^0 = 1$	log 1 = 0		

14-2 Characteristic and Mantissa

The integral part of a logarithm is called the *characteristic;* the decimal part is the *mantissa.* The characteristic is determined mentally by the following two rules.

RULE

The characteristic of the logarithm of a number greater than 1 is 1 less than the number of digits to the left of the decimal point in the given number.

Thus the characteristic of the logarithm of 352.6 is 2 because the integral part of the number contains three places.

RULE

The characteristic of the logarithm of a number less than 1 is a negative number whose absolute value is 1 more than the number of zeros between the decimal point and the first nonzero digit.

Thus, the characteristic of 0.00642 is $\bar{3}$ because there are two zeros between the decimal point and the 6. Note that the negative sign is placed *above* the characteristic to show that it only is negative, whereas *the mantissa is always positive.*

The mantissa is determined by the following rule.

RULE

The mantissa is found in a table of common logarithms and is the same for all numbers made up of the same digits, no matter where the decimal point is located.

To find the logarithm of a number, determine the characteristic and annex to it the mantissa for the digits as found in Table 14-1. (Section 14-3 will further explain how to read the table.) Thus:

log 32,500 = 4.5119	log 325 = 2.5119
log 3,250 = 3.5119	log 32.5 = 1.5119

$$\log 3.25 = 0.5119 \qquad \log 0.00325 = \overline{3}.5119$$
$$\log 0.325 = \overline{1}.5119 \qquad \log 0.000325 = \overline{4}.5119$$
$$\log 0.0325 = \overline{2}.5119 \qquad \log 0.0000325 = \overline{5}.5119$$

Note that in each instance the mantissa is that given in the table for the numeral 325.

TABLE 14-1
TABLE OF COMMON LOGARITHMS

	0	1	2	3	4	5	6	7	8	9
1	0000	0414	0792	1139	1461	1761	2041	2304	2553	2788
2	3010	3222	3424	3617	3802	3979	4150	4314	4472	4624
3	4771	4914	5051	5185	5315	5441	5563	5682	5798	5911
4	6021	6128	6232	6335	6435	6532	6628	6721	6812	6902
5	6990	7076	7160	7243	7324	7404	7482	7559	7634	7709
6	7782	7853	7924	7993	8062	8129	8195	8261	8325	8388
7	8451	8513	8573	8633	8692	8751	8808	8865	8921	8976
8	9031	9085	9138	9191	9243	9294	9345	9395	9445	9494
9	9542	9590	9638	9685	9731	9777	9823	9868	9912	9956
10	0000	0043	0086	0128	0170	0212	0253	0294	0334	0374
11	0414	0453	0492	0531	0569	0607	0645	0682	0719	0755
12	0792	0828	0864	0899	0934	0969	1004	1038	1072	1106
13	1139	1173	1206	1239	1271	1303	1335	1367	1399	1430
14	1461	1492	1523	1553	1584	1614	1644	1673	1703	1732
15	1761	1790	1818	1847	1875	1903	1931	1959	1987	2014
16	2041	2068	2095	2122	2148	2175	2201	2227	2253	2279
17	2304	2330	2355	2380	2405	2430	2455	2480	2504	2529
18	2553	2577	2601	2625	2648	2672	2695	2718	2742	2765
19	2788	2810	2833	2856	2878	2900	2923	2945	2967	2989
20	3010	3032	3054	3075	3096	3118	3139	3160	3181	3201
21	3222	3243	3263	3284	3304	3324	3345	3365	3385	3404
22	3424	3444	3464	3483	3502	3522	3541	3560	3579	3598
23	3617	3636	3655	3674	3692	3711	3729	3747	3766	3784
24	3802	3820	3838	3856	3874	3892	3909	3927	3945	3962
25	3979	3997	4014	4031	4048	4065	4082	4099	4116	4133
26	4150	4166	4183	4200	4216	4232	4249	4265	4281	4298
27	4314	4330	4346	4362	4378	4393	4409	4425	4440	4456
28	4472	4487	4502	4518	4533	4548	4564	4579	4594	4609
29	4624	4639	4654	4669	4683	4698	4713	4728	4742	4757
30	4771	4786	4800	4814	4829	4843	4857	4871	4886	4900
31	4914	4928	4942	4955	4969	4983	4997	5011	5024	5038
32	5051	5065	5079	5092	5105	5119	5132	5145	5159	5172
33	5185	5198	5211	5224	5237	5250	5263	5276	5289	5302
34	5315	5328	5340	5353	5366	5378	5391	5403	5416	5428
35	5441	5453	5465	5478	5490	5502	5514	5527	5539	5551
36	5563	5575	5587	5599	5611	5623	5635	5647	5658	5670
37	5682	5694	5705	5717	5729	5740	5752	5763	5775	5786
38	5798	5809	5821	5832	5843	5855	5866	5877	5888	5899
39	5911	5922	5933	5944	5955	5966	5977	5988	5999	6010
40	6021	6031	6042	6053	6064	6075	6085	6096	6107	6117
41	6128	6138	6149	6160	6170	6180	6191	6201	6212	6222
42	6232	6243	6253	6263	6274	6284	6294	6304	6314	6325
43	6335	6345	6355	6365	6375	6385	6395	6405	6415	6425
44	6435	6444	6454	6464	6474	6484	6493	6503	6513	6522
45	6532	6542	6551	6561	6571	6580	6590	6599	6609	6618
46	6628	6637	6646	6656	6665	6675	6684	6693	6702	6712
47	6721	6730	6739	6749	6758	6767	6776	6785	6794	6803
48	6812	6821	6830	6839	6848	6857	6866	6875	6884	6893
49	6902	6911	6920	6928	6937	6946	6955	6964	6972	6981

TABLE 14-1 (continued)

	0	1	2	3	4	5	6	7	8	9
50	6990	6998	7007	7016	7024	7033	7042	7050	7059	7067
51	7076	7084	7093	7101	7110	7118	7126	7135	7143	7152
52	7160	7168	7177	7185	7193	7202	7210	7218	7226	7235
53	7243	7251	7259	7267	7275	7284	7292	7300	7308	7316
54	7324	7332	7340	7348	7356	7364	7372	7380	7388	7396
55	7404	7412	7419	7427	7435	7443	7451	7459	7466	7474
56	7482	7490	7497	7505	7513	7520	7528	7536	7543	7551
57	7559	7566	7574	7582	7589	7597	7604	7612	7619	7627
58	7634	7642	7649	7657	7664	7672	7679	7686	7694	7701
59	7709	7716	7723	7731	7738	7745	7752	7760	7767	7774
60	7782	7789	7796	7803	7810	7818	7825	7832	7839	7846
61	7853	7860	7868	7875	7882	7889	7896	7903	7910	7917
62	7924	7931	7938	7945	7952	7959	7966	7973	7980	7987
63	7993	8000	8007	8014	8021	8028	8035	8041	8048	8055
64	8062	8069	8075	8082	8089	8096	8102	8109	8116	8122
65	8129	8136	8142	8149	8156	8162	8169	8176	8182	8189
66	8195	8202	8209	8215	8222	8228	8235	8241	8248	8254
67	8261	8267	8274	8280	8287	8293	8299	8306	8312	8319
68	8325	8331	8338	8344	8351	8357	8363	8370	8376	8382
69	8388	8395	8401	8407	8414	8420	8426	8432	8439	8445
70	8451	8457	8463	8470	8476	8482	8488	8494	8500	8506
71	8513	8519	8525	8531	8537	8543	8549	8555	8561	8567
72	8573	8579	8585	8591	8597	8603	8609	8615	8621	8627
73	8633	8639	8645	8651	8657	8663	8669	8675	8681	8686
74	8692	8698	8704	8710	8716	8722	8727	8733	8739	8745
75	8751	8756	8762	8768	8774	8779	8785	8791	8797	8802
76	8808	8814	8820	8825	8831	8837	8842	8848	8854	8859
77	8865	8871	8876	8882	8887	8893	8899	8904	8910	8915
78	8921	8927	8932	8938	8943	8949	8954	8960	8965	8971
79	8976	8982	8987	8993	8998	9004	9009	9015	9020	9025
80	9031	9036	9042	9047	9053	9058	9063	9069	9074	9079
81	9085	9090	9096	9101	9106	9112	9117	9122	9128	9133
82	9138	9143	9149	9154	9159	9165	9170	9175	9180	9186
83	9191	9196	9201	9206	9212	9217	9222	9227	9232	9238
84	9243	9248	9253	9258	9263	9269	9274	9279	9284	9289
85	9294	9299	9304	9309	9315	9320	9325	9330	9335	9340
86	9345	9350	9355	9360	9365	9370	9375	9380	9385	9390
87	9395	9400	9405	9410	9415	9420	9425	9430	9435	9440
88	9445	9450	9455	9460	9465	9469	9474	9479	9484	9489
89	9494	9499	9504	9509	9513	9518	9523	9528	9533	9538
90	9542	9547	9552	9557	9562	9566	9571	9576	9581	9586
91	9590	9595	9600	9605	9609	9614	9619	9624	9628	9633
92	9638	9643	9647	9652	9657	9661	9666	9671	9675	9680
93	9685	9689	9694	9699	9703	9708	9713	9717	9722	9727
94	9731	9736	9741	9745	9750	9754	9759	9763	9768	9773
95	9777	9782	9786	9791	9795	9800	9805	9809	9814	9818
96	9823	9827	9832	9836	9841	9845	9850	9854	9859	9863
97	9868	9872	9877	9881	9886	9890	9894	9899	9903	9908
98	9912	9917	9921	9926	9930	9934	9939	9943	9948	9952
99	9956	9961	9965	9969	9974	9978	9983	9987	9991	9996

Problems 14-2

What is the characteristic of the logarithm of each of the following numbers?

1. 64	2. 300	3. 5600	4. 24.3	5. 8
6. 7.02	7. 5,006	8. 400,000	9. 32.78	10. 400.6

11. 1.609	12. 236.84	13. 300.9	14. 64,200	15. 27.624
16. 5.18	17. 518.6	18. 518.64	19. 51.86	20. 5.186
21. 9	22. 1	23. 1001	24. 12.01	25. 120.1

How many figures must there be before the decimal point in the numbers if the characteristics of their logarithms are as follows?

| 26. 0 | 27. 4 | 28. 6 | 29. 5 | 30. 2 |
| 31. 1 | 32. 3 | 33. 0 | 34. 2 | 35. 3 |

Challenge   What is the characteristic of the logarithm of each of the following numbers?

36. 0.4	37. 0.04	38. 0.004	39. 0.0004
40. 0.0135	41. 0.06002	42. 0.548	43. 0.0063
44. 0.25	45. 0.025	46. 0.0306	47. 0.81
48. 0.8	49. 0.008	50. 0.245	51. 0.00085
52. 0.00009	53. 0.604	54. 0.825	55. 0.3264

14-3   Use of Tables

Finding the Logarithm of a Number

Example 1
Find the logarithm of 5.

SOLUTION   log 5 = 0.6990

EXPLANATION   Since the number consists of one integer, 5, the characteristic is 1 less than 1, namely 0. To find the mantissa, look for the number 5 in the first column of Table 14-1 and since there is no decimal fraction after the integer 5, take the mantissa 6990 which is found in the 0 column. Placing the decimal point before the mantissa, we get log 5 = 0.6990.

Example 2
Find the logarithm of 728.

SOLUTION   log 728 = 2.8621

EXPLANATION   Since there are three integers in the number, the characteristic is 2. To find the mantissa, look for 72 in the first column of Table 14-1 and reading across the page find the number 8621 under 8 at the top. That is, 8621 is the mantissa for 728. Placing the decimal point between the characteristic and the mantissa, we get log 728 = 2.8621.

Example 3
Find the logarithm of 57.4.

SOLUTION   log 57.4 = 1.7589

EXPLANATION   Since there are two figures in the integral part of the number, the characteristic is one less than 2, namely 1. To find the mantissa, look in the first column of the table for the number 57, and reading across the page, find the number 7589 under 4 at the top, that is, 7589 is the mantissa of 574. Place the decimal point between the characteristic and the mantissa, giving the log 57.4 = 1.7589.

Example 4
Find the logarithm of 2,648.

SOLUTION   log 2,648 = 3.4229

EXPLANATION   Set down the characteristic 3 because the number contains four figures before the decimal point. To find the mantissa, imagine the decimal point after the third figure. This would make the number 264.8. The mantissa for 2,648 is the same as for 264.8. To obtain the mantissa for 264.8, proceed as follows:

Mantissa of 264 = 4216
Mantissa of 265 = 4232
Difference = 16

Since 264.8 lies $\frac{8}{10}$ of the way from 264 to 265, the mantissa of 264.8 will lie $\frac{8}{10}$ of the way from 4216 to 4232; that is, the mantissa of 264.8 is 4216 + ($\frac{8}{10}$ × 16) = 4216 + 13 = 4229. Hence the logarithm of 2,648 = 3.4229.

The process of computing an intermediate number between two numbers in a table is known as *interpolation.*

Finding the Logarithm of a Decimal Fraction

Example 5
Find the logarithm of 0.05834.

SOLUTION   log 0.05834 = $\bar{2}$.7660

EXPLANATION   The characteristic is $\bar{2}$ because there is one zero between the decimal point and the first significant figure. To find the mantissa, proceed as if 5,834 were an integral number, as follows:

Mantissa of 583 = 7657
Mantissa of 584 = 7664
Difference = 7
Mantissa of 583.4 = 7657 + ($\frac{4}{10}$ × 7) = 7657 + 3 = 7660

Therefore,

$$\log 0.05834 = \bar{2}.7660.$$

Find the logarithms of the following numbers.

1. 9	2. 28	3. 70	4. 120
5. 400	6. 245	7. 248.6	8. 2,486
9. 0.0524	10. 0.345	11. 2,600	12. 6,735
13. 0.000382	14. 0.004965	15. 234.56	16. 0.4008
17. 12.3	18. 4.207	19. 0.8935	20. 0.000089
21. 149,200	22. 80.06	23. 0.538	24. 0.0538
25. 53.8			

Finding the Number Corresponding to a Given Logarithm

Example 1

Find the number whose logarithm is 1.5051.

SOLUTION   The number whose logarithm is 1.5051 is 32.

EXPLANATION   In the table of logs find that the mantissa 5051 corresponds to the number 32. Since the characteristic is 1, there must be two figures before the decimal point. The number whose log is 1.5051 is 32.

Example 2

Find the number whose logarithm is 2.8609.

SOLUTION   The number whose logarithm is 2.8609 is 726.

EXPLANATION   In the table, find the mantissa 8609 on line 72 in the column under 6. Therefore the number corresponding to the mantissa 8609 is 726. Since the characteristic is 2, there must be three figures before the decimal point. The number whose log is 2.8609 is 726.

Example 3

Find the number whose logarithm is 3.3927.

SOLUTION   The number whose log is 3.3927 is 2,470.

EXPLANATION   In the table of logs, find that the mantissa 3927 corresponds to the number 247. Since the characteristic is 3, there must be four integral figures; annex a zero to 247 and get the number 2,470.

Similarly,

> The number whose log is 2.3927 is 247.
> The number whose log is 1.3927 is 24.7.
> The number whose log is 0.3927 is 2.47.
> The number whose log is $\overline{1}.3927$ is 0.247.
> The number whose log is $\overline{2}.3927$ is 0.0247.

Example 4
Find the number whose log is $\overline{3}.5595$.

SOLUTION The number whose log is $\overline{3}.5595$ is 0.003627.

EXPLANATION The mantissa 5595 cannot be found in the table. The nearest mantissa *below* 5595 is 5587, which is the log of 362. Since the given mantissa lies between 5587 and 5599, the number will be between 362 and 363. The difference between the mantissas for 362 and 363 is $5599 - 5587 = 12$; the difference between the given mantissa and that of 362 is $5595 - 5587 = 8$. The number corresponding to mantissa 5595 is $362 + \frac{8}{12} = 362.7$. But the characteristic $\overline{3}$ tells us there must be two zeros between the decimal point in the number and the first significant figure; hence the number corresponding to the logarithm $\overline{3}.5595$ is 0.003627.

Problems 14-3B

Find the numbers whose logarithms are the following.

1. 2.9212	2. 1.9212	3. 0.9212	4. $\overline{1}.9212$	5. 4.9212
6. 2.6390	7. 3.3862	8. 0.8761	9. 0.9559	10. $\overline{1}.8348$
11. 3.7057	12. 2.7378	13. 2.0152	14. 1.0099	15. $\overline{2}.8951$
16. $\overline{2}.9225$	17. $\overline{3}.9530$	18. $\overline{1}.8239$	19. 4.9488	20. 4.9488
21. 2.8879	22. 1.7870	23. 3.5385	24. $\overline{1}.4594$	25. 0.6284

14-4 Multiplication by the Use of Logarithms

RULE
To find the product of two or more numbers, add their logarithms; the sum is the log of the product.

Example 1
Multiply 32.4 by 71.6

SOLUTION
$$\log 32.4 = 1.5105$$
$$\log 71.6 = +1.8549$$
$$\text{log of product} = 3.3654 \therefore \text{ the product is } 2{,}319.4$$

Multiply 32.4 by 7.16 and get 2,319.8. The discrepancy between the exact product and the one obtained by logarithms is due to the fact that only four-place log tables are used. Greater accuracy can be obtained by the use of a six-place table.

Example 2
Find the value of 6.45 × 25.92 × 0.088.

SOLUTION   log 6.45 =   0.8096
                      log 25.92 =   1.4136
                      log 0.088 = +$\bar{2}$.9445
          log of product =   1.1677 ∴ the product is 14.71

Problems 14-4

Find the approximate value of the following by means of logs.

1. 2 × 7 × 5   2. 0.6 × 8.5 × 40
3. 124 × 532   4. 788 × 436
5. 42.5 × 684   6. 232 × 456
7. 14.8 × 12.7   8. 46.87 × 4.24
9. 0.444 × 3.25   10. 3.1416 × 0.063
11. 5,280 × 16,400   12. 0.085 × 0.006254
13. 6,053 × 286.7   14. 2.005 × 82.695
15. 788.1 × 0.00342   16. 72 × 320 × 0.654
17. 34.8 × 28.6 × 14.2   18. 1,765 × 5.5 × 0.075 × 23.8
19. 144 × 6.25 × 0.875   20. 24.5 × 24.5 × 0.7854
21. 5.4 × 8 × 48.56 × 0.085   22. 9,876 × 0.00325 × 2.59
23. 320.8 × 2.31 × 4.42   24. 2,004 × 0.00375 × 543
25. 12.22 × 5.26 × 0.125

14-5   Division by the Use of Logarithms

RULE

To find the quotient of two numbers, subtract the log of the divisor from the log of the dividend; the difference is the log of the quotient.

Example 1
Divide 864 by 16.

SOLUTION   log 864 =   2.9365
                      log 16 = − 1.2041
          log of quotient =   1.7324 ∴ the quotient is 54

Example 2
  Divide 1.82 by 1.3.

SOLUTION   log 1.82 =   0.2601
                        log 1.3 = −0.1139
           log of quotient =   0.1462 ∴ the quotient is 1.403

Example 3
  Divide 872 by 93.

SOLUTION   log 872 =   2.9405
                         log 93 = −1.9685
           log of quotient =   0.9720 ∴ the quotient is 9.376

Example 4
  Divide 42.75 by 752.3.

SOLUTION   log 42.75 =   1.6309 = 11.6309 − 10 (adding 10 and
            log 752.3 = −2.8764 = −2.8764   subtracting 10)
                  log of quotient =   8.7545 − 10
                  log of quotient = $\overline{2}$.7545 ∴ the quotient is 0.05682

14-6 Cologarithms

The *cologarithm* of a number is found by subtracting the log of the number from 10 and then annexing −10 to the remainder. This is done mentally by beginning at the left and subtracting each figure from 9 until the last significant figure is reached, which must be subtracted from 10. Thus we have

$$\log 234 = 2.3692$$
$$\text{colog } 234 = 7.6308 - 10$$

Instead of *subtracting* the log of the divisor from the log of the dividend, the same result is obtained if we *add* the colog of the divisor to the log of the dividend. This method greatly simplifies and expedites the work.

Example
  42.75 ÷ 752.3

SOLUTION   log 42.75 =   1.6309
                    colog 752.3 = +7.1236 − 10
            log of quotient =   8.7545 − 10 = $\overline{2}$.7545

  ∴ the quotient is 0.05682

Find the cologarithms corresponding to the following logarithms.

1. 1.6284	2. $\overline{0}$.8149	3. $\overline{2}$.8762	4. 1.3909	5. $\overline{0}$.6990
6. 3.4524	7. $\overline{3}$.4524	8. $\overline{4}$.7356	9. 2.4728	10. $\overline{2}$.4728
11. 1.8069	12. $\overline{1}$.8998	13. 4.5391	14. 0.6096	15. 0.7723
16. 2.6628	17. 3.5944	18. 1.4757	19. 0.5955	20. $\overline{2}$.6484
21. 1.8156	22. 2.8704	23. 3.3856	24. $\overline{1}$.5416	25. 1.1614

Find the cologarithms of the following numbers.

26. 8	27. 35	28. 864	29. 5,762	30. 400
31. 6.2	32. 9.54	33. 12.36	34. 64.582	35. 0.06
36. 0.044	37. 0.0069	38. 0.000428	39. 382.6	40. 42.9
41. 1.602	42. 83.4	43. 932	44. 105	45. 600
46. 124	47. 23.64	48. 8.92	49. 6.02	50. 58.9

Find the value of the following by means of logarithms.

51. 7856 ÷ 2.34	52. 0.0653 ÷ 0.005809	53. 55,400 ÷ 325
54. 0.00448 ÷ 0.06925	55. 4.62 ÷ 889.6	56. 0.988 ÷ 245
57. 82 ÷ 0.0743	58. 62.3 ÷ 15.7	59. 3,266 ÷ 9,452
60. 0.0287 ÷ 6.3	61. 234 ÷ 18	62. 652 ÷ 2.4
63. 59.6 ÷ 1.2	64. 842.8 ÷ 2.86	65. 700 ÷ 82.4
66. 6.592 ÷ 56.88	67. 416 ÷ 69	68. 2,702 ÷ 3.14
69. 3.1 ÷ 0.026	70. 42.4 ÷ 0.00892	

14-7   Combined Multiplication and Division

The following example will illustrate the method used in solving problems involving both multiplication and division.

Example

Find the value of $\dfrac{236 \times 8.92}{455 \times 0.00264}$

SOLUTION

$$
\begin{aligned}
\log 236 &= 2.3729 \\
\log 8.92 &= 0.9504 \\
\text{colog } 455 &= 7.3420 - 10 \\
\text{colog } 0.00264 &= +12.5784 - 10 \\
\hline
\log \text{ of quotient} &= 23.2437 - 20 = 3.2437
\end{aligned}
$$

∴ the quotient is 1,752.8

EXPLANATION   Instead of *subtracting* the *logs* of 455 and 0.00264 from the logs of 236 and 8.92, *add* the *cologs* of the numbers in the divisor to the logs of the numbers in the dividend.

Problems 14-7

Find the value of the following by means of logarithms and cologarithms.

1. $\dfrac{24.6 \times 563}{32}$

2. $\dfrac{2.35 \times 0.0084}{9.2}$

3. $\dfrac{62.44 \times 0.056}{3.14}$

4. $\dfrac{681.4 \times 0.065}{8.66}$

5. $\dfrac{24 \times 37}{4.3 \times 61}$

6. $\dfrac{72}{83.6 \times 54.9}$

7. $\dfrac{5}{18.3 \times 0.064}$

8. $\dfrac{42.1 \times 26.5}{52.4}$

9. $\dfrac{10.3 \times 50.9}{30.6 \times 16.8}$

10. $\dfrac{0.076 \times 0.0048}{1.09}$

11. $\dfrac{2.4 \times 3.6}{1.8 \times 4.5}$

12. $\dfrac{16.2 \times 12.8}{45.6 \times 1.16 \times 32.3}$

13. $\dfrac{18 \times 42}{16 \times 13}$

14. $\dfrac{196 \times 2.4}{0.56 \times 0.885}$

15. $\dfrac{0.0726 \times 0.625}{0.375 \times 0.00543}$

16. $\dfrac{432 \times 8.25}{396 \times 0.055}$

17. $\dfrac{2.32 \times 4546 \times 722}{682.5 \times 300.5 \times 0.042}$

18. $\dfrac{399 \times 7600 \times 814}{483 \times 8,856 \times 212}$

19. $\dfrac{3.26 \times 8.98 \times 6.33}{16.8 \times 9.61 \times 4.52}$

20. $\dfrac{1.23 \times 3.45 \times 4.56 \times 5.67}{0.321 \times 5.43 \times 6.45 \times 7.56}$

14-8   Raising a Number to Any Power

RULE

To find the power of a number, multiply the log of the number by the exponent of the power; the result is the log of the power.

Example 1
Find the value of 8.4^2.

SOLUTION $\log 8.4 = 0.9243$
$$\underline{\times 2}$$
$$\log 8.4^2 = 1.8486 \therefore 8.4^2 = 70.55$$

Example 2
Find the value of 8.654^3.

SOLUTION $\log 8.654 = 0.9372$
$$\underline{\times 3}$$
$$\log 8.654^3 = 2.8116 \therefore 8.654^3 = 648$$

Problems 14-8

Find the value of the following by means of logarithms.

1. 18^2	2. 3.5^2	3. 93^2
4. 84.5^2	5. 324.6^2	6. 235^2
7. 36.92^2	8. 0.463^2	9. 6.52^3
10. 0.00656^2	11. 3.412^2	12. 1.692^2

13. $0.866 \times (1.463)^2$ 14. $(0.854)^2 \times (0.692)^2$
15. $(3.14)^3 \times (1.75)^2$ 16. $11\frac{1}{4} \times (6\frac{1}{4})^2$
17. $3.14 \times (1\frac{1}{8})^3$ 18. $0.707 \times (7\frac{3}{4})^2$
19. $4.832 \times (3\frac{1}{2})^2$ 20. $0.577 \times (7.22)^2$
21. $2.598 \times (3\frac{5}{8})^2$ 22. $0.828 \times (11\frac{1}{2})^2$
23. $1.57 \times (3\frac{3}{4})^3$ 24. $0.828 \times (6\frac{1}{4})^2$
25. $(1\frac{1}{4})^2 \times (2\frac{1}{2})^3$

14-9   Extracting Any Root of a Number

RULE

To find the root of a number, divide the log of the number by the index of the root; the result is the log of the root.

Example 1
   Find $\sqrt{256}$.

SOLUTION $\log 256 = 2.4082$

$$\log \sqrt{256} = \tfrac{1}{2} \log 256 = \frac{2.4082}{2} = 1.2041$$

$\therefore \sqrt{256} = 16$

Example 2
   Find $\sqrt{1.89}$.

SOLUTION $\log 1.89 = 0.2765$

$$\log \sqrt{1.89} = \tfrac{1}{2} \log 1.89 = \frac{0.2765}{2} = 0.1382$$

$\therefore \sqrt{1.89} = 1.375$

Example 3

   Find $\sqrt[3]{343}$.

SOLUTION $\log 343 = 2.5353$

$$\log \sqrt[3]{343} = \tfrac{1}{3} \log 343 = \frac{2.5353}{3} = 0.8451$$

$\therefore \sqrt[3]{343} = 7$

Example 4
Find $\sqrt{.7652}$.

SOLUTION $\quad \log 0.7652 = \overline{1}.8838$

$$= 9.8838 - 10$$

$$\tfrac{1}{2} \log 0.7652 = 4.9419 - 4 = \overline{1}.9419$$

$\therefore \sqrt{0.7652} = 0.8748$

Before dividing by 2, adding 20 and subtracting 20 to the logarithm will eliminate the negative 1 of the characteristic. Similarly, in extracting the cube root of a decimal, adding and subtracting 30 will make it easy to divide by 3. The principle to bear in mind is that the number added and subtracted is selected to make the arithmetic of logarithms easier.

Example 5
Find the value of $\sqrt{\dfrac{6.62 \times 48.75 \times 12.8}{18.4 \times 33.42}}$

SOLUTION

$$\log 6.62 = 0.8209$$

$$\log 48.75 = 1.6880$$

$$\log 12.8 = 1.1072$$

$$\text{colog } 18.4 = 8.7352 - 10$$

$$\text{colog } 33.42 = 8.4761 - 10$$

$$\log \frac{6.62 \times 48.75 \times 12.8}{18.4 \times 33.42} = 20.8274 - 20 = 0.8274$$

$$\log \sqrt{\frac{6.62 \times 48.75 \times 12.8}{18.4 \times 33.42}} = \frac{1}{2} \log \frac{6.62 \times 48.75 \times 12.8}{18.4 \times 33.42}$$

$$= \frac{0.8274}{2} = 0.4137, \text{ the log of the answer}$$

$$\therefore \sqrt{\frac{6.62 \times 48.75 \times 12.8}{18.4 \times 33.42}} = 2.594$$

Problems 14-9

Solve the following problems by means of logarithms and cologarithms.

1. $\sqrt{196}$ 2. $\sqrt{2.89}$

3. $\sqrt{361}$

4. $\sqrt{0.0324}$

5. $\sqrt{0.000225}$

6. $\sqrt{6954}$

7. $\sqrt{7.216}$

8. $\sqrt{0.8143}$

9. $\sqrt[3]{15}$

10. $\sqrt[3]{82.16}$

11. $\sqrt{0.06952}$

12. $\sqrt{0.0004882}$

13. $\sqrt[3]{28.69}$

14. $\sqrt{116.7}$

15. $\sqrt{8.72}$

16. $\sqrt{9.82} \times \sqrt[3]{4.12}$

17. $\sqrt{5.12 \times 7.64 \times 8.19}$

18. $\sqrt{0.00986 \times 5280}$

19. $\sqrt{0.0587 \times 6032}$

20. $4 \times 3.14 \times \sqrt{36.24}$

21. $21.92 \times \sqrt{\dfrac{278.45}{44.18}}$

22. $\sqrt{\dfrac{1029 \times 672 \times 408}{1852}}$

23. $\sqrt{\dfrac{8.84 \times 36.45 \times 15.6}{19.7 \times 44.52}}$

24. $\dfrac{11.76 \times \sqrt{516.9}}{146.5}$

25. $6 \times 2.73 \times \sqrt{86.66}$

26. The volume of a sphere $= \frac{4}{3}\pi r^3$. Find the volume of a sphere whose radius is 14.8″.

27. Find the radius of a sphere whose volume is 1,000 cu in.

$$r = \sqrt[3]{\dfrac{3 \times V}{4\pi}}$$

28. The formula for the surface area of a sphere is $4\pi r^2$. Find the surface area of a sphere whose radius is 12.5″.

29. The volume of a cylinder is $\pi r^2 h$ where h is the altitude. Find the volume of a cylinder 16.5″ high if the radius of the base is 9.25″.

30. Find the radius of a cylinder 15.5″ high, having a volume of 2,400 cu in.

$$r = \sqrt{\dfrac{V}{\pi h}}$$

14-10   Logarithms and the Calculator

Examine your calculator to see if it has a log key. If it does, then it probably has a 10^x key, which is an anti-log key. The two functions are inverses of each other. Practice using the two keys. First, enter 10 and press the log key. Did you get a 1? Press the 10^x key. Did you get back the 10? Now enter 15000 and press the log key. Did you get 4.176091259? Press the 10^x key. You should get back the 15000.

This exercise should show you that log is really an exponent. The base number in the log system is 10. There is probably an *ln* function on your calculator. This is also a logarithm system. The base of this system is *e*, a number that never ends or repeats; it starts out with the numbers 2.7182818 . . . . Your calculator has the inverse of this system if it has an e^x key. This logarithm system is used in advanced mathematics.

Problems 14-10

Use your calculator to do all the problems in Chapter 14 again.

Self-Test

Use the Table of Common Logarithms on page 340 to solve the following problems. Avoid arithmetical errors by organizing, arranging, and labeling your work. Estimate the size of the answer before solving the problem. Compare your answers with the answers in the back of the book.

1. To what power must 10 be raised to produce 1,000?
2. What is the characteristic of the logarithm of 456.8?
3. What is the characteristic of the logarithm of 0.0457?
4. Find the logarithm of 45.8.
5. Find the logarithm of 0.0555.
6. Use logarithms to find the value of 34.5 × 678.
7. Use logarithms to find the value of $\sqrt[3]{87.8}$.
8. Use logarithms to find the value of $\dfrac{(87.5)^3}{(4.55)^2}$.
9. Find the logarithm of 5.652.
10. Use logarithms to find the value of $(89.76 \times 876.5) \div (78.44)^2$.

Chapter Test

Use the Table of Common Logarithms on page 340 to solve the following problems. Avoid arithmetical errors by organizing, arranging, and labeling your work. Estimate the size of your answer before solving the problem.

1. To what power of 10 must 10 be raised to produce 10,000?
2. To what power of 10 must 10 be raised to produce 0.001?
3. What is the characteristic of the logarithm of 362.3?
4. What is the characteristic of the logarithm of 7?
5. What is the characteristic of the logarithm of 0.5625?
6. A logarithm of a number is actually a(n) _____.
7. Find the logarithm of 32.
8. Find the logarithm of 462.
9. Find the logarithm of 3,694. (Interpolate and round off to the nearest four-place mantissa.)
10. Find the logarithm of 5.618. (Interpolate and round off to the nearest four-place mantissa.)
11. Find the logarithm of 0.03162. (Interpolate and round off to the nearest four-place mantissa.)
12. Use logarithms to multiply 47.2 × 35.8.
13. Use logarithms to find the value of 83.6 × 43.8 × 39.2.
14. Use logarithms to divide 3.66 by 2.47. (Round off the quotient to the nearest hundredth.)

15. Use logarithms and cologarithms to find the logarithm of the reciprocal of 37.2.
16. Use logarithms and cologarithms to solve (36.2 × 43.7) ÷ 12.7.
17. Use logarithms to find the square of 572.
18. Use logarithms to find the cube of 428.
19. Use logarithms to find the square root of 74.2.
20. Use logarithms to find the cube root of 937.

Solve the following problems by using logarithms and cologarithms.

21. 56.31 × 49.27

22. $\dfrac{63.81 \times 14.92}{63.72}$

23. 5.678 × 4.929 ÷ 3.728
24. 0.0631 ÷ 0.2392
25. 0.003125 × 0.001562 × 0.02125

26. $\dfrac{27.6 \times 43.81 \times 529}{3.142}$

27. 4.919^2
28. 3.663^3
29. $\sqrt[3]{6398}$

30. $\sqrt{\dfrac{673.2 \times 4.7}{28.6}}$

15

Essentials of Trigonometry

15-1   The Right Triangle

Every triangle has three sides and three angles. If any three parts are known, at least one of which is a side, then the triangle can be constructed. In Fig. 15-1, triangle ABC is a right triangle in which angle C is the right angle. The sum of the three angles of any triangle is 180°. Since angle C is a right angle, that is, 90°, then the sum of angle A and angle B is 90°. Therefore, angles A and B are *acute* angles. An acute angle is an angle less than 90°.

For convenience let a represent the side opposite the acute angle $A;$ b the side opposite the angle $B;$ and c the side opposite the right angle C. The side c, opposite the right angle, is called the *hypotenuse;* the side a, opposite the angle A, is *adjacent to the angle B;* the side b, opposite the angle B, is *adjacent to the angle A.*

In Fig. 15-2, the measure of side a (the side opposite angle A) is $\frac{3}{4}$ in. and the measure of side b (the side opposite angle B) is $1\frac{1}{2}$ in. The ratio of side a to side b, that is $\frac{3}{4}:1\frac{1}{2}$, is 0.5. In Fig. 15-2, if side b were extended to point E so that the measure of AE is $2\frac{1}{2}$ in. and side c (the side opposite angle C) were extended to D, a point that meets the perpendicular erected at point E,

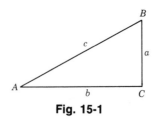

Fig. 15-1

another right triangle would be formed, triangle ADE. The measure of side DE is $1\frac{1}{4}$ in. Notice that the ratio of side DE to side AE, that is, $1\frac{1}{4}:2\frac{1}{2}$, is again 0.5. Notice also that angle A is common to both of these triangles and the ratios used the side opposite angle A and the side adjacent to angle A. The slant side is called the hypotenuse of the right triangle and is not considered an adjacent side. The value of the ratio of the side opposite the angle A to the side adjacent to the angle A will be found to be the same as long as the angle A remains unchanged. If the angle A were to be increased, the ratio of the opposite side to the adjacent side would no longer be 0.5 but some larger number. If the angle A were to be decreased, the ratio of the two sides would be less than 0.5. Since the size of the angle determines the value of the ratio of the two sides, the ratio may be taken as a measure of the angle.

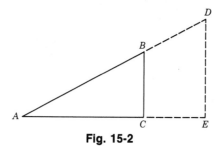

Fig. 15-2

The ratio obtained by dividing the side opposite the acute angle A by the side adjacent to the angle A is known as the **tangent** to the angle A and is written tan A. Tables in the back of this book give the value of the natural tangent for every angle from $0°0'$ to $90°$ in $10'$ increments. Refer to this table of natural tangents and find the angle whose tangent is 0.5 is between $26°30'$ and $26°40'$. Actually, the angle whose tangent is 0.5 is very close to $26°34'$.

Again referring to Fig. 15-1, suppose that the angle $A = 26°34'$ and the side $b = 6$ in., and it is required to find the length of side a. Proceed as follows: Express the ratio of the side a to the side b thus, $a/b = a/6$. This ratio of the side opposite A to the side adjacent to A is the tangent of A. From a table of tangents find that tan $26°34' = 0.5$. Therefore, write $a/6 = 0.5$, so that $a = 6 \times 0.5 = 3$ in. Similarly, $A = 26°34'$ and $a = 3$ in., to find b proceed thusly:

$$\tan A = \frac{\text{opposite side}}{\text{adjacent side}}$$

$$\frac{a}{b} = \frac{3}{b} = 0.5$$

Therefore,

$$b = \frac{3}{0.5} = 6 \text{ in.}$$

Summary

1. Given a and b, find A.

$$\frac{a}{b} = \tan A \qquad \therefore A = \text{angle whose tangent is } \frac{a}{b}$$

2. Given A and b, find a.

$$\frac{a}{b} = \tan A \qquad \therefore a = b \times \tan A$$

3. Given A and a, find b.

$$\frac{a}{b} = \tan A \qquad \therefore b = \frac{a}{\tan A}$$

15-2   Trigonometric Functions

Instead of determining the value of the angle from the ratio of side a to side b, divide any side by either of the remaining two and use the value of the ratio thus obtained as a measure of angle A. There are six possible ratios or *trigonometric functions* of each angle (see Fig. 15-3).

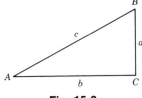

Fig. 15-3

$$\textit{sine of angle } A = \frac{\text{opposite side}}{\text{hypotenuse}} = \frac{a}{c}; \text{ written } \sin A = \frac{a}{c}$$

$$\textit{cosine of angle } A = \frac{\text{adjacent side}}{\text{hypotenuse}} = \frac{b}{c}; \text{ written } \cos A = \frac{b}{c}$$

$$\textit{tangent of angle } A = \frac{\text{opposite side}}{\text{adjacent side}} = \frac{a}{b}; \text{ written } \tan A = \frac{a}{b}$$

$$\textit{cotangent of angle } A = \frac{\text{adjacent side}}{\text{opposite side}} = \frac{b}{a}; \text{ written } \cot A = \frac{b}{a}$$

$$\textit{secant of angle } A = \frac{\text{hypotenuse}}{\text{adjacent side}} = \frac{c}{b}; \text{ written } \sec A = \frac{c}{b}$$

$$cosecant \ of \ angle \ A = \frac{\text{hypotenuse}}{\text{opposite side}} = \frac{c}{a}; \text{ written cosec } A = \frac{c}{a}$$

For angle B the six functions are as follows:

$$\sin B = \frac{b}{c} \qquad \tan B = \frac{b}{a} \qquad \sec B = \frac{c}{a}$$

$$\cos B = \frac{a}{c} \qquad \cot B = \frac{a}{b} \qquad \text{cosec } B = \frac{c}{b}$$

15-3   Use of Tables

This section refers to Appendix Table 2, Natural Trigonometric Functions, which begins on page 540.

To find the functions of an angle less than 45°, look for the degrees and minutes in the left-hand column of the table and look for the function at the top of the page. Go down the function column until the row which contains the desired degrees is reached.

Example 1
Find the sine of 39°50′.

PROCEDURE   On page 544 find 39° in heavy type on the left-hand side of the page. Go down further to the 50′ row. This is the row for 39°50′. Read at the top of the page the word "sin" (the usual abbreviation for sine). Go down the "sin" column until the row for 39°50′ is reached. The 0.6406 found there is the sine of 39°50′.

   Therefore, sin 39°50′ = 0.6406.

Example 2
Find the tangent of 84°10′.

PROCEDURE   On page 540 find 84° in the right-hand column of the table. (You will note that all degrees greater than 45° and less than 90° are in the right-hand column of the table.) Now go *up* the column until 84°10′ is reached. This is the row for 84°10′. Now read at the bottom of the page the word "tan" (the usual abbreviation for tangent). Go up the "tan" column until the row for 84°10′ is reached. The 9.788 found there is the tangent of 84°10′.

   Therefore, tan 84°10′ = 9.788.

Example 3
Find the cosine of 23°15′.

PROCEDURE   23°15′ is not in the table, since the table is in degrees and 10′ increments. Find the value for 23°15′ by the method known as interpolation. Since 23°15′ is halfway between 23°10′ and 23°20′, assume the tabular values for 23°15′ are also halfway between the values of 23°10′ and 23°20′. We find cosine 23°10′ in the table as 0.9194 and the cosine of 23°20′ in the table as 0.9182. (Note that the cosine ratio is a decreasing ratio.) Halfway between 0.9194 and 0.9182 is 0.9189.

Therefore, cos 23°15′ is 0.9189.

Problems 15-3

Find the value of each of the following from the tables in the back of this book.

1. sin 20°	2. sin 30°	3. sin 25°	4. cos 30°
5. cos 60°	6. cos 45°	7. tan 45°	8. tan 30°
9. tan 60°	10. sin 45°	11. cos 40°10′	12. tan 30°20′
13. sin 45°30′	14. tan 25°10′	15. cos 56°20′	16. cos 70°30′
17. tan 20°40′	18. sin 47°50′	19. cos 37°20′	20. cos 21°20′

Find the value of each of the following by interpolation.

21. sin 30°15′	22. sin 40°25′	23. sin 60°35′	24. cos 28°25′
25. cos 35°35′	26. cos 62°45′	27. tan 20°45′	28. tan 40°55′
29. tan 65°25′	30. sin 80°22′	31. cos 80°42′	32. tan 37°36′
33. sin 75°34′	34. tan 62°27′	35. cos 47°53′	36. tan 27°55′
37. sin 68°31′	38. cos 43°43′	39. tan 35°12′	40. cos 26°5′

15-4   Sin, Cos, and Tan on the Calculator

Examine your calculator and see if it has trigonometry function keys. You had to use degrees and decimal parts of a degree to do the previous problems. You can get your displayed answers converted to degrees, minutes, and seconds with the →D.MS key, but you cannot key in degrees, minutes, and seconds. Try some of the problems in this manner.

Example
What is the sin 47°35′?

SOLUTION

The display shows

Turn on the calculator	0.
Enter 35	35.
Press ÷	35.
Enter 60	60.
Press +	0.583333333
Enter 47	47.
Press =	47.58333333
Press sin	0.738259162 (answer)

This is the trig ratio of the sin 47°35′. Round off to 0.7383.

Use your calculator to do the trig ratios in Problem Set 15-3. Check the answers with those that you got from looking in the tables. Practice will increase your skills.

15-5 Finding an Angle Corresponding to a Given Function

Example 1

Find the angle whose sine is 0.5373.

PROCEDURE Turn to the trigonometry table and search the "sin" column until you find 0.5373. Read across to the left column headed "Degrees" and read 32°30′.

Therefore, the angle whose sine is 0.5373 is 32°30′.

Example 2

Find the angle whose tangent is 0.9340.

PROCEDURE Turn to the trigonometry table and search the tangent column until you find 0.9340. You will note that 0.9340 is not in our table. You will have to interpolate to find the angle whose tangent is 0.9340.

$$
\begin{array}{c}
\left.\begin{array}{c}
\left.\begin{array}{c}
0.9325 = \tan 43°0′ \\
0.0015 \left[0.9340 = \tan 43°x′ \right. \\
\end{array} \right] x' \\
0.0055 \\
0.9380 = \tan 43°10′
\end{array} \right] 10′
\end{array}
$$

By proportion,

$$\frac{0.0015}{0.0055} = \frac{x'}{10'}$$

$$\frac{15}{55} = \frac{x'}{10'}$$

$$x \approx 3'$$

Therefore, 0.9340 is $\frac{15}{55}$ of the way from 43°0′ to 43°10′ in our tangent table, or $\frac{15}{55} \times 10 = \frac{30}{21} = 2\frac{8}{11}$ or almost 3.

Therefore, the angle whose tangent is 0.9340 is 43°3′.

Problems 15-5

Find the value of angle A in each of the following.

1. $\sin A = 0.5225$	2. $\sin A = 0.0987$
3. $\cos A = 0.7826$	4. $\cos A = 0.9377$
5. $\tan A = 0.5206$	6. $\tan A = 0.8952$
7. $\sin A = 0.7071$	8. $\cos A = 0.9689$
9. $\tan A = 1.117$	10. $\sin A = 0.9100$
11. $\sin A = 0.2840$	12. $\cos A = 0.4147$
13. $\tan A = 1.540$	14. $\cos A = 0.2164$
15. $\tan A = 14.30$	16. $\sin A = 0.3393$
17. $\cos A = 0.6428$	18. $\tan A = 0.4557$
19. $\cos A = 0.7509$	20. $\sin A = 0.4410$

Find the value of angle B in each of the following. Interpolation will be necessary. Round off to the nearest minute.

21. $\sin B = 0.4710$	22. $\sin B = 0.2907$
23. $\cos B = 0.9894$	24. $\cos B = 0.1968$
25. $\tan B = 0.5867$	26. $\tan B = 0.2978$
27. $\sin B = 0.6788$	28. $\cos B = 0.9166$
29. $\tan B = 1.649$	30. $\cos B = 0.3902$
31. $\sin B = 0.7067$	32. $\cos B = 0.8531$
33. $\tan B = 1.108$	34. $\cos B = 0.4904$
35. $\sin B = 0.7161$	36. $\sin B = 0.7731$
37. $\cos B = 0.7707$	38. $\tan B = 1.047$
39. $\sin B = 0.9801$	40. $\cos B = 0.6873$

15-6 Using the Calculator to Find the Angle, Given the Function

If your calculator has the sin key, then it also has the arc sin key. The arc sin means "find an angle whose sine is x." On the calculator the key is usually the second function (2nd F) of the sin key and is denoted $\sin^{-1}$. The cos and tan keys likewise have arc cos and arc tan keys. The cotangent, secant, and cosecant of an angle are not on most calculators. These three functions and their inverses (arc cotangent, etc.) are reciprocals of the functions on the calculator and can easily be evaluated by using the reciprocal key $(1/x)$.

Example
Find the angle whose sine is 0.5373.

SOLUTION

The display shows

Turn on the calculator	$0.$
Enter 0.5373	0.5373
Press 2nd F	0.5373
Press $\sin^{-1}$	32.50002661
Press 2nd F	32.50002661 (decimal answer)
Press →D.MS	32.300009 (answer)

This answer is in degrees, minutes, and seconds in the calculator display. The answer is read 32 degrees, 30 minutes, and 00.09 seconds. Round off this answer to 32°30′. The 00.09 seconds is the result of the calculator's using a nine-place decimal (0.537300000) when we keyed in a four-place decimal.

Be careful with the results you get on the calculator display. Your answers may be in degrees and decimal parts of a degree (32.50002661 in our example) or in degrees, minutes, and seconds, which looks like the decimal mode but is read *degrees* for all the digits to the left of the decimal point, *minutes* for the next two places to the right of the decimal point, and *seconds* and *hundredths of a second* for the last four places in the display.

Problems 15-6

Using a calculator, do the problems from Problem Set 15-5 again.

15-7   Solution of Right Triangles

The following examples will illustrate the method to be followed in the solution of right triangles.

Example 1

Given $A = 36°30′$ and $b = 14$ in., find B, a, and c (Fig. 15-4).

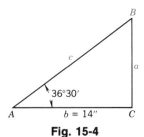

Fig. 15-4

SOLUTION   Since $A + B = 90°$

$$B = 90° - A$$
$$= 89°60′ - 36°30′$$
$$= 53°30′$$

To find a, write a function of the given angle A, employing for that purpose that given side b and the required side a. That function is $a/b = \tan A$.

$$\therefore a = b \times \tan A$$

That is, the length of side a is found by multiplying b by the value of the tangent of A, which is given in the table of natural tangents.

$$a = b \times \tan A = 14 \times \tan 36°30'$$
$$= 14 \times 0.7400 = 10.3600 \text{ in.}$$

Similarly, to find c, write a function of angle A, employing for that purpose the given side b and the required side c. That function is $b/c = \cos A$, from which $c = b/\cos A$. That is, the length of side c is found by dividing b by the value of the cosine of A given in the table of natural cosines.

$$c = \frac{b}{\cos A} = \frac{14}{\cos 36°30'} = \frac{14}{0.8039} = 17.4 \text{ in.}$$

A satisfactory way of arranging the work in the preceding example is as follows:

Given $A = 36°30'$, $b = 14$ in., find B, a, and c (Fig. 15-5).

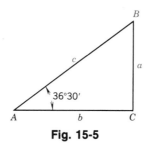

Fig. 15-5

To find B: $B = 90 - A = 89°60' - 36°30'$
$$= 53°30'$$

To find a: $a/b = \tan A \therefore a = b \times \tan A$
$$= 14 \times \tan 36°30'$$
$$= 14 \times 0.7400$$
$$= 10.3600 \text{ in.}$$

To find c: $b/c = \cos A \therefore c = \dfrac{b}{\cos A}$

$$= \frac{14}{\cos 36°30'}$$

$$= \frac{14}{0.8039} = 17.4 \text{ in.}$$

EXPLANATION   The method of solving a right triangle when an acute angle and one of the sides are given is as follows:

1. To find the unknown acute angle, subtract the given acute angle from 90°.
2. To find an unknown side, write a function of the given angle, employing for that purpose the given side and the required side. From that function obtain an expression for the required side in terms of the given side and the function of the angle. For example:

I.   *Given A and a.*

To find *b*: $\dfrac{b}{a} = \cot A$ $\therefore b = a \times \cot A$

To find *c*: $\dfrac{a}{c} = \sin A$ $\therefore c = \dfrac{a}{\sin A}$

II.   *Given A and c.*

To find *a*: $\dfrac{a}{c} = \sin A$ $\therefore a = c \times \sin A$

To find *b*: $\dfrac{b}{c} = \cos A$ $\therefore b = c \times \cos A$

III.   *Given A and b.*

To find *a*: $\dfrac{a}{b} = \tan A$ $\therefore a = b \times \tan A$

To find *c*: $\dfrac{b}{c} = \cos A$ $\therefore c = \dfrac{b}{\cos A}$

If angle *B* is given, derive angle *A* by subtracting *B* from 90° and use the formulas *I, II,* and *III* to find the missing sides.

Example 2

Given *a* = 8 in. and *b* = 15 in. Find *A, B,* and *c* (Fig. 15-6).

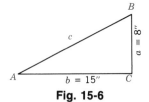

Fig. 15-6

SOLUTION
To find *A:*

$$\tan A = \frac{a}{b} = \frac{8}{15} = 0.5333 \quad \therefore A = 28°04'$$

To find B:

$$B = 90° - A = 90° - 28°04' = 61°56'$$

To find c:

$$c = \sqrt{a^2 + b^2} = \sqrt{8^2 + 15^2} = 17.00 \text{ in.}$$

EXPLANATION Since side a (the side opposite angle A) and side b (the side opposite angle B) are known, then the tangent of angle A (opposite side divided by the adjacent side) can be found, namely $\frac{8}{15}$. Reducing this to a decimal gives $\tan A = 0.5333$. Referring to a table of natural tangents, find by interpolation that the angle whose tangent is 0.5333 is $28°04'$.

Having found angle A, find angle B by subtracting angle A from $90°$. To find c, write the sine function of angle A, thus:

$$\frac{a}{c} = \sin A \quad \therefore c = \frac{a}{\sin A}$$

or

$$\frac{b}{c} = \cos A \quad \therefore c = \frac{b}{\cos A}$$

Side c can also be found by using the Pythagorean Theorem, that is, the square of the hypotenuse is equal to the sum of the squares of the other two sides.

Example 3
Given $a = 4.5$ in. and $c = 7.25$ in., find A, B, and b (Fig. 15-7).

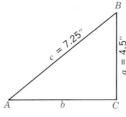

Fig. 15-7

SOLUTION
To find A:

$$\sin A = \frac{a}{c} = \frac{4.5}{7.25} = 0.6207 \quad \therefore A = 38°22'$$

To find B:

$$B = 90° - A = 90° - 38°22' = 51°38'$$

To find b:

$$b = \sqrt{c^2 - a^2} = \sqrt{7.25^2 - 4.5^2} = 5.68 \text{ in.}$$

Problems 15-7

Find the missing parts of the following right triangles.

1. Given $A = 24°$ and $a = 7''$, find B, b, and c.
2. Given $B = 30°40'$ and $a = 15''$, find A, b, and c.
3. Given $A = 70°40'$ and $a = 11.4''$, find B, b, and c.
4. Given $A = 48°10'$ and $b = 54.8''$, find B, a, and c.
5. Given $A = 8°$ and $b = 45''$, find B, a, and c.
6. Given $B = 16°10'$ and $b = 90$ m, find A, a, and c.
7. Given $B = 66°20'$ and $c = 60.55$ cm, find A, a, and b.
8. Given $A = 53°28'$ and $c = 16.5$ mm, find B, a, and b.
9. Given $A = 31°56'$ and $c = 21.7$ m, find B, a, and b.
10. Given $B = 45°45'$ and $a = 32.5$ cm, find A, b, and c.

Find the missing parts of the following right triangles.

11. Given $a = 7.5''$ and $b = 8.75''$, find A, B, and c.
12. Given $a = 9.4''$ and $b = 12.8''$, find A, B, and c.
13. Given $a = 12.8''$ and $b = 16''$, find A, B, and c.
14. Given $a = 6.6''$ and $b = 14.48''$, find A, B, and c.
15. Given $a = 40'$ and $b = 28'$, find A, B, and c.
16. Given $a = 38.3$ cm and $b = 45.6$ cm, find A, B, and c.
17. Given $a = 20$ mm and $b = 28$ mm, find A, B, and c.
18. Given $a = 102$ m and $b = 126$ m, find A, B, and c.
19. Given $a = 16.75$ cm and $b = 24.5$ cm, find A, B, and c.
20. Given $a = 34$ cm and $b = 40$ cm, find A, B, and c.

Find the missing parts of the following right triangles.

21. Given $a = 3.8''$ and $c = 5.2''$, find A, B, and b.
22. Given $b = 18''$ and $c = 22''$, find A, B, and a.
23. Given $b = 92'$ and $c = 124''$, find A, B, and a.
24. Given $a = 15''$ and $c = 25''$, find A, B, and b.
25. Given $b = 7.1''$ and $c = 15.84''$, find A, B, and a.
26. Given $a = 21$ cm and $c = 38$ cm, find A, B, and b.
27. Given $a = 14.3$ cm and $c = 23.4$ cm, find A, B, and b.
28. Given $b = 10.6$ m and $c = 16.2$ m, find A, B, and a.
29. Given $a = 45$ m and $c = 70$ m, find A, B, and b.
30. Given $b = 25.5$ cm and $c = 40$ cm, find A, B, and a.

15-8   Isosceles Triangles

A perpendicular dropped from the vertex of an isosceles triangle to the base divides the isosceles triangle into two equal right triangles that can be solved by the methods shown in the preceding section (see Fig. 15-8).

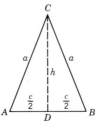

Fig. 15-8

Example 1
Given $a = 16$ in. and $c = 12$ in., find A, B, C, and h (Fig. 15-9).

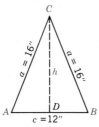

Fig. 15-9

SOLUTION See Fig. 15-10.
To find A:

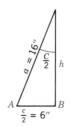

Fig. 15-10

$$\cos A = \frac{c/2}{a} = \frac{6}{16} = 0.3750 \quad \therefore A = 67°59'$$

$$B = A = 67°59'$$

To find C, first find $C/2$:

$$C/2 = 90° - A = 90° - 67°59' = 22°01'$$
$$C = 2 \times 22°01' = 44°02'$$

or

$$C = 180° - (A + B)$$
$$= 180° - (67°59' + 67°59')$$
$$= 180° - 135°58' = 44°02'$$

To find h:

$$h = \sqrt{a^2 - (c/2)^2} = \sqrt{16^2 - 6^2} = 14.83 \text{ in.}$$

Example 2

Given $C = 54°42'$ and $a = 25$ in., find A, B, c, and h (Fig. 15-11).

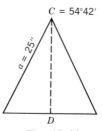

Fig. 15-11

SOLUTION   See Fig. 15-12.

To find A:

$$A = 90° - C/2 = 90° - 27°21' = 62°39'$$
$$B = A = 62°39'$$

To find c, first find $c/2$:

$$\frac{c/2}{a} = \sin 27°21'$$

$$c/2 = a \times \sin 27°21'$$
$$= 25 \times 0.4594 = 11.48 \text{ in.}$$
$$\therefore c = 2 \times 11.48 \text{ in.} = 22.96 \text{ in.}$$

To find h:

$$h/a = \cos 27°21'$$
$$\therefore h = a \times \cos 27°21'$$
$$= 25 \times 0.8882 = 22.20 \text{ in.}$$

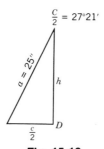

Fig. 15-12

Problems 15-8

Compute the missing dimensions of the isosceles triangles in the following chart.

ISOSCELES TRIANGLES

	Base	Sides	Altitude	Base Angles	Vertex Angle
1.	12″	10″	_____	_____	_____
2.	10″	13″	_____	_____	_____
3.	8″	_____	_____	60°	_____
4.	_____	15″	_____	30°	_____
5.	18 m	_____	_____	_____	90°
6.	_____	_____	10 cm	45°	_____
7.	_____	_____	8 cm	_____	120°
8.	_____	16 mm	_____	_____	60°
9.	_____	10.8″	_____	64°15′	_____
10.	_____	14.6″	_____	_____	85°30′
11.	18.2″	_____	_____	_____	71°40′
12.	16″	_____	_____	69°45′	_____
13.	24.5 cm	_____	15.6 cm	_____	_____
14.	_____	_____	28 cm	82°30′	_____
15.	_____	32.8 cm	25.4 cm	_____	_____
16.	_____	_____	19.3 mm	_____	100°

Self-Test

Solve the following problems. Use Table 2, Natural Trigonometric Functions, in the back of this book. Check your answers with the answers given in the back of the book.

1. What is the sin 30°?

Use the figure of the right triangle below to do Problems 2 to 5.

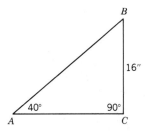

2. How large is angle *B*?
3. What is the sine of angle *A*?
4. How long is side *AB*?
5. How long is side *AC*?

Use the figure of the isosceles triangle below to do Problems 6 to 10.

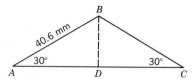

6. How long is side BC?
7. How long is the altitude of triangle ABC?
8. What is the size of the vertex angle B?
9. How long is the base AC?
10. What is the area of triangle ABC?

Chapter Test

Solve the following problems involving triangles. Make a sketch of each triangle and label all the known parts. Organize your work and avoid arithmetical errors. Use the tables in the back of the book.

1. Find the tangent of $35°10'$.
2. Find the sine of $25°25'$.
3. Find the angle whose cotangent is 3.867.
4. Find the angle whose cosine is 0.2565.
5. In right triangle ABC, angle $A = 25°$, angle $C = 90°$, and side $a = 15''$. Find side c.
6. In right triangle ABC, angle $B = 42°20'$, angle $C = 90°$, and side $b = 18''$. Find side a.
7. In right triangle ABC, angle $C = 90°$, side $a = 24''$, and side $b = 20''$. Find angle A.
8. Find side c in Problem 7.
9. In right triangle ABC, side $a = 19''$, side $c = 28''$, and angle $C = 90°$. Find side b.
10. Find angle A in Problem 9.
11. The two legs of a right triangle are 7 mm and 24 mm, respectively. How long is the hypotenuse?
12. In isosceles triangle ABC, angle $A =$ angle B, angle $C = 100°$, and side $c = 16''$. Find side a.
13. In isosceles triangle ABC, angle $A =$ angle B, side $c = 26$ cm, and side $b =$ side $a = 16$ cm. Find angle B.
14. Find angle A in Problem 12.
15. Find the area of a triangle whose sides are 20 cm, 20 cm, and 24 cm. (Use Hero's formula in section 7-12.)

Miscellaneous Problems Using Trigonometry

The following problems can be solved by using the mathematics you have learned so far. The key to solving these problems is organizing your work.

Make a neat sketch and label the known parts. Study your sketch and determine what you need to know to find the solution to the problem. Avoid arithmetical errors. Estimate the answer before solving the problem.

1. The equal sides of an isosceles triangle are 25″ long and the altitude is 14″. Compute the base and the angles.
2. An isosceles triangle has sides of 13″, 13″, and 5″. Compute the angles and the altitude.
3. An isosceles triangle has a base of 4.34 cm and an altitude of 10 cm. Compute the angles and the length of the equal sides.
4. The base angles of an isosceles triangle are each 72°46′ and the altitude is 7.5″. Compute the vertex angle, the base, and the length of the equal sides.
5. The vertex angle of an isosceles triangle is 38°22′ and the equal sides are each 2.5″ long. Find the base angles, the base, and the altitude.
6. Find the perpendicular distance from the center to the side of a hexagon with 5″ sides. (*Suggestion:* Draw *OA*, Fig. 15-13. Angle *OAB* = 60°. Solve for *OB*.)

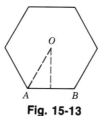

Fig. 15-13

7. Find the included taper angle in a piece of work having a taper of 6.25 cm per meter.
8. Find the included taper angle in a piece of work having a taper of 0.6″ per foot.
9. Figure 15-14 shows a 29° Acme thread. Find the bottom flat of a 1″ pitch thread if the depth is $\frac{1}{2}$ pitch + 0.01″.

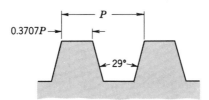

Fig. 15-14

10. In the Whitworth thread in Fig. 15-15, the sides make an angle of 55° with each other. The depth of the thread is two-thirds of the depth of the triangle formed by prolonging the sides. Find the depth of a thread of 1″ pitch.

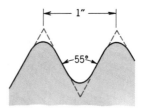

Fig. 15-15

11. A road has a rise of 4 m in 100 m. What is the angle of slope with the horizontal?
12. A road has a slope of 1°50′ with the horizontal. What is the rise in 100 m?
13. What is the largest square that can be milled from a circular disk 6″ in diameter?
14. Find the radius of a circle circumscribed about an octagon with 10″ sides.
15. Find the angle of slope of the rafter in Fig. 15-16.

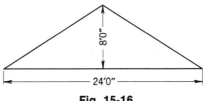

Fig. 15-16

16. A chord of a $2\frac{1}{2}″$ circle is 1.028″ from the center. Find its length.
17. The chord of a 19.5-cm circle is 11.25 cm long. Find the central angle and the distance from the center.
18. Lay out an angle of 12°40′ by means of the value of its natural tangent.
19. Compute the missing dimension in the sketch in Fig. 15-17.

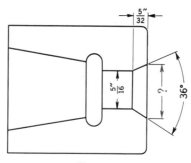

Fig. 15-17

20. Compute the missing dimension in the flat-headed screw shown in Fig. 15-18.

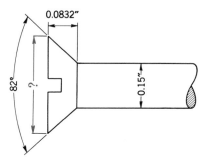

Fig. 15-18

21. Figure 15-19 shows a taper gib with a taper of $\frac{3}{16}$" per foot. Compute the following dimensions:

    (*a*) Thickness at the large end.
    (*b*) Dimension *A* at the small end.
    (*c*) Width *W* at the large end.
    (*d*) Width *w* at the small end.

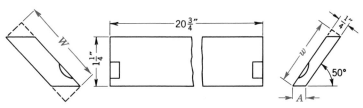

Fig. 15-19

22. Twelve bolt holes are to be drilled in a cylinder head on the circumference of a 36-cm circle. What is the straight line distance between centers of holes?

23. If 15 holes are drilled on the circumference of a 27" circle, find the distance center to center of holes.

24. Nine holes are to be drilled on the circumference of a circle 16.5 cm in diameter. What should be the distance center to center of holes?

25. Find the length of the chord joining two points 119° apart on the circumference of a $3\frac{1}{2}$" circle.

26. Find the distance between two points 37°30' apart on the circumference of a circle 2.37 cm in diameter.

27. What should be the check measurement over a pair of $\frac{1}{2}$" wires on the dovetail slide shown in Fig. 15-20, both inside and outside? The sides form an angle of $62\frac{1}{2}$° with the horizontal. (*Suggestion:* The dashed lines indicate the solution.)

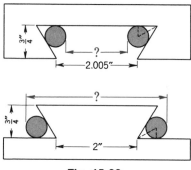

Fig. 15-20

28. Figure 15-21*a* shows the arrangement for gaging screw threads by the three-wire system. Compute the measurement over the wires for the following sharp V-threads:

   (*a*) $\frac{1}{4}''$ diameter, 20 threads per inch, wires 0.035'' diameter
   (*b*) $\frac{9}{16}''$ diameter, 12 threads per inch, wires 0.050'' diameter

   (*Suggestion:* The solution is indicated on Fig. 15-21*b*. Add twice the distance *AB* to the root diameter and add half a wire diameter on each side.)

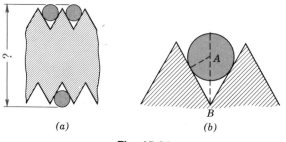

(*a*) (*b*)

Fig. 15-21

29. Figure the check measurements for the same two screws as in Problem 28 but for American National threads. Use the same wires.
30. Compute the height of a $7\frac{1}{2}''$ sine bar for an angle of 27°55'.
31. A sine bar 10'' long has a difference in elevation of 4.873'' between the two ends. What angle is indicated?
32. A cylindrical gasoline tank 14'' in diameter and 3'1'' long is lying in a horizontal position. A gage inserted through the top indicates a depth of gasoline of $11\frac{1}{4}''$. How many gallons of gasoline are there in the tank? (Fig. 15-22).

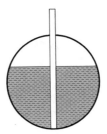

Fig. 15-22

33. Two roads cross each other at an angle of 70° (Fig. 15-23). The arc *AB* is drawn tangent to the main curb lines with a radius of 3 m. Find the length of the piece of curbing *AB*. If the roadways are 20 m wide, how many square meters of pavement are required for their intersection?

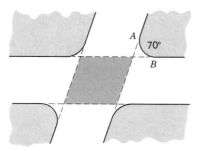

Fig. 15-23

34. In a Brown and Sharpe 29° worm thread, find the diameter of the wire laid in the groove that will lie exactly flush with the top of the thread: (*a*) 4 threads per inch; (*b*) 5 threads per inch; (*c*) 3 threads per inch. See Fig. 15-24.

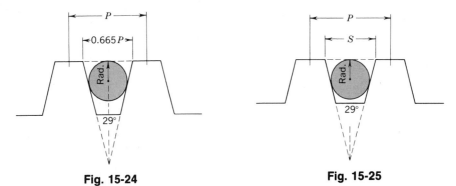

Fig. 15-24　　　　　　　　　　　**Fig. 15-25**

35. In the 29° Acme tap thread the space $S = 0.6293P + 0.0052''$. Find the diameter of a wire that will lie in the groove flush with the top of the thread if the pitch is (*a*) $\frac{1}{8}''$; (*b*) $\frac{1}{6}''$; (*c*) $\frac{1}{10}''$. See Fig. 15-25.

16

Strength
of
Materials

16-1 Stress and Strain

When a load of 100 lb is suspended by a wire, the load tends to stretch or elongate the wire. This tendency of the load to *distort* or *change* the shape of the wire is called **strain.** The wire resists the tendency of the load to pull it apart and in so doing exerts a force opposite to the load. The internal force by which the wire resists the tendency of the outside load to change its shape is called **stress.**

Similarly, a load on a column tends to compress or crush the column. The column reacts against the tendency of the load to crush it and exerts a force opposite to the load. The tendency of the outside load to change the shape of the column is called *strain,* and the internal force by which the column resists the tendency of the outside load to change its shape is called *stress.*

16-2 Kinds of Stresses

Three kinds of direct stress may be developed in a body by the application of an outside force or load: *tension, compression,* and *shear.*

376

Tension

A weight suspended from a rod tends to pull the rod apart. The stress developed in the rod to resist being *pulled* apart is called ***tensile stress.*** The rod is said to be in ***tension.***

Compression

A load on a column tends to crush it. The stress the column develops to resist being *crushed* is called ***compressive stress.*** The column is said to be in ***compression.***

Shear

A rivet connecting two plates will sometimes give way because the plates push so hard on the rivet in opposite directions as to cause one half of the rivet to slide past the other; that is, the two halves of the rivet are separated as if the rivet were cut by a pair of shears (see Fig. 16-1). The stress the rivet develops to resist being cut apart is called ***shearing stress.*** The rivet is said to be in ***shear.***

   Bending stress is a combination of tension and compression.

   Twisting or ***torsion*** is a form of shearing stress.

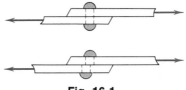

Fig. 16-1

16-3   Unit Stress

The ***unit stress*** is the stress per unit of area and is usually expressed in pounds per square inch. Thus if a column whose cross-section area is 6 sq in. supports a load of 60,000 lb, the unit stress is $\frac{60,000}{6}$ or 10,000 lb per square inch.

16-4   Elastic Limit

If a suspended rod with a cross-section area of 1 sq in. is loaded with a weight, the latter stretches the rod a certain amount, depending upon the length of the rod and the load applied. If too great a load is not used, the rod will spring back to its original length when the load is removed. When a sufficiently heavy load is hung from the rod, the rod will not spring back to its original length when the load is removed. There remains a permanent set. So the ***elastic limit*** is the unit stress beyond which the body will not return to its original shape when the load is removed.

   Thus, if a steel rod of 1 sq in. cross-section area from which a load of 35,000 lb is suspended springs back to its original length when the load is removed but acquires a permanent set when a load of 36,000 lb is applied, we say the elastic limit of that rod is 35,000 lb.

16-5 Ultimate Strength

By the *ultimate* or *breaking strength* of a body is meant the unit stress that causes that body to break. Thus it takes almost 50,000 lb to pull apart a wrought iron bar with a cross-section area of 1 sq in. We therefore say that the ultimate tensile strength of wrought iron is 50,000 lb per square inch.

It is necessary for the engineer to know the strength of the materials that enter into the machine or structure under design. To determine the ultimate strength of the materials in common use, specimens are subjected to tensile, compressive, or shearing stresses in testing machines, and the breaking point in each instance is noted. In this way the average values of the elastic limits and ultimate strengths in pounds per square inch were obtained for the following materials.

ELASTIC LIMIT AND ULTIMATE STRENGTH (POUNDS PER SQUARE INCH)

Material	Elastic Limit	Tension	Compression	Shear
Timber	3,000	8,000	8,000	500 (with grain) 3,000 (across grain)
Stone			6,000	1,500
Brick		250–300	3,000	1,000
Cast iron	6,000 (tension) 20,000 (comp.)	25,000	90,000	18,000
Wrought iron	28,000	50,000	55,000	40,000
Medium steel	36,000	60,000	60,000	50,000
Concrete, Portland 1:2:4		100–150	2,400	300–500

The column group "Ultimate Strength" spans Tension, Compression, Shear.

16-6 Factor of Safety

Although the ultimate compressive strength of steel is 60,000 lb per square inch, in designing a machine or structure the engineer will design each member on the assumption that the material can develop a compressive stress of only 15,000 lb per square inch. In other words, he will make each member of the machine or structure four times as strong as is required in order to keep it from breaking down when loaded. The ratio of the ultimate strength to the actual stress is called the *safety factor.* In the example just cited, the safety factor is $\dfrac{60,000}{15,000} = 4$. The safety factor ranges from 4 for steel, designed for steady stress, to 35 for brick, designed to withstand shock. A larger safety factor is assumed in the design of a railroad bridge than in the design of an office building because the moving load on a railroad bridge causes the structure to vibrate, whereas for the building, the load is steady.

16-7 Working Unit Stresses

As a result of investigation and experience, the following values have been adopted as safe working stresses for the materials most commonly employed. Notice that in each instance the working unit stress is well within the elastic limit.

SAFE WORKING STRESSES (POUNDS PER SQUARE INCH)			
Material	Tension	Compression	Shear
Timber	750	750	500
Cast iron	4,000	15,000.	3,000
Wrought iron	12,000	12,000	10,000
Medium steel	15,000	15,000	10,000

Example 1

How many pounds can be safely carried in tension by a steel rod whose cross section is a rectangle 2 in. $\times \frac{3}{4}$ in.? (From the preceding table we note that steel has a safe working stress of 15,000 lb per sq inch when in tension.)

SOLUTION   Area of cross section $= 2 \times \frac{3}{4} = 1\frac{1}{2}$ sq in.
Carrying capacity in tension $= 1\frac{1}{2} \times 15,000 = 22,500$ lb

Example 2

Allowing a tensile stress of 15,000 lb per square inch, how many pounds can be safely carried by a steel rod $\frac{7}{8}$ in. in diameter?

SOLUTION   Area of cross section $= 0.7854 \times 0.875^2 = 0.6013$ sq in.
Safe carrying capacity $= 15,000 \times 0.6013 = 9,020$ lb

Example 3

Allowing a tensile stress of 15,000 lb per square inch, find the diameter of a steel rod to carry 100,000 lb.

SOLUTION   Required area of cross section $= \dfrac{100,000}{15,000} = 6.6667$ sq in.

$$\text{Diameter} = 2 \sqrt{\frac{\text{area}}{\pi}} \text{ (see Chapter 8 for areas of circles)}$$

$$= 2 \sqrt{\frac{6.6667}{3.1416}} = 2 \sqrt{2.1221}$$

$$= 2 \times 1.46 = 2.92 \text{ in.}$$

Problems 16-7

Answer the following questions involving working stresses. Organize and label all work.

1. A steel rod has a rectangular cross section $3\frac{1}{2}'' \times 1\frac{3}{4}''$. Allowing a tensile stress of 15,000 lb per square inch, how many pounds can it carry?
2. How many pounds can be safely carried in tension by a steel rod $1\frac{7}{8}''$ in diameter?
3. What must be the diameter of a steel bar to carry safely a tensile force of 65,000 lb?
4. What should be the diameter of a steel bar to carry a load of 60,000 lb, allowing a tensile stress of 16,000 lb per square inch?
5. What should be the diameter of a steel bar to carry safely a load of 72,000 lb in tension, allowing 15,000 lb per square inch?
6. A balcony is supported from the ceiling by means of six wrought iron rods. Allowing a tensile stress of 12,000 lb per square inch, what size rods must be used if the total weight is 80,000 lb?
7. Allowing a stress of 10,000 lb per square inch, what is the shearing strength of an iron bar $2\frac{1}{2}'' \times \frac{7}{16}''$ in cross section?
8. The ends of a floor beam carrying a total load of 6,400 lb uniformly distributed rest on cast iron plates placed on the brick. If the plates are $8'' \times 10''$, what is the pressure per square inch on the brick?
9. A loading platform is supported by six square sticks of timber. What must be the size of the supports if the platform is to carry 120,000 lb and the allowable compressive stress is 500 lb per square inch?
10. What is the ultimate compressive strength in pounds per square inch of a brick $8'' \times 4'' \times 2''$ if it gave way in the testing machine under a load of 56,000 lb when lying flat?

16-8 Pressure in Pipes

The tendency of steam or water pressure in a pipe is to burst the pipe longitudinally. The material of the pipe is therefore under tension when the pipe is under pressure. The total pressure on a diametral plane is equal to

$$P = p \times L \times D$$

where P = total pressure
p = pressure per square inch
L = length of pipe in inches
D = diameter of pipe in inches

The tendency of the steam or water pressure to burst the pipe is resisted by the tension in the walls of the pipe (see Fig. 16-2). The total resisting stress in the pipe is equal to

$$S = 2 \times s \times L \times t$$

where S = total resisting stress
s = unit tensile stress
L = length of pipe in inches
t = thickness of pipe

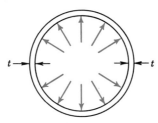

Fig. 16-2

For the pipe to withstand bursting, the resisting stress must equal the total pressure. We have therefore

$$2 \times s \times L \times t = p \times L \times D$$

$$\therefore t = \frac{p \times L \times D}{2 \times s \times L} = \frac{p \times D}{2 \times s}$$

This formula enables us to find the required thickness of pipe to withstand a given pressure.

To find the safe internal unit pressure in a pipe of given thickness, the preceding formula is solved for p, giving,

$$p = \frac{2 \times s \times t}{D}$$

Example
Allowing a tensile stress of 5,000 lb per square inch, how thick should a 20-in. wrought iron steam pipe be to withstand a pressure of 300 lb per square inch?

SOLUTION $\quad t = \dfrac{p \times D}{2 \times s} = \dfrac{300 \times 20}{2 \times 5,000} = 0.6$ in.

This is the minimum thickness of pipe required. The next higher standard thickness would be used.

Problems 16-8

Use the formulas in Section 16-8 to solve the following problems.

1. How thick must a 15″ steam pipe be to withstand a steam pressure of 250 lb per square inch with a unit tensile stress of 5,000 lb per square inch?
2. What may be the maximum pressure per square inch in a steam pipe $\frac{3}{4}$″ thick if its diameter is 16″? Allow a unit tensile stress of 5,500 lb per square inch.
3. A cast iron water pipe 30″ in diameter is to withstand a pressure of 75 lb per square inch. Allowing a tensile stress of 2,000 lb per square inch, how thick must the pipe be?
4. Allowing a unit tensile stress of 2,000 lb per square inch, what is the maximum allowable pressure per square inch in a cast iron pipe 15″ in diameter and $\frac{5}{8}$″ thick?
5. Find the maximum allowable pressure per square inch in a steam pipe $\frac{7}{8}$″ thick and 20″ in diameter, allowing a tensile stress of 2,500 lb per square inch.
6. What must be the thickness of a steam pipe 18″ in diameter to withstand a pressure of 100 lb per square inch if the unit tensile stress is 3,500 lb per square inch?

16-9   Riveted Joints

A riveted joint may fail in one of two ways: either the rivets may be sheared off or the plates may be crushed by the pressure of the rivets. In the former, the area to be sheared is the cross-section area of the rivet. The safe load on the rivet for this type of stress is called the *shearing value* of the rivet.

In order to fail by tearing through the plate, the rivet must crush an area that is a rectangle formed by the thickness of the plate and the diameter of the rivet. The safe load per rivet for this type of stress is called the *bearing value* of the rivet. The common unit stresses allowed for steel plates and rivets are: 10,000 lb per square inch for shearing and 15,000 lb per square inch for bearing.

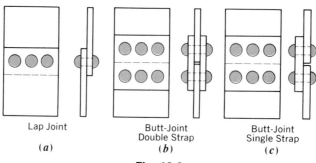

Lap Joint

(a)

Butt-Joint
Double Strap

(b)

Butt-Joint
Single Strap

(c)

Fig. 16-3

Figure 16-3 shows three types of riveted joints. In (*a*) and (*c*) the rivets are said to be in ***single shear*** because they may be destroyed by shearing at one place only. In (*b*) the rivets are in ***double shear*** because they must be sheared off in two places at the same time in order to fail.

Example 1
Figure 16-4*a* shows two plates each $\frac{1}{2}$ in. thick connected by one rivet $\frac{7}{8}$ in. in diameter. Find the strength of the joint.

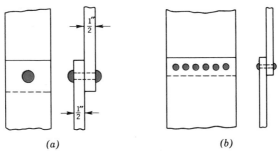

(*a*)   (*b*)

Fig. 16-4

SOLUTION   Cross-section area of rivet $= 3.14 \times \frac{7}{16} \times \frac{7}{16} = 0.601$ sq in.
Shearing value of rivet $= 0.601 \times 10,000 = 6,010$ lb
Bearing area of rivet $= \frac{7}{8} \times \frac{1}{2} = \frac{7}{16} = 0.438$ sq in.
Bearing value of rivet $= 0.438 \times 15,000 = 6,570$ lb

The smallest value determines the strength of the joint, which is 6,010 lb.

Example 2
If the plates in Fig. 16-4*b* are connected by six rivets each $\frac{7}{8}$ in. in diameter, what is the strength of the joint?

SOLUTION $6 \times 6,010 = 36,060$

EXPLANATION   Since one rivet is good for 6,010 lb, six rivets will carry $6 \times 6,010$, or 36,060 lb.

Example 3
How many $\frac{7}{8}$-in. rivets are required to carry a stress of 100,000 lb through the two plates in Fig. 16-4*b*?

SOLUTION $\dfrac{100,000}{6,010} = 16.64$ or 17 rivets

EXPLANATION   The number of rivets required is found by dividing the total stress to be carried by the lowest value of one rivet. You must always round the answer *up* to a whole number of rivets.

Example 4

In Fig. 16-5 the inside plate is $\frac{3}{4}$ in. thick and the outside plates are each $\frac{1}{2}$ in. thick. The rivet is $\frac{7}{8}$ in. in diameter. What is the strength of the joint?

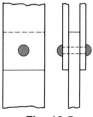

Fig. 16-5

SOLUTION

$$\text{Cross-section area of rivet} = 3.14 \times \tfrac{7}{16} \times \tfrac{7}{16} = 0.601 \text{ sq in.}$$
$$\text{Shearing value of rivet (double shear)} = 2 \times 0.601 \times 10,000 = 12,020 \text{ lb}$$
$$\text{Bearing area of rivet} = \tfrac{7}{8} \times \tfrac{3}{4} = \tfrac{21}{32} = 0.656 \text{ sq in.}$$
$$\text{Bearing value of rivet} = 0.656 \times 15,000 = 9,840 \text{ lb}$$

The strength of the joint, as determined by the lowest value of the rivet, is 9,840 lb.

CROSS-SECTION AREA OF RIVETS			
Diameter (Inches)	Area (Square Inches)	Diameter (Inches)	Area (Square Inches)
$\frac{3}{8}$	0.1104	$\frac{3}{4}$	0.4418
$\frac{1}{2}$	0.1963	$\frac{7}{8}$	0.6013
$\frac{5}{8}$	0.3068	1	0.7854

Problems 16-9

Find the solutions to the following problems involving strength of rivets. Organize your work. Use the example problems as a guide.

1. Two $\frac{5}{8}''$ plates are connected by a lap joint with $\frac{7}{8}''$ rivets. What is the strength of the joint if 12 rivets are used?
2. Find the strength of the joint shown in Fig. 16-6 if eight $\frac{3}{4}''$ rivets are used.
3. How many $\frac{5}{8}''$ rivets will be required on each side of the joint shown in Fig. 16-7 if all plates are $\frac{9}{16}''$ thick and the stress to be developed is 60,000 lb?
4. Find the strength of the joint shown in Fig. 16-8 if ten $\frac{1}{2}''$ rivets are used on each side of the butt joint. All plates are $\frac{1}{2}''$ thick.
5. Find the strength of the joint shown in Fig. 16-9 if the plates are $\frac{1}{2}''$ thick and the straps $\frac{9}{16}''$ thick. Eight $\frac{3}{4}''$ rivets are used on each side of the butt joint.

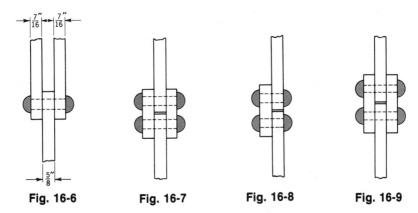

Fig. 16-6   Fig. 16-7   Fig. 16-8   Fig. 16-9

6. Compute the strength of the joint in Problem 1 if: (*a*) eighteen $\frac{1}{2}$" rivets are used; (*b*) sixteen $\frac{5}{8}$" rivets; (*c*) fourteen $\frac{3}{4}$" rivets; (*d*) ten 1" rivets.

7. (*a*) How many $\frac{7}{8}$" rivets are required to transmit a stress of 15,000 lb through a lap joint of $\frac{3}{4}$" plates? (*b*) How many $\frac{1}{2}$" rivets? (*c*) How many $\frac{5}{8}$" rivets? (*d*) How many $\frac{3}{8}$" rivets? (*e*) How many 1" rivets?

8. A single strap butt joint of two $\frac{1}{2}$" plates is required to transmit 75,000 lb. (*a*) How many $\frac{1}{2}$" rivets are required if the strap is $\frac{1}{2}$" thick? (*b*) How many $\frac{5}{8}$" rivets if the strap is $\frac{9}{16}$" thick? (*c*) How many $\frac{3}{4}$" rivets with a $\frac{1}{2}$" strap? (*d*) How many $\frac{7}{8}$" rivets with a $\frac{7}{16}$" strap? (*e*) How many 1" rivets with a $\frac{1}{2}$" strap?

9. Two $\frac{11}{16}$" plates are connected by a double strap butt joint with $\frac{1}{2}$" straps. (*a*) What is the strength of such a joint with eight $\frac{7}{8}$" rivets on each side of the butt joint? (*b*) Compute the strength of this joint with twenty-four $\frac{1}{2}$" rivets on each side of the butt joint. (*c*) Twenty $\frac{5}{8}$" rivets on each side. (*d*) Eight 1" rivets on each side.

10. (*a*) How many $\frac{3}{4}$" rivets are required on each side of the butt joint in Problem 9 if a stress of 90,000 lb is to be transmitted? (*b*) How many $\frac{1}{2}$" rivets? (*c*) How many $\frac{5}{8}$" rivets? (*d*) How many 1" rivets?

17

Work and Power

17-1 Work and Power

When we lift a weight of 1 lb a height of 1 ft, we do 1 *foot-pound* of work. If we exert a pull of 50 lb in dragging a load a distance of 10 ft, we do 10×50, or 500 foot-pounds of work.

A *foot-pound* (abbreviated as ft-lb) is the unit of work. It is the work done in raising 1 lb a height of 1 ft, or it is the pressure of 1 lb exerted over a distance of 1 ft in any direction.

A freight elevator lifting 2,000 lb a height of 40 ft does 80,000 ft-lb of work whether it takes 2 minutes or 4 minutes to do it. But an elevator that would lift the load in 2 minutes would have twice the *power* of one that would take 4 minutes to do the same amount of work.

Power is the rate at which work is done. It is measured in foot-pounds per minute. Thus, if an elevator did 80,000 ft-lb of work in 2 minutes, its power would be $\frac{80,000}{2}$ or 40,000 ft-lb per minute. If it took 4 minutes to do it, its power would be $\frac{80,000}{4}$ or 20,000 ft-lb per minute.

The unit of power is the *horsepower* (abbreviated as hp), which is equal to 33,000 ft-lb per minute. It is interesting to note that 33,000 ft-lb per minute is considered more than the work a horse can do. The horse used in the original tests was an unusually strong horse, so three-quarters of the figure used for a horsepower more nearly approximates the work a horse can do.

Example
An engine does 120,000 ft-lb of work in 3 minutes. Compute the horsepower.

SOLUTION $\text{Horsepower} = \dfrac{120,000}{3 \times 33,000} = 1.2$

EXPLANATION   Since the engine takes 3 minutes to do 120,000 ft-lb of work, in 1 minute it will do $\frac{1}{3}$ of 120,000 or 40,000 ft-lb of work.

Since $33,000 \dfrac{\text{ft-lb}}{\text{min}} = 1$ hp, divide by 33,000.

Problems 17-1

Find the amount of work done in the following problems. Use foot-pounds as the unit of work. Convert this work to horsepower when asked to do so.

1. A man carries a load of 50 lb up a flight of stairs 10 ft high. Compute the work done in foot-pounds.
2. Compute the horsepower required for an elevator to lift 2,400 lb a height of 120 ft in $1\frac{3}{4}$ minutes.
3. Find the horsepower of an engine that pumps 30 cu ft of water per minute from a depth of 320 ft. Water weighs 62.5 lb per cubic foot.
4. What must be the horsepower of an engine to lift a girder weighing 7 tons a height of 40 ft in 5 minutes?
5. A man pushing a wheelbarrow a distance of 450 ft exerts a force of 45 lb. How much work does he do?
6. An engine does 84,000 ft-lb of work in $2\frac{1}{2}$ minutes. Compute the horsepower.

17-2   Horsepower of a Steam Engine

In speaking of the horsepower of a steam engine, the *indicated horsepower* is meant. The formula for finding the indicated horsepower is

$$\text{hp} = \frac{P \times L \times A \times N}{33,000}$$

where hp = indicated horsepower
   P = mean effective pressure on the piston in pounds per square inch (P is found by means of an indicator card)
   L = length of stroke in feet
   A = area of piston in square inches
   N = number of strokes per minute = 2 × rpm for a double-acting engine

Example

Find the horsepower of a 26 in. × 40 in. double-acting engine running at 80 rpm under a mean effective pressure of 75 lb per square inch.

NOTE.   In specifying the dimensions of a cylinder, the first number gives the diameter or bore; the second number, the stroke.

PROCEDURE

1. Find the area of the piston in square inches (use the formula $A = 0.7854d^2$).
2. Find the length of the stroke in feet.
3. If the engine is a double-acting engine multiply the rpm by 2.
4. Substitute the values in the formula for horsepower of steam engines and solve.

SOLUTION $$hp = \frac{P \times L \times A \times N}{33,000}$$

$P = 75$

$L = 40 \text{ in.} = 3\frac{1}{3} \text{ ft}$

$A = 26^2 \times 0.785 = 530.66 \text{ sq in.}$

$N = 80 \times 2 = 160$

$$hp = \frac{75 \times 10 \times 530.66 \times 160}{3 \times 33,000} = 643.2 \text{ hp}$$

Problems 17-2

Find the horsepower of the following steam engines.

1. What is the indicated horsepower of an 18″ × 28″ double-acting steam engine making 100 rpm with a mean effective pressure of 48 lb?
2. Compute the horsepower of a 32″ × 50″ double-acting steam engine running at 75 rpm with a mean effective pressure of 55 lb?
3. A 24″ × 40″ double-acting engine is running at 80 rpm under a mean effective pressure of 50 lb. Find the horsepower.
4. Find the horsepower of a 22″ × 26″ double-acting steam engine making 75 rpm under a mean pressure of 110 lb.
5. Compute the horsepower of a double-acting steam engine with two 16″ × 30″ cylinders, if the engine is making 140 rpm under a mean effective pressure of 50 lb.

17-3   Horsepower of Gas Engines

There are many formulas for computing the horsepower of a gas engine. The standard formula for the rated horsepower of a 4-cycle gas engine is that of the Society of Automotive Engineers, formerly the American Licensed Automobile Manufacturers. This formula, now known as the N.A.C.C. (National Automobile Chamber of Commerce) formula, is as follows:

$$\text{hp} = \frac{D^2 \times N}{2.5} = 0.4 \times D^2 \times N$$

where D = diameter of cylinders, in inches
$\quad\quad N$ = number of cylinders

This formula is based on a piston speed of 1,000 ft per minute and a mean effective pressure of 90 lb per square inch.

Example
Find the horsepower of a $4\frac{1}{2}'' \times 6''$, 4-cycle, 4-cylinder gas engine.

SOLUTION $\text{hp} = 0.4 \times D^2 \times N = 0.4 \times 4.5^2 \times 4 = 32.4$ hp

Problems 17-3

Use the N.A.C.C. formula to find the horsepower of the following gasoline engines.

1. Find the rated horsepower of a 6-cylinder automobile engine if the cylinders have a $3\frac{5}{16}''$ bore and a $3\frac{1}{2}''$ stroke.
2. A gas engine with $3\frac{1}{4}''$ bore and $4\frac{1}{2}''$ stroke has 8 cylinders. What is the rated or taxable horsepower?
3. Find the rated horsepower of a 12-cylinder engine with a bore of $3\frac{1}{8}''$.
4. A certain V-8 engine has cylinders $3\frac{1}{16}'' \times 3\frac{3}{4}''$. What is the rated horsepower?
5. If the bore of the engine in Problem 4 is increased by $\frac{1}{8}''$, what increase in horsepower will result?
6. Find the horsepower of a 16-cylinder $3'' \times 4''$ engine.
7. Find the horsepower of a 4-cylinder $3\frac{1}{8}'' \times 4\frac{3}{8}''$ engine.
8. How much additional horsepower will be developed by the engine in Problem 7 if the bore is increased by $\frac{3}{16}''$?

17-4   Brake Horsepower

The *brake horsepower* (bhp) is the power delivered by an engine at its flywheel, or the power actually available for use. The brake horsepower of an engine may be determined by the use of a device known as the *Prony brake* (Fig. 17-1). This consists of a steel or leather belt carrying a number of wooden blocks or shoes clamped around the pulley by means of the bolt B. The bolt enables us to vary the friction between the shoes and the rim of the pulley. The pull developed by this friction at a distance L from the shaft is measured by the platform scale on which the lever rests. Allowance must be

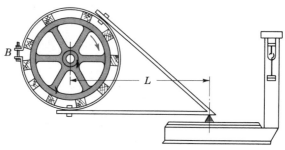

Fig. 17-1

made for the weight of the lever itself. The formula for determining brake horsepower is

$$\text{bhp} = \frac{2 \times \pi \times L \times N \times W}{33,000}$$

where L = length of lever arm in feet
      N = number of revolutions per minute
      W = net force = scale reading minus weight of lever

Example

Find the brake horsepower of an engine determined by a Prony brake where L is 4 ft, the engine is running at 400 rpm, and the net force on the scale is 30 lb.

SOLUTION $\text{bhp} = \dfrac{2 \times \pi \times L \times N \times W}{33,000}$

$$= \frac{2 \times 3.1416 \times 4 \times 400 \times 30}{33,000} = 9.14 \text{ hp}$$

Problems 17-4

Find the brake horsepower of the following engines.

1. Calculate the brake horsepower of a gas engine running at 450 rpm if the net scale reading is 25 lb and the lever arm is 4'6" long.
2. What is the brake horsepower of a steam engine making 120 rpm if the Prony brake lever is 5'0" long and the net force on the scale is 28 lb?
3. The brake horsepower of an engine making 60 rpm was determined by a Prony brake having a lever arm of 3'4". Compute the brake horsepower if the net force on the scale was 40 lb.
4. Compute the brake horsepower of a gas engine as determined by a Prony brake having a lever arm of 3'3", a net force on the scale of 20 lb, and a speed of 400 rpm.
5. An engine making 80 rpm is tested with a Prony brake having a lever arm of 3'6". What is the brake horsepower if the net force on the scale is 28 lb?

17-5   Electrical Power

The unit of electrical power is the *watt*. A *watt* is the power used when 1 volt causes 1 ampere of current to flow. The watts consumed in any circuit are found by multiplying the amperes (current) in the circuit by the volts (pressure) impressed on the circuit.

Kilowatt

Because the watt is too small a unit for convenience in computation, we use the *kilowatt* (kw), which is equal to 1,000 watts. In electrical power machines 1 horsepower is equal to 746 watts, or in round numbers, 750 watts, which is equal to $\frac{3}{4}$ of a kilowatt.

Electrical Horsepower

Electric tools such as hand electric drills and saws always have the voltage and amperage ratings listed. The mechanic is interested in the amount of power the tool can deliver. This is easily deduced using the formula that converts watts to horsepower.

The formula for the horsepower of an electric motor is

$$hp = \frac{amperes \times volts}{750}$$

Example 1

A generator delivers a current of 50 amperes at a pressure of 110 volts. What power in kilowatts does it supply?

SOLUTION $kw = \dfrac{amperes \times volts}{1000} = \dfrac{50 \times 110}{1000} = 5.5 \text{ kw}$

Example 2

Compute the horsepower input of a motor running on a 110-volt line and taking 20 amperes.

SOLUTION $hp = \dfrac{amperes \times volts}{750} = \dfrac{20 \times 110}{750} = 2.93 \text{ hp}$

Problems 17-5

Solve the following problems about electric devices.

1. An electric drill requires 110 volts and uses 2.3 amperes of current. What is its horsepower?
2. An electric saw requires 110 volts and uses 10 amperes of current. What is its horsepower?
3. An electric jigsaw requires 110 volts and uses 2.7 amperes of current. What is its horsepower?

4. An electric router requires 110 volts and uses 3.7 amperes of current. What is its horsepower?

5. If two electric drills have the same voltage requirements and one drill uses twice as much current as the other, how would the horsepower of the two compare?

6. A 100-watt bulb requires 110 volts. How many amperes of current are used by the bulb?

7. A motor taking 5 amperes is running on a 220-volt line. How much power does the motor receive?

8. Find the kilowatt capacity of a generator delivering 12 amperes at 220 volts.

9. How many watts are consumed by an electric iron taking 6.4 amperes on a 110-volt line?

10. What must be the horsepower of a motor for a trolley car if the car requires a current of 50 amperes and a voltage of 550 volts?

11. A generator delivers to a motor 200 amperes at 400 volts. What is the horsepower input of the motor?

12. Ten percent of the power delivered to a certain motor is lost by friction. What horsepower will this motor deliver if it receives 25 amperes at 110 volts?

17-6   Mechanical Efficiency of Machines

Because of friction losses, and for other reasons, no machine gives out all the power it receives. The power put into a machine is called the *input*. The power which the machine delivers is called the *output*. The *efficiency* of a machine is the ratio of the output to the input and is always expressed in percent.

$$\text{Efficiency} = \frac{\text{output}}{\text{input}}$$

Thus, if a motor receives 10 hp and gives out only 8 hp, its efficiency is $\frac{8}{10} = 0.80$ or 80%.

The efficiency of a steam engine is the ratio of brake horsepower to indicated horsepower. If the indicated horsepower is 7.5 and the brake horsepower is only 6, the efficiency is 6/7.5 = 0.80 or 80%.

Example 1

A motor receiving 2,500 watts has an efficiency of 90%. What horsepower will it deliver?

SOLUTION   Input = 2,500 watts = 2.5 kw
Output = 90% of input = 0.90 × 2.5 = 2.25 kw
hp = kw ÷ $\frac{3}{4}$ = 2.25 ÷ $\frac{3}{4}$ = 2.25 × $\frac{4}{3}$ = 3 hp

EXPLANATION   One kilowatt = 1,000 watts; therefore 2,500 watts = $\frac{2500}{1000}$ = 2.5 kw. Since the efficiency is 90%, the motor gives out 90% or 0.90 of what it gets. Therefore, the output = 0.90 × 2.5 or 2.25 kw. One horsepower = $\frac{3}{4}$ kw; therefore, to convert kilowatt to horsepower, we divide kw by $\frac{3}{4}$ or multiply by $\frac{4}{3}$.

Example 2
What current does a 5-hp 220-volt motor take if its efficiency is 80%?

SOLUTION $\text{Input} = \dfrac{\text{output}}{\text{efficiency}} = \dfrac{5}{0.80} = 6.25 \text{ hp}$

$\text{Watts} = \text{hp} \times 750 = 6.25 \times 750 = 4{,}687.50$

$\text{Current} = \dfrac{\text{watts}}{\text{volts}} = \dfrac{4{,}687.5}{220} = 21.31 \text{ amperes}$

EXPLANATION   If the motor is a 5-hp motor with an efficiency of 80%, it means that 5 hp is 80% of the input. Hence, to find the input, we divide the output by the efficiency, that is, 5/0.80 = 6.25 hp.

To convert the horsepower to watts, we multiply by 750. We find the current by dividing watts by volts.

Example 3
What is the efficiency of a generator that delivers 60 amperes at 110 volts if the input is 10 hp?

SOLUTION $\text{Efficiency} = \dfrac{\text{output}}{\text{input}}$

$\text{Output} = \dfrac{60 \times 110}{1{,}000} = 6.6 \text{ kw}$

$\text{Input} = 10 \times \tfrac{3}{4} = 7.5 \text{ kw}$

$\text{Efficiency} = \dfrac{6.6}{7.5} = 0.88 = 88\%$

EXPLANATION   Multiplying 60 amperes by 110 volts gives the output in watts. Dividing watts by 1,000 gives the output in kilowatts. To change the horsepower input to kilowatts, we multiply by $\frac{3}{4}$. We then divide the kilowatts output by the kilowatts input to get the efficiency.

NOTE. Output and input must be in the same units; that is, to find efficiency, we divide kilowatt output by kilowatt input, or horsepower output by horsepower input.

Supply the missing information about the following motors.

1. A motor having an efficiency of 75% receives 6 kw. What horsepower will it deliver?
2. A steam engine whose indicated horsepower is 8.5 hp has an efficiency of 70%. What will be the brake horsepower of the engine?
3. A 110-volt motor taking 20 amperes has an efficiency of 85%. What horsepower will it deliver?
4. An engine supplies 120 hp to a generator delivering 125 amperes at 550 volts. What is the efficiency of the generator?
5. Find the efficiency of a steam engine whose indicated horsepower is 6.2 hp and whose brake horsepower is 4.8 hp.
6. What is the efficiency of a 4.5-hp motor if it takes 32 amperes on a 110-volt line?
7. On what voltage must a 10-hp motor be run if it takes a current of 37.5 amperes and is 86% efficient?
8. What current does a 10-hp, 220-volt motor take if it is 88% efficient?
9. A 12-hp, 110-volt motor has an efficiency of 90%. What current does it require?
10. A generator having an efficiency of 80% receives 60 hp from the driving engine. What current will it deliver at 110 volts?
11. What is the efficiency of a 12-hp motor if it requires 90 amperes at 120 volts?

Solve the following problems involving work and power. Refer to the formulas in the text. Check your answers with the answers in the back of the book.

1. A machine does 64,000 ft-lb of work in four minutes. Compute its horsepower.
2. A roofer carries an 80-lb bundle of shingles 12 ft up a ladder. How many foot-pounds of work does he do?
3. A 20″ × 36″ double-acting steam engine is running at 100 rpm under a mean effective pressure of 40 lb. Find the horsepower.
4. Find the brake horsepower of an engine determined by a Prony brake where the lever length is 4 ft, the engine is running at 4,000 rpm, and the net force on the scale is 60 lb.
5. A generator delivers 10 amperes at 120 volts. What power in kilowatts does it supply?
6. A sander draws 3.4 amperes of current at 120 volts. What is its horsepower?
7. How many amperes of current will twelve 100-watt light bulbs draw on a 220-volt line?

8. A 10-hp motor delivers 8.7 hp. What is its efficiency?
9. A motor delivers $\frac{3}{4}$ horsepower and draws 5.9 amperes on a 110-volt line. What is its input horsepower?
10. A shaper requires 110 volts and uses 7.5 amperes. What is its horse-power?

Chapter Test

Solve the following problems involving work and power. Refer to the formulas in the text. Organize your work and label all parts of the problem.

1. A man moves one ton of coal 12 ft. How much work does he do in foot-pounds?
2. An engine can do 48,000 ft-lb of work in $2\frac{1}{2}$ minutes. Compute its horse-power.
3. What must be the horsepower of a machine to lift a 2-ton girder to a height of 30 ft in 2 minutes?
4. What is the indicated horsepower of a 16″ × 28″ double-acting steam engine making 100 rpm with a mean effective pressure of 54 lb?
5. Find the rated horsepower of a 6-cylinder automobile engine if the cylinders have $3\frac{1}{4}$″ bore and a $3\frac{1}{2}$″ stroke.
6. If the bore in Problem 5 were increased to $3\frac{5}{16}$″, how much additional horsepower would be developed?
7. Calculate the brake horsepower of a gasoline engine running at 1,000 rpm, if the net scale reading is 80 lb and the lever arm is 4 ft long.
8. An electric drill requires 110 volts and uses 2.7 amperes of current. What is its horsepower?
9. An electric saw requires 110 volts and is rated at one horsepower. How much current does it draw?
10. An electric jigsaw requires 110 volts and uses 1.8 amperes of current. What is its horsepower?
11. A 150-watt electric bulb requires 110 volts. How many amperes of current are used by the bulb?
12. A motor having an efficiency of 75% receives 5 kw. What horsepower will it deliver?
13. A 110-volt motor taking 5.5 amperes of current has an efficiency of 90%. What horsepower will it deliver?
14. A motor has an input of 5 hp and delivers 4.7 hp. What is its efficiency?
15. What is the efficiency of a 1-hp motor if it takes 9.0 amperes on a 110-volt line?

18

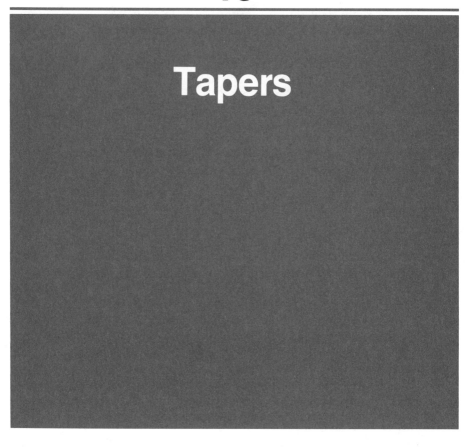

Tapers

18-1 Definitions

A piece of work that decreases gradually in diameter so that it assumes a conical shape is said to be *tapered*. Figure 18-1 shows a piece of tapered work.

When we speak of the **taper** on a piece of work, we mean the *difference between the diameters* at the ends of the piece. If, as in Fig. 18-1, the large diameter is $1\frac{1}{2}$ in. and the small diameter 1 in., the taper, or difference between the diameters, is $\frac{1}{2}$ in. Therefore, when we say that a piece of work has a taper of $\frac{1}{2}$ in., we mean that the diameter at the large end is $\frac{1}{2}$ in. greater than the diameter at the small end.

Taper is usually designated as a *fraction of an inch per foot of length;* the length is measured along the axis or center line of the piece of work. When we say, for example, that a piece of work has a taper of $\frac{5}{8}$ in. per foot, we mean that there is a difference of $\frac{5}{8}$ in. between the diameters 1 ft apart.

Taper may also be specified as a fraction of an inch per inch of length of the work. Thus we may say that a piece of work has a taper of $\frac{1}{16}$ in. per inch. This means that the diameters 1 in. apart differ by $\frac{1}{16}$ in.

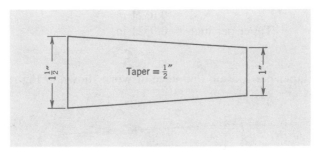

Fig. 18-1

18-2 Computing Taper and Diameter

In Fig. 18-2, the taper in 2 in. is twice as much as the taper in 1 in., the taper in 3 in. is three times as much as the taper in 1 in., and so on. Hence if we know the taper in a 1-in. length of a piece of work, we can find the taper in any other length by multiplying the taper per inch by the number of inches in the given length.

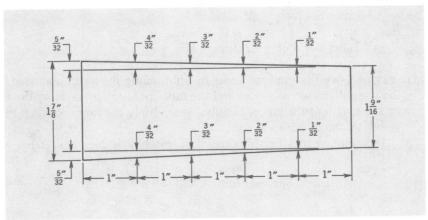

Fig. 18-2

Example 1
The taper per inch on a piece of work is 0.05 in.; what is the taper in 8 in.?

SOLUTION   Taper = 8 × 0.05 = 0.4 in.

If, on the other hand, we are told that the taper in a foot length of a piece of work is $\frac{1}{2}$ in., and we are required to find the taper in a 1-in. length, we divide by 12 because in 1 in. there is $\frac{1}{12}$ as much taper as there is in 12 in.

Example 2
Find the taper per inch if the taper per foot is $\frac{5}{8}$ in.

SOLUTION $\frac{5}{8} \div 12 = \frac{5}{8} \times \frac{1}{12} = \frac{5}{96} = 0.0521$ in.

   Taper per inch $= 0.0521$ in.

Example 3

Find the taper per foot on the piece of work shown in Fig. 18-3.

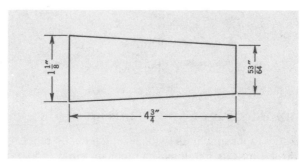

Fig. 18-3

SOLUTION   Taper in piece $= 1\frac{1}{8}$ in. $- \frac{53}{64}$ in. $= \frac{19}{64}$ in.

   Taper per inch $= \frac{19}{64} \div 4\frac{3}{4}$

   $= \frac{19}{64} \times \frac{4}{19} = \frac{1}{16}$ in.

   Taper per foot $= 12 \times \frac{1}{16} = \frac{12}{16} = \frac{3}{4}$ in.

EXPLANATION   We first find the taper by subtracting the small diameter from the large diameter. We then find the taper per inch by dividing the taper by the length of the piece. Finally, we multiply the taper per inch by 12, giving the taper per foot, $\frac{3}{4}$ in.

   The solution may be abbreviated by combining the steps, thus

$$\text{Taper} = 1\frac{1}{8} - \frac{53}{64} = \frac{19}{64} \text{ in.}$$
$$\text{Taper per foot} = \frac{19}{64} \div 4\frac{3}{4} \times 12$$
$$= \frac{19}{64} \times \frac{4}{19} \times 12 = \frac{3}{4} \text{ in.}$$

Example 4

The piece of work shown in Fig. 18-4 has a taper of 0.6 in. per foot. Find the small diameter.

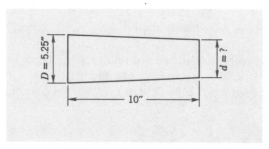

Fig. 18-4

SOLUTION   Taper per inch = $\frac{1}{12}$ × 0.6 = 0.05 in.
             Taper in 10 in. = 10 × 0.05 = 0.50 in.
             Small diameter = large diameter − taper
                           = 5.25 − 0.50 = 4.75 in.

EXPLANATION   To find the small diameter, we must subtract the taper in the piece of work from the large diameter. Dividing the taper per foot by 12 gives 0.05 in., the taper per inch. Multiplying this by 10 gives 0.50 in., the taper. Subtracting the taper from the large diameter gives the small diameter, 4.75 in.

Example 5

Find the large diameter of a piece of work $7\frac{3}{4}''$ long, having a small diameter of 0.658 in. and a taper of $\frac{1}{16}$ in. per inch (Fig. 18-5).

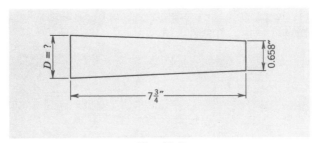

Fig. 18-5

SOLUTION   Taper = $7\frac{3}{4} \times \frac{1}{16} = \frac{31}{4} \times \frac{1}{16}$
                            = $\frac{31}{64}$ = 0.485 in.
             Large diameter = small diameter + taper
                           = 0.658 + 0.485 = 1.143 in.

EXPLANATION   The large diameter is equal to the small diameter plus the taper. The taper in the piece which is $7\frac{3}{4}$ in. long is $7\frac{3}{4}$ times the taper per inch, or $7\frac{3}{4} \times \frac{1}{16}$ = 0.485 in. The large diameter is therefore equal to 0.658 + 0.485 = 1.143 in.

18-3 American Standard Self-Holding (Slow) Taper Series

The taper series most commonly used are the following:

1. Brown and Sharpe
2. Morse
3. $\frac{3}{4}$ in. per ft series

The Brown and Sharpe taper is $\frac{1}{2}$ in. per ft.
The Morse taper is nominally $\frac{5}{8}$ in. per ft. The exact taper for each size is as follows:

No. 1: 0.600 in. per ft
No. 2: 0.600 in. per ft
No. 3: 0.602 in. per ft
No. 4: 0.623 in. per ft
No. 4½: 0.623 in. per ft
No. 5: 0.630 in. per ft

In the ¾ in. per ft series, the taper in all cases is ¾ in. per ft.

A less commonly used taper is the *steep taper*. The most common steep taper has a taper of 3½ in. per ft.

<hr/>

Problems 18-3

<hr/>

Find the following tapers.

1. What is the taper in the piece of work in Fig. 18-6? What is the taper per inch? What is the taper per foot?

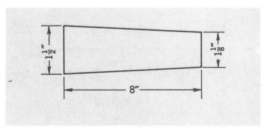

Fig. 18-6

2. What is the taper per inch in the Brown and Sharpe? Morse No. 1? ¾ in. per foot series?

3. Compute the small diameter of the lathe center in Fig. 18-7. The taper is Morse No. 2.

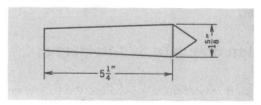

Fig. 18-7

4. What is the diameter 4 in. from the small end of the lathe center in Problem 3?

5. Compute the taper per foot on the taper socket in Fig. 18-8.

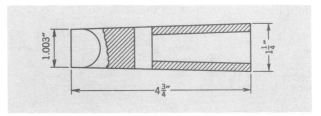

Fig. 18-8

6. Compute the taper per foot on the bushing in Fig. 18-9.

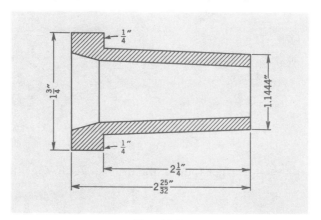

Fig. 18-9

7. Compute the taper per foot on the reamer in Fig. 18-10.

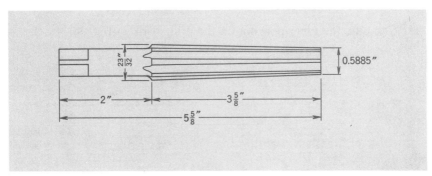

Fig. 18-10

8. Compute the taper per foot on the collet in Fig. 18-11.

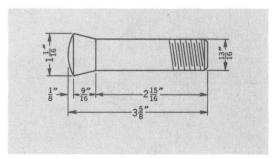

Fig. 18-11

9. Compute the taper per foot on the milling machine arbor in Fig. 18-12.

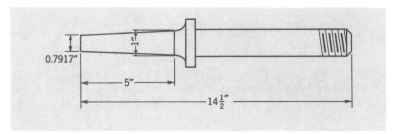

Fig. 18-12

10. Compute the taper per foot on the drill socket in Fig. 18-13.

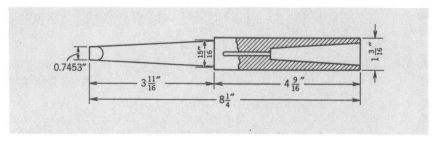

Fig. 18-13

11. Compute the taper per foot on the drill shank in Fig. 18-14.

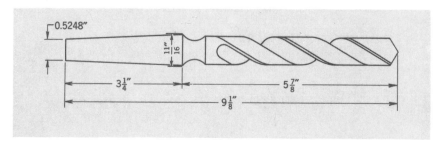

Fig. 18-14

12. Compute the taper per foot on the mandrel in Fig. 18-15.

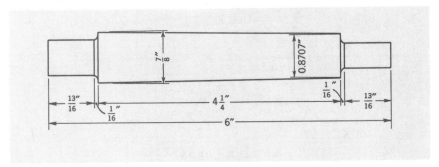

Fig. 18-15

13. Compute the taper per foot at both ends of the piston rod in Fig. 18-16.

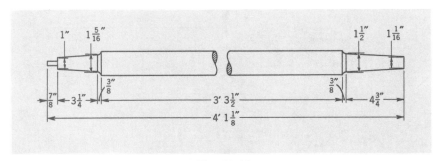

Fig. 18-16

14. The inside taper of the taper socket in Fig. 18-17 is Morse No. 5. Compute the small diameter.

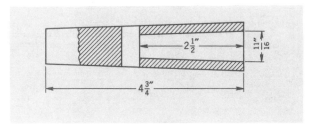

Fig. 18-17

15. Compute the large diameter of the inside taper (Brown and Sharpe) of the taper bushing in Fig. 18-18.

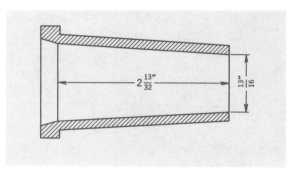

Fig. 18-18

16. Compute the small end diameter of the inside taper (Brown and Sharpe) of the drill socket in Fig. 18-19.

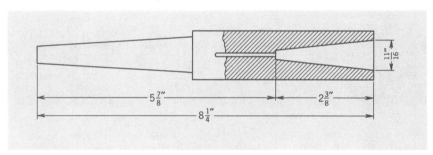

Fig. 18-19

17. Find the diameter at the large end of the taper pin reamer in Fig. 18-20. The taper is $\frac{3}{8}''$ per foot.

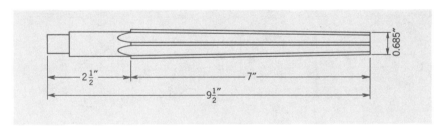

Fig. 18-20

18. The crank pin in Fig. 18-21 is to be cut with Morse No. 4 taper. Find the diameter at the small end of the tapered position.

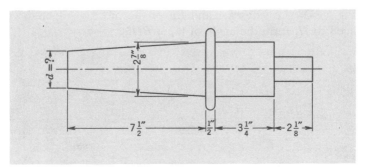

Fig. 18-21

19. A piece of work $6\frac{1}{2}''$ long has a taper of $\frac{3}{4}''$ per foot. Find the small diameter if the large diameter is $\frac{1}{2}''$.
20. Find the large diameter of a piece of tapered work $4\frac{1}{8}''$ long, having a taper of $\frac{5}{8}''$ per foot and a small diameter of 0.572″.
21. Find the small diameter of the shaft in Fig. 18-22. The taper is Brown and Sharpe.

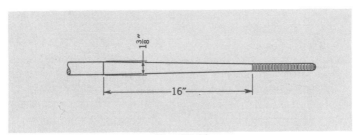

Fig. 18-22

18-4   Taper Angle

The taper on a piece of work is sometimes specified by giving *the angle included between the sides* or the prolongation of the sides of the piece of work. Thus in Fig. 18-23 the sides *AB* and *CB* meet at *B*, forming the angle *ABC*, which is the *included angle* or the *taper angle*.

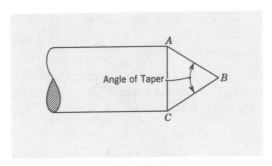

Fig. 18-23

In Fig. 18-24 the sides *BA* and *EF* of the piece of tapered work, when prolonged to *H*, form the angle of taper *BHE*.

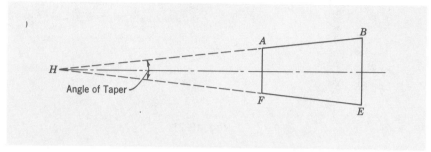

Fig. 18-24

Example 1

Compute the taper angle in a piece of work 10 in. long, whose small diameter is 2 in. and large diameter $3\frac{1}{2}$ in. (see Fig. 18-25).

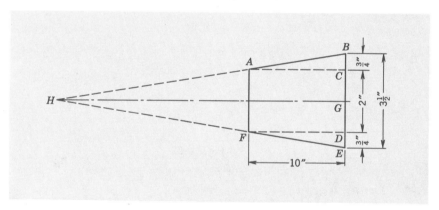

Fig. 18-25

SOLUTION   Draw *AC* parallel to the center line of the piece of work, forming the right triangle *ABC*. Angle *BAC* = *BHG* = half of the taper angle. From trigonometry we know that the tangent of an angle $= \dfrac{\text{side opposite}}{\text{side adjacent}}$.

Therefore tangent $BAC = \dfrac{BC}{AC} = \dfrac{\text{half of taper}}{\text{length}} = \dfrac{0.75}{10} = 0.0750$. By referring to a table of natural functions, we find that this tangent corresponds to an angle of 4°17′. Since the angle *DFE* = angle *BAC*, the included angle at *H*, which is twice *BAC*, equals 2 × 4°17′ = 8°34′.

Example 2

Find the angle of taper in a piece of work having a taper of $\frac{5}{8}$ in. per foot (see Fig. 18-26).

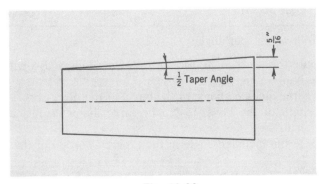

Fig. 18-26

SOLUTION   Tangent of $\frac{1}{2}$ taper angle $= \dfrac{\frac{1}{2} \text{ of } \frac{5}{8}}{12}$

$$= \tfrac{1}{2} \times \tfrac{5}{8} \times \tfrac{1}{12} = \tfrac{5}{192} = 0.02604$$

$$\tfrac{1}{2} \text{ taper angle } = 1°29\tfrac{1}{2}'$$

$$\text{Taper angle } = 2 \times 1°29\tfrac{1}{2}' = 2°59'$$

Example 3

Find the taper per foot if the included angle is 12°40′ (see Fig. 18-27).

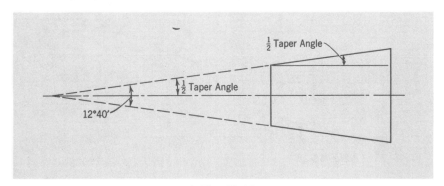

Fig. 18-27

SOLUTION $\frac{1}{2}$ of the taper angle $= \dfrac{12°40'}{2} = 6°20'$

$$\text{Tangent } 6°20' = 0.1110$$

$$\text{Taper per inch } = 2 \times 0.1110 = 0.2220$$

$$\text{Taper per foot } = 12 \times 0.2220 = 2.6640 \text{ in.}$$

EXPLANATION   From a table of natural functions obtain the tangent of half the taper angle. This gives half the taper in one inch of length. To get the taper per inch, multiply this result by 2. The taper per foot is 12 times the taper per inch.

Problems 18-4

Find the following taper angles.

1. Compute the taper angle for the following tapers: Brown and Sharpe, Morse No. 5, $\frac{3}{4}$ in. per foot series.
2. What is the included angle of the bevel of the piece shown in Fig. 18-9? The length of the bevel is $\frac{3}{8}''$, the large diameter $1\frac{1}{8}''$, and the small diameter 0.9253".
3. Compute the included angle in the tapered portion of the collet in Fig. 18-11.
4. What is the angle of taper of the left-hand end of the piston rod in Fig. 18-16?
5. Find the taper per foot if the taper angle is 5°.
6. Find the taper per inch if the taper angle is $3\frac{1}{2}°$.
7. The included angle on a piece of tapered work is 7°. What is the taper per foot?
8. What standard taper has a taper angle of 2°59'?
9. The angle with the center line in a piece of tapered work is 2°23'. What is the taper per foot?
10. Compute the included angle in the bevel gear blank shown in Fig. 18-28.

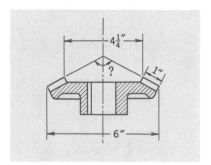

Fig. 18-28

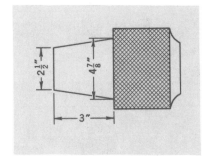

Fig. 18-29

11. Find the included or taper angle in the drill chuck in Fig. 18-29.
12. Compute the included angle in the point of the countersink in Fig. 18-30.

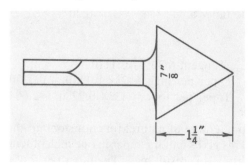

Fig. 18-30

18-5   Taper Turning by Offsetting the Tail Stock

There are three methods of turning a taper in a lathe. The first method, by offsetting the tail stock, can be used for outside tapers only and when the taper to be turned is not too great. The effect of offsetting the tail stock will be seen from a study of Fig. 18-31.

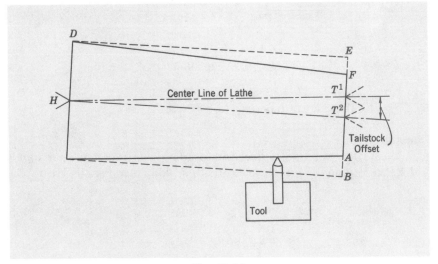

Fig. 18-31

For cylindrical turning, the tail stock center T_1 is on the center line of the lathe, that is, directly in line with the head stock center H. When the tail stock is offset for cutting a taper, the center assumes the position T_2 and the piece of work assumes the position $DEBC$. The tool, however, travels in a straight line parallel to the center line of the machine from C to A. The radius of the work is therefore reduced at the tail end by the amount AB, which is equal to the tail stock offset from T_1 to T_2. But, since the work revolves, an equal amount is taken off all around. In other words the diameter at the tail end is less than the diameter at the head end by an amount equal to twice the offset of the tail stock. The tail stock offset is therefore equal to half the total taper.

RULE

The amount of set-over of the tail stock is equal to one-half the total length of the work in inches multiplied by the taper per inch.

Example 1

Calculate the tail stock offset for turning the taper on the piece in Fig. 18-32.

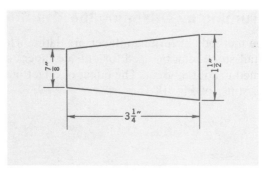

Fig. 18-32

SOLUTION   Taper = $1\frac{1}{2} - \frac{7}{8} = \frac{5}{8}$ in.
Offset = $\frac{1}{2}$ of taper = $\frac{1}{2} \times \frac{5}{8} = \frac{5}{16}$ in.

Example 2
Calculate the tail stock offset for turning a taper of $\frac{3}{4}$ in. per foot for a distance of 4 in. on a piece of work 10 in. long (see Fig. 18-33).

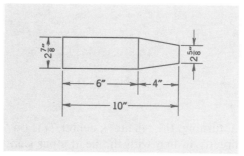

Fig. 18-33

SOLUTION
$$\text{Taper per inch} = \frac{1}{12} \times \frac{3}{4} = \frac{1}{16} \text{ in.}$$
$$\text{Taper in whole length of piece of work} = 10 \times \frac{1}{16} = \frac{5}{8} \text{ in.}$$
$$\text{Tail stock offset} = \frac{1}{2} \text{ of } \frac{5}{8} = \frac{5}{16} \text{ in.}$$

Notice that even though the tapered part is only 4 in. long, the taper for the whole length of the piece is found and the offset is computed as one-half of this overall taper.

Example 3
Compute the tail stock offset for turning tapers at both ends of the piston rod in Fig. 18-34.

SOLUTION   To turn the left end taper at A, we must compute the taper for the whole length of the rod as if it were tapered the same amount per inch as the part at A. The tail stock will then be offset half of the total taper so computed.

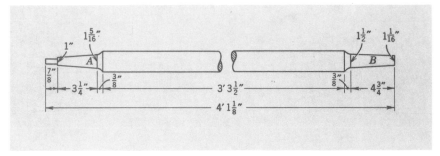

Fig. 18-34

Taper in part $A = 1\frac{5}{16} - 1 = \frac{5}{16}$ in.
Taper per inch $= \frac{5}{16} \div 3\frac{1}{4} = \frac{5}{16} \times \frac{4}{13} = \frac{5}{52}$ in.
Total length $= 4$ ft $1\frac{1}{8}$ in. $= 49\frac{1}{8}$ in.
Taper in whole length of rod $= 49\frac{1}{8} \times \frac{5}{52} = 4.724$ in.
Tail stock offset $= \frac{1}{2}$ of $4.724 = 2.362$ in. or $2\frac{23}{64}$ in.

To turn the taper at the right end, we again find the taper in the entire length of the rod as if it were tapered the same amount per inch as the right end B. The tail stock offset will be one-half of the total taper.

Taper in part $B = 1\frac{1}{2} - 1\frac{1}{16} = \frac{7}{16}$ in.
Taper per inch $= \frac{7}{16} \div 4\frac{3}{4} = \frac{7}{16} \times \frac{4}{19} = \frac{7}{76}$ in.
Overall taper $= 49\frac{1}{8} \times \frac{7}{76} = 4.525$ in.
Tail stock offset $= \frac{1}{2}$ of $4.525 = 2.263$ in. or $2\frac{17}{64}$ in.

Problems 18-5

Compute the following tail stock offsets.

1. Compute the tail stock offset for turning the taper on the piece of work in Fig. 18-35.

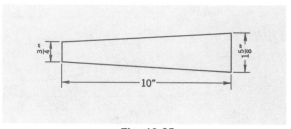

Fig. 18-35

2–9. Compute the tail stock offsets for cutting the tapers in Problems 5 through 12 from Problem Set 18-3.
10. Compute the tail stock offsets for turning the tapers at the ends of the piston rod in Fig. 18-36.

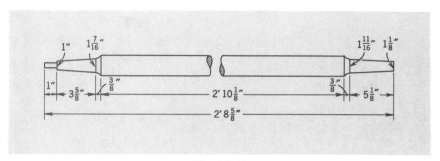

Fig. 18-36

18-6   Taper Turning by Using the Compound Rest

This method is used where short tapers or tapers having a considerable angle are to be turned. To set the compound rest for cutting a taper we must know the included angle, that is, the taper angle (see Fig. 18-37).

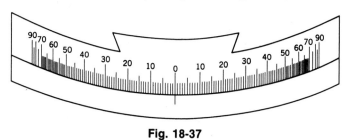

Fig. 18-37

Example 1

Compute the setting of the compound rest for cutting the taper on the piece in Fig. 18-38.

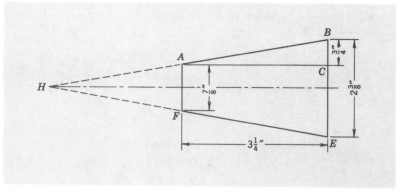

Fig. 18-38

SOLUTION

$$\text{Tangent } BAC = \text{tangent of } \tfrac{1}{2} \text{ included angle} = \frac{BC}{AC} = \frac{\tfrac{3}{4}}{3\tfrac{1}{4}}$$

$$= \tfrac{3}{4} \times \tfrac{4}{13} = \tfrac{3}{13} = 0.23077$$
Angle $BAC = \tfrac{1}{2}$ taper angle $= 13°0'$
Included angle $= 2 \times 13°0' = 26°0'$

To turn this taper, the compound rest is not set to read the included angle, 26°, but to *90° minus half the included angle;* 90° − $\tfrac{1}{2}$ of 26° = 90° − 13° − 77°0'. See Fig. 18-39.

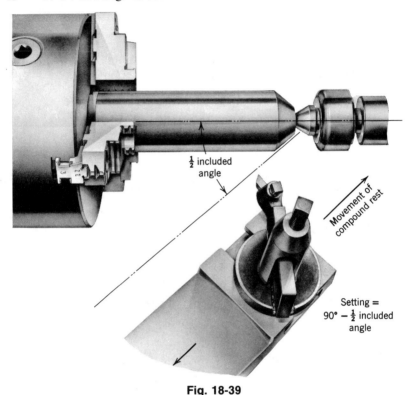

$\tfrac{1}{2}$ included
angle

Movement of
compound rest

Setting =
90° − $\tfrac{1}{2}$ included
angle

Fig. 18-39

The *complement* of an angle is equal to 90° minus the given angle. For example, the complement of 30° = 90° − 30° = 60°; the complement of 6°50' is 90° − 6°50' = 83°10'.

We can therefore state the rule for setting the compound rest as follows:

RULE _____

Set the compound rest to the complement of one-half the included angle.

Example 2

To what angle must the compound rest be set to turn the lathe center, which has an included angle of 60°? See Fig. 18-40.

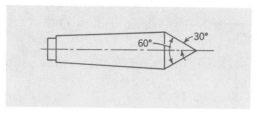

Fig. 18-40

SOLUTION   Included angle = 60°
                        $\frac{1}{2}$ included angle = $\frac{1}{2}$ of 60° = 30°
                Setting of compound rest = 90° − 30° = 60°

Problems 18-6

Answer the following questions involving compound rest settings.

1. To what angle must the compound rest be set to cut the tapers and bevels shown in Fig. 18-41?

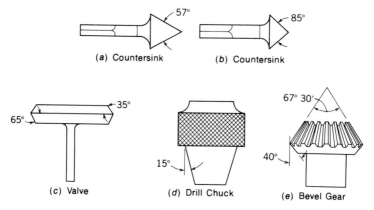

Fig. 18-41

2. Compute the angle for cutting the following tapers with the compound rest: Brown and Sharpe, Morse No. 5, $\frac{3}{4}$ in. per foot series.

3. Figure the compound rest setting for cutting the inside bevel on the taper bushing in Fig. 18-42.

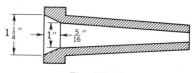

Fig. 18-42

4. Figure the compound rest setting for cutting the taper on the collet in Fig. 18-43.

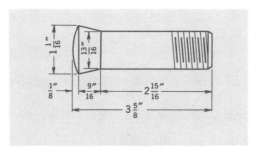

Fig. 18-43

5. Compute the compound rest setting for cutting the tapers at both ends of the piston rod in Fig. 18-44.

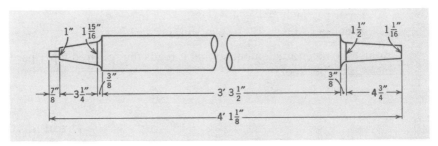

Fig. 18-44

6. How would you set the compound rest for cutting the following included angles: $5°$; $3\frac{1}{2}°$; $7°30'$; $8°20'$; $16°50'$?

18-7 Taper Turning by Using the Taper Attachment

To turn a taper by means of the taper attachment, we compute the taper per foot, disregarding the length of the tapered portion and the overall length of the work. In one type of taper attachment, the swivel bar that controls the taper is graduated at one end in inches per foot of taper and at the other end in degrees. The attachment can be set for any taper up to 3 in. per foot, which corresponds to a taper angle of $14°15'$ (see Fig. 18-45). In another type of attachment, the graduations at one end give the taper in eighths of an inch per foot, and the other end gives it in tenths of an inch per foot.

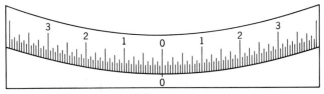

Fig. 18-45

Example

Compute the setting of the taper attachment for turning the taper shown in Fig. 18-46.

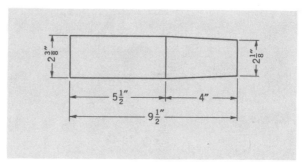

Fig. 18-46

SOLUTION   Taper in 4 in. $= 2\frac{3}{8} - 2\frac{1}{8} = \frac{1}{4}$ in.

Taper per inch $= \frac{1}{4} \div 4 = \frac{1}{4} \times \frac{1}{4} = \frac{1}{16}$ in.

Taper per foot $= 12 \times \frac{1}{16} = \frac{3}{4}$ in. or $\frac{6}{8}$ in.

Adjust the taper attachment so that it reads the required taper per foot, namely, $\frac{6}{8}$ in. per foot.

Problems 18-7

Find the setting of the taper attachment for cutting the following tapers.

1. How would you set the taper attachment for turning a piece of work having a Brown and Sharpe taper? Morse No. 1? $\frac{3}{4}$ in. per foot?
2. How should the taper attachment be set to turn tapers having the following included angles: 4°30′; 3°00′; 5°30′; 7°00′; 9°40′?
3. Compute the setting of the taper attachment for turning the taper on the taper socket in Fig. 18-8.
4. Compute the setting of the taper attachment for turning the outside taper on the bushing in Fig. 18-9.
5. Compute the setting of the taper attachment for turning the taper on the reamer in Fig. 18-10.
6. Compute the setting of the taper attachment for turning the taper on the collet in Fig. 18-11.
7. Compute the setting of the taper attachment for turning the taper on the milling machine arbor in Fig. 18-12.
8. Compute the setting of the taper attachment for turning the outside taper on the drill socket in Fig. 18-13.
9. Compute the setting of the taper attachment for turning the taper on the drill shank in Fig. 18-14.
10. Compute the setting of the taper attachment for turning the taper on the mandrel in Fig. 18-15.

11. Compute the setting of the taper attachment for cutting the tapers at both ends of the piston rod shown in Fig. 18-16.

Self-Test

Solve the following problems involving tapers. Check your answers with the answers in the back of the book.

1. The taper per inch on a piece of work is 0.02 in. What is its taper in 12 in.?
2. The taper per foot on a piece of work is 0.6 in. What is its taper in 8 in.?
3. What is the taper per inch if the taper per foot is $\frac{3}{4}$ in.?
4. What is the taper per foot if the taper per inch is $\frac{1}{32}$ in.?
5. Find the taper on a piece of work 9 in. long if the diameter of the large end is 2 in. and the diameter of the small end is $1\frac{3}{4}$ in.
6. What is the taper per inch of the piece of work in Problem 5?
7. What is the taper per foot of the piece of work in Problem 5?
8. A tapered piece of work is 14 in. long. What is its taper per inch if the diameter of the large end is 3.562 in. and the diameter of the small end is 3.182 in.?
9. What is the taper angle on a piece of work having a taper of 0.75 in. per foot?
10. What is the taper angle on a piece of work having a taper of 0.0156 in. per inch if the piece is 2 ft long?

Chapter Test

Solve the following problems involving tapers.

1. The taper per inch on a piece of work is 0.06"; what is the taper in 10"?
2. What is the taper per inch if the taper per foot is $\frac{7}{8}''$?
3. Find the taper per foot on a piece of work if the diameter of the large end is $1\frac{1}{4}''$, the diameter of the small end is $1\frac{7}{64}''$, and the piece is $4\frac{1}{2}''$ long.
4. A piece of work has a taper of 0.55" per foot. What is its diameter 6" away from the large end that has a diameter of 2.37"?
5. Find the large diameter of a piece of work $8\frac{3}{4}''$ long, having a small diameter of 0.625 and a taper of $\frac{1}{16}''$ per inch.
6. The Brown and Sharpe taper is $\frac{1}{2}''$ per foot. What is the taper per inch?
7. Compute the taper angle on a piece of work 20" long whose small diameter is 2" and large diameter is $3\frac{1}{4}''$.
8. Find the taper angle in a piece of work having a taper of $\frac{1}{2}''$ per foot.
9. The compound rest is set to the _____ of one-half the included angle.
10. The amount of set-over of the tail stock is equal to _____ (the total length of the work in inches multiplied by the taper per inch).

19

Speed Ratios of Pulleys and Gears

19-1 Gear Trains

In trains of gears, the gears that transmit power are called ***driving gears,*** whereas the others are ***followers*** or ***driven gears.***

Two meshing gears, such as A and B (Fig. 19-1), with the same number of teeth, will revolve at the same rate of speed. Assume A to be the driving gear and B the driven gear. If A revolves *clockwise,* that is, in the direction of the hands of a clock, as shown by the arrow, it will cause B to revolve in the opposite direction, or counterclockwise.

Let each gear have 20 teeth. With A revolves, every tooth on it passes the point C and engages a tooth on B. When A has made one complete turn, that is, when its 20 teeth have passed the point C, they will have engaged 20 teeth on gear B, thus causing B also to make one complete turn.

If, however, A has 20 teeth and B 40 teeth (Fig. 19-2), in one complete turn of A its 20 teeth will pass C and will engage 20 teeth on gear B. But since gear B has 40 teeth, gear A will have to make two complete revolutions in order to cause B to make one complete turn. In this instance the gears are said to

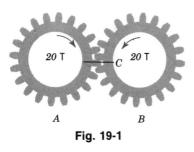

Fig. 19-1

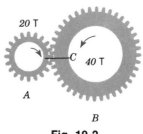

Fig. 19-2

have a speed ratio of 1 to 2; that is, one revolution of *B* corresponds to two revolutions of *A*. If *A* had 20 teeth and *B* 60 teeth, *A* would make three revolutions while *B* made one. The gears would then have a speed ratio of 1 to 3. In each case the *smaller* gear is the *faster* and the *larger* gear the *slower*.

The relation existing between the speeds of gears and their numbers of teeth is called an **inverse ratio**; that is, the speeds of gears vary not in the same order as the number of their teeth but in the inverse order. In other words, instead of the speed of a gear increasing with an increase in its number of teeth, it diminishes.

The rule governing the speeds of gears can be expressed as follows:

$$\frac{\text{Teeth on driven}}{\text{Teeth on driver}} = \frac{\text{Revolutions of driver}}{\text{Revolutions of driven}}$$

RULE _____

To find the speed of either gear, multiply the size and speed of the other gear and divide by the size of the gear whose speed is sought.

Example 1

Two meshing gears have 40 and 60 teeth, respectively. If the small gear makes 120 rpm, how many revolutions per minute will the larger gear make?

SOLUTION   rpm of large gear $= \dfrac{\text{size} \times \text{rpm of small gear}}{\text{size of large gear}}$

$$= \frac{40 \times 120}{60} = 80 \text{ rpm}$$

Example 2

In Fig. 19-3, *A* has 50 teeth and *B* 80 teeth. If *B* makes 100 rpm, how many revolutions per minute will *A* make?

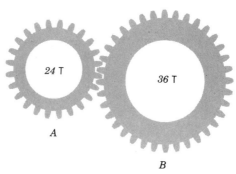

Fig. 19-3

SOLUTION   rpm of $A = \dfrac{\text{teeth} \times \text{rpm of } B}{\text{teeth of } A} = \dfrac{80 \times 100}{50} = 160$ rpm

RULE _____

To find the number of teeth on either gear, multiply the size and speed of the other gear and divide by the speed of the one whose number of teeth is sought.

Example 3

The driving gear has 24 teeth and makes 120 rpm. How many teeth must the driven gear have to make 80 rpm?

SOLUTION   teeth of driven $= \dfrac{\text{teeth} \times \text{rpm of driver}}{\text{rpm of driven}}$

$$= \dfrac{24 \times 120}{80} = 36 \text{ teeth}$$

Problems 19-1

Solve the following problems involving driving and driven gears. Make a sketch and label the known parts.

1. A gear having 120 teeth meshes with one having 24 teeth. If the large gear makes one revolution, how many revolutions will the small gear make?
2. In Fig. 19-4, gear A has 36 teeth and B has 96 teeth. If gear A makes 75 rpm, how many revolutions per minute will gear B make?

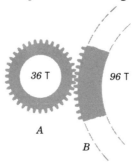

Fig. 19-4

3. In Fig. 19-5, gear A has 56 teeth and gear B has 80 teeth. If gear A makes 210 rpm, how many revolutions per minute will gear B make?

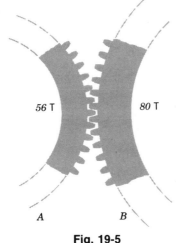

Fig. 19-5

4. In Fig. 19-6, gear A has 40 teeth and makes 180 rpm. If gear B is to make 120 rpm, how many teeth must it have?

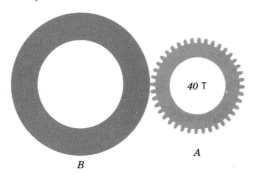

Fig. 19-6

5. A 48-tooth gear running 320 rpm is required to drive another gear at 240 rpm. How many teeth must the driven gear have?

19-2 Idlers

If we want to transmit power from shaft A to shaft B (Fig. 19-7), and the distance between the shafts is so great that it would require unusually large gears to bridge the gap, we must resort to idler gears. Gear C in Fig. 19-7 is such an idler.

Assume A has 20 teeth, C 40 teeth, and B 80 teeth. When A revolves once, C makes $\frac{1}{2}$ a revolution; and, while C revolves once, B makes $\frac{1}{2}$ a revolution; so that for one revolution of A the gear B makes $\frac{1}{4}$ of a revolution. That is, the speed of B is the same as if it were in direct mesh with gear A. Hence the idler has no effect on the relative speeds of the gears between which it is placed and it is omitted from all computations.

The only effect of the idler is to reverse the direction of the motion of B. Without the idler, it was seen in Figs. 19-3 and 19-4 that *B* turned counter-clockwise while *A* turned clockwise. With the idler interposed between *A* and *B*, when *A* turns clockwise *B* also turns clockwise, as can readily be seen by a study of Fig. 19-7.

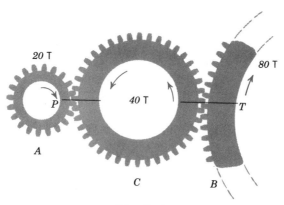

Fig. 19-7

In the case of two idlers such as *C* and *D* in Fig. 19-8, a study of the sketch will show that the direction of rotation of *B* is the same as when there are no idlers. That is, idler *D* neutralizes the effect of idler *C*. In general, the effect of idlers on the direction of motion of a follower or driven gear may be stated thus: *An odd number of idlers between two gears will cause the driven gear to rotate in the same direction as the driver; an even number of idlers will cause the driven gear to rotate in the direction opposite to that of the driver.*

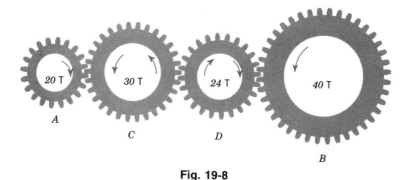

Fig. 19-8

19-3   Finding the Number of Teeth for a Given Speed Ratio

The method of computing the number of teeth in gears that will give a desired speed ratio is illustrated by the following example.

Example

Find two suitable gears that will give a speed ratio between driver and driven of 2 to 3.

SOLUTION $\dfrac{2}{3} = \dfrac{2 \times 12}{3 \times 12} = \dfrac{24}{36} \begin{array}{l} \text{teeth on follower} \\ \text{teeth on driver} \end{array}$

EXPLANATION Express the desired ratio as a fraction and multiply both terms of the fraction by any convenient multiplier that will give an equivalent fraction whose numerator and denominator will represent available gears. In this instance 12 was chosen as a multiplier giving the equivalent fraction $\frac{24}{36}$. Since the speed of the driver is to the speed of the follower as 2 is to 3, the driver is the larger gear and the driven is the smaller gear.

Problems 19-3

Solve the following problems involving gear trains. Make a sketch of the train and label all the known parts.

1. The speeds of two gears are in the ratio of 1 to 3. If the faster one makes 180 rpm, find the speed of the slower one.
2. The speed ratio of two gears is 1 to 4. The slower one makes 45 rpm. How many revolutions per minute does the faster one make?
3. Two gears are to have a speed ratio of 2.5 to 3. If the larger gear has 72 teeth, how many teeth must the smaller one have?
4. Find two suitable gears with a speed ratio of 3 to 4.
5. Find two suitable gears with a speed ratio of 3 to $5\frac{1}{2}$.
6. In Fig. 19-9, *A* has 24 teeth, *B* has 36 teeth, and *C* has 40 teeth. If gear *A* makes 200 rpm, how many revolutions per minute will gear *C* make?

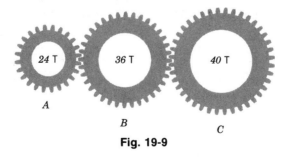

A

B

C

Fig. 19-9

7. In Fig. 19-10, *A* has 36 teeth, *B* has 60 teeth, *C* has 24 teeth, and *D* has 72 teeth. How many revolutions per minute will gear *D* make if gear *A* makes 175 rpm?

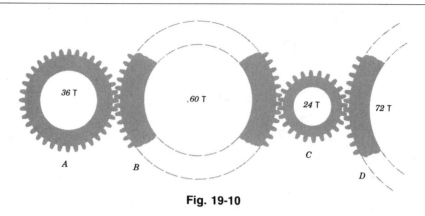

Fig. 19-10

19-4   Compound Gearing

Figure 19-11 shows a train of compound gears. The driver *A* is in mesh with *B*, making *B* a driven gear. Gear *C*, keyed to the same shaft as *B*, is also a driver and meshes with *D*, so that *D* is a driven gear. Gears *B* and *C*, being keyed to the same shaft, revolve at the same rate of speed.

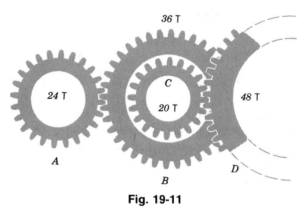

Fig. 19-11

The following example illustrates the method of finding the speed of a driven gear of a compound train.

Example 1
In Fig. 19-11, if *A* makes 72 rpm, how many revolutions per minute will gear *D* make?

SOLUTION $72 \times \frac{24}{36} \times \frac{20}{48} = 20$ rpm

EXPLANATION   Since *B* has 36 teeth and *A* has 24 teeth, *B* will make $\frac{24}{36}$ as many revolutions as *A*. That is, $72 \times \frac{24}{36}$ represents the speed of *B*. But *C*, being keyed to the same shaft as *B*, revolves at the same rate of speed; hence when *A* makes 72 revolutions, *C* will make $72 \times \frac{24}{36}$ revolutions. Since *D* has 48 teeth, whereas C has 20 teeth, D will make $\frac{20}{48}$ as many revolutions as C, or $72 \times \frac{24}{36} \times \frac{20}{48} = 20$ rpm.

A similar analysis is applicable where the number of teeth of a driven gear is wanted, or the speed or the number of teeth of a driver is desired.

In the solution of the preceding example it will be noticed that all the numbers in the numerator applied to the drivers, whereas those in the denominator applied to the driven gears. This enables us to state the following rule.

RULE

To find the revolutions per minute or number of teeth of any driven gear in a train, divide the continued product of the given factors of the driving gears by the continued product of the given factors of the driven gears.

Example 2

In Fig. 19-11, if A makes 72 rpm, how many teeth must D have if it is to make 20 rpm?

SOLUTION $\quad \dfrac{\text{Product of driver data}}{\text{Product of driven data}} = \dfrac{72 \times 24 \times 20}{36 \times 20} = 48$ teeth

To find the number of teeth or revolutions per minute of a driving gear, divide the continued product of all that is known about the driven gears by the continued product of all that is known about the driving gears.

Example 3

In Fig. 19-11, how many revolutions per minute must gear A make if gear D is required to make 20 rpm?

SOLUTION $\quad \dfrac{\text{Product of driven data}}{\text{Product of driver data}} = \dfrac{36 \times 48 \times 20}{24 \times 20} = 72$ rpm

If a certain speed ratio is required between the first driver and the last driven in a compound gear train, we proceed as in the following example.

Example 4

Find four suitable gears that will transmit motion from A to D at the ratio of 3 to 5 (see Fig. 19-12).

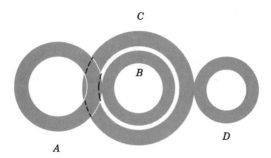

Fig. 19-12

SOLUTION

$$\frac{3}{5} = \frac{2}{2\frac{1}{2}} \times \frac{1\frac{1}{2}}{2} = \frac{2 \times 16}{2\frac{1}{2} \times 16} \times \frac{1\frac{1}{2} \times 24}{2 \times 24} = \frac{\overset{B}{32} \times \overset{D}{36}}{\underset{A}{40} \times \underset{C}{48}} = \begin{matrix}\text{followers} \\ \text{drivers}\end{matrix}$$

EXPLANATION   Write the desired speed ratio as a fraction $\frac{3}{5}$. Split each term into two factors, namely, $3 = 2 \times 1\frac{1}{2}$ and $5 = 2\frac{1}{2} \times 2$, giving the two fractions $\frac{2}{2\frac{1}{2}}$ and $\frac{1\frac{1}{2}}{2}$. Treat each of these fractions separately as in the case of a simple gear train; that is, multiply each term by some convenient multiplier which will give an equivalent fraction whose terms will express available gears. The terms of the fraction $\frac{2}{2\frac{1}{2}}$ are multiplied by 16, giving the gears 32 and 40; and the terms of the fraction $\frac{1\frac{1}{2}}{2}$ are multiplied by 24, giving the gears 36 and 48. Since speeds of gears are inversely proportional to the number of their teeth, we have the proportion:

$$\frac{\text{Speed of driver}}{\text{Speed of driven}} = \frac{\text{teeth of driven}}{\text{teeth of driver}}$$

Hence the two numbers in the numerator of the result represent driven gears in the train and the two numbers in the denominator represent drivers. In mesh they would appear as in Fig. 19-12, where A drives B and C drives D.

19-5   Worm and Gear

To find the revolutions per minute of a worm wheel when we know the revolutions per minute of the worm and also whether it is single or double threaded, we proceed as in the following example.

Example

If the worm A is triple-threaded and makes 100 rpm, how many revolutions per minute will the worm gear B make if it has 50 teeth? (See Fig. 19-13.)

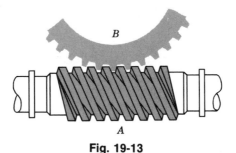

Fig. 19-13

SOLUTION $\dfrac{100 \times 3}{50} = 6$ rpm

EXPLANATION   The worm thread is regarded as a gear with one tooth if single-threaded, two teeth if double-threaded, and so on. Hence to find the revolutions per minute of the worm gear, multiply the revolutions per minute and the number of threads of the worm and divide by the number of teeth on the wheel.

19-6   Trains of Spur, Bevel, and Worm Gearing

Problems involving speed ratios of bevel gears are solved exactly like problems with spur gears.

Example

If gear F in Fig. 19-14 makes 240 rpm and has 48 teeth, E has 32 teeth, D has 20 teeth, C has 24 teeth, worm A is double-threaded, and worm gear B has 40 teeth, how many revolutions per minute will gear B make?

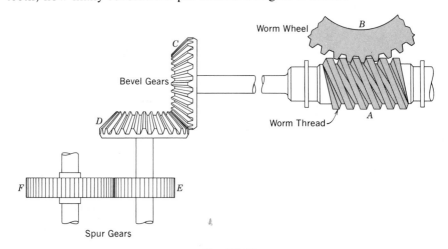

Fig. 19-14

SOLUTION   rpm of gear $B = \dfrac{\text{product of driver data}}{\text{product of driven data}}$

$$= \dfrac{240 \times 48 \times 20 \times 2}{32 \times 24 \times 40}$$

$$= 15 \text{ rpm}$$

19-7   Pulley Trains

The formulas used for gear trains are applicable also to pulley trains, except that for pulleys the diameter is used instead of the number of teeth.

Example 1

Pulley A in Fig. 19-15 is 8 in. in diameter and makes 120 rpm. How many revolutions per minute does pulley B make if its diameter is 12 in.?

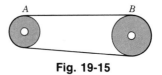

Fig. 19-15

SOLUTION   rpm of $B = \dfrac{8 \times 120}{12} = 80$

Example 2

In Fig. 19-16, pulley A is 10 cm in diameter and makes 140 rpm. What must be the diameter of B if it is required to make 100 rpm?

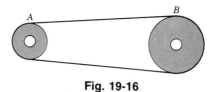

Fig. 19-16

SOLUTION   Diameter of $B = \dfrac{10 \times 140}{100} = 14 \text{ cm}$

Example 3

In Fig. 19-17, A is 8 in. in diameter, B is 20 in., C is 10 in., and D is 16 in. in diameter. If A makes 200 rpm, how many revolutions per minute will D make?

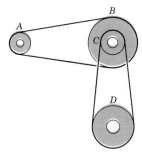

Fig. 19-17

SOLUTION   Speed of $D = \dfrac{\text{product of all driver data}}{\text{product of all driven data}}$

$$= \frac{200 \times 8 \times 10}{20 \times 16} = 50 \text{ rpm}$$

Problems 19-7

Solve the following problems involving gear trains and pulley trains. Sketch the train described and label the parts with the given information.

1. In Fig. 19-18, gear A has 28 teeth, B has 48 teeth, C has 24 teeth, and D has 32 teeth. If gear A makes 180 rpm, how many revolutions per minute will gear D make?

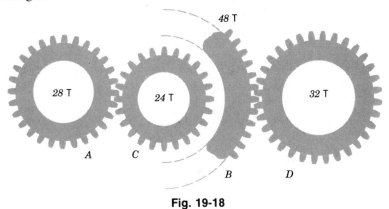

Fig. 19-18

2. Figure 19-19 shows a compound gear train. Find the speed ratio between shaft A and shaft D.

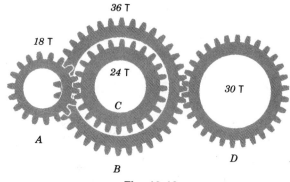

Fig. 19-19

3. If A in Fig. 19-20 makes 42 rpm, how many teeth must D have in order to make 105 rpm?

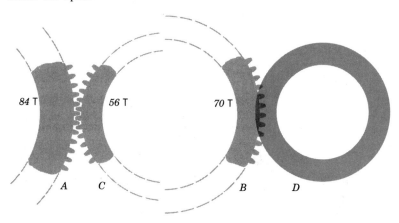

Fig. 19-20

4. In Fig. 19-21, if A makes one revolution, how many revolutions will D make?

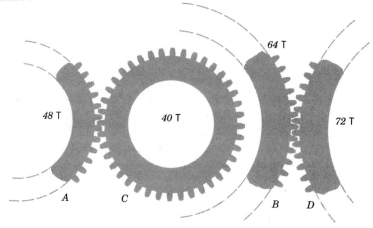

Fig. 19-21

5. In Problem 4, what must be the speed of *A* in order to cause *D* to make 50 rpm?

6. In Fig. 19-22, how many teeth must gear *C* have if gear *A* makes 100 rpm and gear *D* makes 180 rpm?

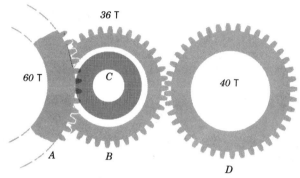

Fig. 19-22

7. In Fig. 19-23, gear *A* has 40 teeth; *B*, 36 teeth; *C*, 60 teeth; *D*, 48 teeth; and *E*, 72 teeth. How many revolutions per minute will *E* make if *A* makes 240 rpm?

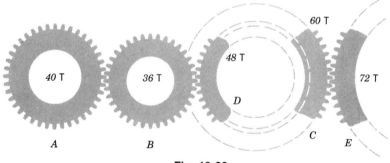

Fig. 19-23

8. Gear *E* in Fig. 19-24 is required to make 100 rpm while *A* makes 75 rpm. How many teeth must *E* have if *A* has 24 teeth, *B*, 64; *C*, 32; and *D*, 40 teeth?

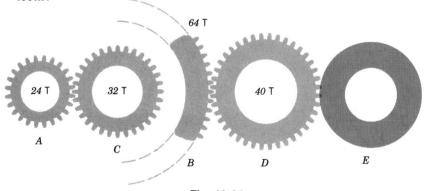

Fig. 19-24

9. If a speed ratio of 2 to 5 is required between *A* and *D* in Fig. 19-25, what gears would you use?

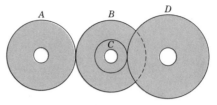

Fig. 19-25

10. The smaller pulley in Fig. 19-26 is 12 cm in diameter and the larger 20 cm. How many revolutions per minute will the larger pulley make if the smaller pulley makes 80 rpm?

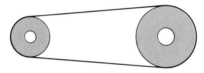

Fig. 19-26

11. If the large pulley in Fig. 19-27 is 18″ in diameter and makes 60 rpm, what must be the diameter of the small pulley to make 90 rpm?

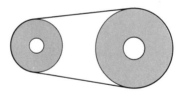

Fig. 19-27

12. In Fig. 19-28, pulley *A* is 14″ in diameter, *B*, 10″; *C*, 16″; and *D*, 12″. If *A* makes 120 rpm, how many revolutions per minute will *D* make?

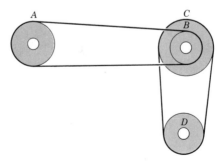

Fig. 19-28

13. The pulleys of the train in Fig. 19-29 have the following diameters: *A*, 12″; *B*, 20″; *C*, 8″; *D*, 16″; *E*, 10″; *F*, 12″. If *A* makes 75 rpm, how many revolutions per minute will *F* make?

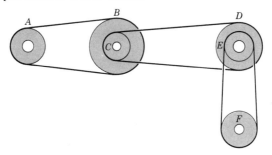

Fig. 19-29

14. It is required to transmit power from *A* to *D* in Fig. 19-30 at a speed ratio of 3 to 2. Find four suitable pulleys for this train.

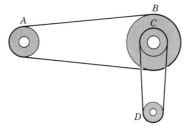

Fig. 19-30

15. A cone pulley having steps of 11″, 9″, and $6\frac{1}{2}$″ in diameter, and running at 140 rpm, drives another cone pulley having diameters of 5″, 7″, and $9\frac{1}{2}$″. What speeds can be obtained?

16. A driving pulley 36″ in diameter on a line shaft making 200 rpm is belted to a 16″ pulley on a countershaft. The countershaft has a pulley 14″ in diameter belted to a 6″ pulley on an emery wheel. How many revolutions per minute will the emery wheel make?

17. A countershaft running at 240 rpm has a cone pulley with steps 11″, 8″, and 5″ in diameter driving a cone on a lathe with steps 6″, 9″, and 12″ in diameter. What range of speeds can be obtained?

18. In Fig. 19-14, the driving gear *F* has 44 teeth and makes 280 rpm. Gear *E* has 28 teeth; *D*, 36 teeth; *C*, 24 teeth; and the worm *A* is triple-threaded. If the worm wheel has 32 teeth, how many revolutions per minute will it make?

19. If the single-threaded worm *A* in Fig. 19-14 makes 140 rpm, how many revolutions per minute will the worm wheel *B* make if it has 48 teeth?

20. In Fig. 19-31, *A* has 60 teeth and makes 160 rpm, *B* has 28 teeth, *C*, 24 teeth; *D*, 20 teeth; *E*, 36 teeth; and *F*, 32 teeth. How many revolutions per minute will *F* make?

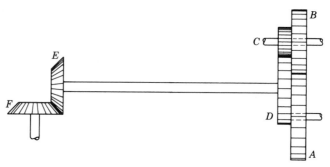

Fig. 19-31

21. In Fig. 19-31, how many teeth must F have if it is required to make 320 rpm and the other conditions are as follows: A makes 120 rpm and has 72 teeth; B has 24 teeth; C, 40 teeth; D, 36 teeth; and E, 32 teeth?

Self-Test

Solve the following problems. Check your answers with the answers in the back of the book.

1. Two gears have 30 and 60 teeth respectively. If the small gear makes 180 rpm what is the rpm of the large gear?
2. If a driving gear has 28 teeth and makes 100 rpm, how many teeth must the driven gear have to make 70 rpm?
3. The speeds of two gears are in the ratio of 1 to 2. If the faster gear makes 180 rpm, what is the speed of the slower gear?
4. The speeds of two gears are in the ratio of 2 to 5. If the slower gear makes 2,800 rpm, what is the speed of the faster gear?
5. If a worm is double threaded and makes 120 rpm, how many revolutions will the worm gear make if it has 80 teeth?
6. Two pulleys are connected by a vee-belt. The pulleys are 6 cm and 8 cm in diameter. What is the rpm of the larger pulley if the rpm of the smaller is 1,750?
7. A sharper must turn about 10,000 rpm. Its spindle pulley is 1.5 in. in diameter and is to be connected by a vee-belt to a 3,400-rpm motor. What size motor pulley diameter is needed?
8. A step pulley has steps of 2 in., $2\frac{1}{2}$ in., and 3 in. in diameter and, running at 1,725 rpm, drives a 6-in. pulley. What speeds can be obtained at the shaft of the 6-in. pulley?
9. An idler of 20 teeth is placed between a driven gear and a driver gear. Will the rpm of the driven gear be changed by the addition of the idler?
10. If the driver in Problem 9 rotated clockwise, what will be the direction of the driven gear?

Chapter Test

Solve the following problems involving gears and pulleys. Make a sketch for each problem. Label all known parts. Estimate the answer before solving.

1. Two meshing gears have 20 and 40 teeth, respectively. If the smaller gear makes 150 rpm, how many revolutions per minute will the larger gear make?

2. If the gears in Problem 1 have 20 and 35 teeth, respectively, and the smaller gear makes 150 rpm, how many rpm will the larger gear now make?

3. If a driving gear has 30 teeth and makes 132 rpm, how many teeth must the driven gear have to make 44 rpm?

4. An idler of 40 teeth is placed between the driven gear and the driver gear in Problem 3. How many rpm will the driven gear now make?

5. The speeds of two gears are in the ratio of 1 to 3. If the faster gear makes 150 rpm, what is the speed of the slower one?

6. In Fig. 19-32, A has 30 teeth, B has 60 teeth, C has 24 teeth, and D has 72 teeth. How many rpm will gear D make if gear A makes 180 rpm?

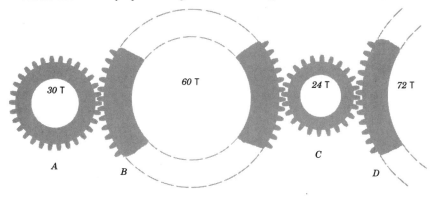

Fig. 19-32

7. In Fig. 19-32, under the conditions as stated in Problem 6, how many rpm does gear C make?

8. In Fig. 19-32, under the conditions as stated in Problem 6, how many rpm does gear B make?

9. In Fig. 19-33, if A makes 60 rpm, how many rpm will gear D make?

10. In Fig. 19-33, what is the speed ratio from A to D?

11. If a worm is double-threaded and makes 100 rpm, how many revolutions per minute will the worm gear make if it has 75 teeth?

12. If two pulleys are connected by a vee-belt, the smaller pulley is 8″ in diameter, the larger pulley is 12″ in diameter, and the larger pulley makes 1,750 rpm, how many rpm does the smaller pulley make?

13. An electric motor makes 1,750 rpm and is to be connected by pulleys and vee-belt to a saw that must make close to 4,000 rpm. Find two pulley sizes for this job.

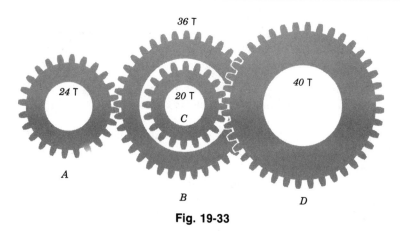

Fig. 19-33

14. A step pulley has steps of 3″, $3\frac{1}{2}$″, and 4″ in diameter, and running at 3,400 rpm drives a 4″ pulley. What speeds can be obtained at the shaft of the 4″ pulley?

15. If instead of the 4″ pulley in Problem 14, another step pulley of 4″, $3\frac{1}{2}$″, and 3″ is the driven pulley, what speeds can be obtained?

20

Screw Threads

20-1   Introduction

Screw threads are used (*a*) to fasten parts together such as with bolts, (*b*) to adjust parts as with levellers on the base of machinery, or (*c*) to transmit power such as is done with a vise. World War II convinced the United States, Britain, and Canada that it would be a great convenience if a standard were adopted for threads. The American Standard Unified Thread was the standard that was adopted.

20-2   Pitch

The *pitch* of a screw thread is the distance between corresponding points of two adjacent threads. In Fig. 20-1, the space marked *P* indicates the pitch for the threads shown.

If, for example, the distance between threads is $\frac{1}{8}$ in., we say that the thread is $\frac{1}{8}$-in. pitch. If the pitch is $\frac{1}{8}$ in., there are eight threads to the inch. Hence a machinist will often speak of an 8-pitch thread instead of a $\frac{1}{8}$-in. pitch thread. In this chapter both methods of designating pitch will be employed, since the meaning in either case is unmistakable.

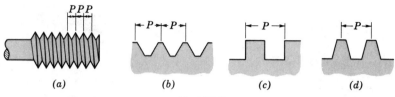

Fig. 20-1

From the definition of pitch it follows that:

$$\text{Pitch} = \frac{1}{\text{number of threads per inch}}$$

and

$$\text{Number of threads per inch} = \frac{1}{\text{pitch}}$$

Example 1
Find the pitch of a thread having 12 threads per inch.

SOLUTION $\text{Pitch} = \dfrac{1}{\text{number of threads per inch}} = \dfrac{1}{12}$ in.

Example 2
Find the pitch of a screw having $4\frac{1}{2}$ threads per inch.

SOLUTION $\text{Pitch} = \dfrac{1}{\text{number of threads per inch}}$

$$= \frac{1}{4\frac{1}{2}} = 1 \div \frac{9}{2} = 1 \times \frac{2}{9} = \frac{2}{9} \text{ in.}$$

Example 3
A screw has a pitch of $\frac{5}{8}$ in. Find the number of threads per inch.

SOLUTION $\text{Number of threads per inch} = \dfrac{1}{\text{pitch}} = \dfrac{1}{\frac{5}{8}} = 1 \times \dfrac{8}{5}$

$$= \tfrac{8}{5} = \text{threads per inch}$$

Problems 20-2

Solve the following pitch problems.

1. What is the pitch of a screw having $3\frac{1}{2}$ threads per inch?
2. A screw having five threads per inch has what pitch?
3. Find the pitch of a screw having $5\frac{1}{2}$ threads per inch.
4. Find the pitch of a screw having 18 threads per inch.

5. Compute the pitch of a screw with $2\frac{3}{4}$ threads per inch.
6. A $\frac{3}{4}''$-pitch screw has how many threads per inch?
7. How many threads per inch are on a screw of $\frac{2}{7}''$ pitch?
8. A screw has a pitch of $\frac{3}{8}''$. Find the number of threads per inch.
9. Find the number of threads per inch in a $\frac{7}{16}''$-pitch screw.
10. The pitch of a thread is $\frac{1}{14}''$. How many threads are there in $1''$?

20-3 Lead

The **lead** of a screw is the distance the nut advances on the screw in one turn, or the distance the screw would advance if given one complete turn.

In a **single-threaded** screw, the lead is *equal* to the pitch; that is, in one complete turn, the screw advances a distance equal to the pitch. In a **double-threaded** screw, the lead is *twice* the pitch; in a **triple-threaded** screw, the lead is *three times* the pitch, and so forth.

In a screw of $\frac{1}{8}$-in. pitch, the lead has the following values:

$$\begin{aligned}
\text{Single-threaded:} \quad &\text{lead} = \text{pitch} = \tfrac{1}{8} \text{ in.} \\
\text{Double-threaded:} \quad &\text{lead} = 2 \times \text{pitch} = 2 \times \tfrac{1}{8} = \tfrac{1}{4} \text{ in.} \\
\text{Triple-threaded:} \quad &\text{lead} = 3 \times \text{pitch} = 3 \times \tfrac{1}{8} = \tfrac{3}{8} \text{ in.}
\end{aligned}$$

Figure 20-2 illustrates the relation between pitch and lead in a single, double, triple, and quadruple square thread.

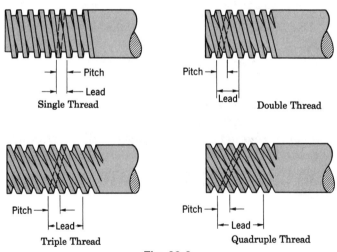

Single Thread   Double Thread

Triple Thread   Quadruple Thread

Fig. 20-2

Problems 20-3

Solve the following lead problems.

1. A single-threaded screw has 12 threads per inch. What is its pitch? What is its lead?

2. What is the lead of an 8-pitch double-threaded screw?
3. What is the lead of a triple 8-pitch thread?
4. The lead of a double-threaded screw is $\frac{3}{8}''$. What is its pitch?
5. What is the lead of a $\frac{3}{16}''$-pitch quadruple thread?
6. Find the pitch of a screw with a triple thread whose lead is $\frac{3}{4}''$.
7. A single thread of $\frac{1}{16}''$-pitch has what lead?
8. What is the pitch of a double-threaded screw having a lead of $\frac{3}{32}''$?

20-4   Definitions Applying to Screw Threads

Figure 20-3 illustrates the following terms:

The top of the thread is called the *point* or *crest.*

The bottom of the thread is called the *root.*

The *depth* of the thread is the vertical distance between the top and the bottom of the thread. Twice the depth is known as the *double depth.*

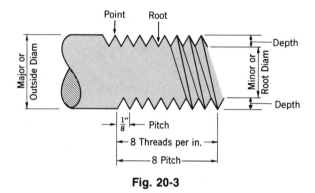

Fig. 20-3

The *major diameter,* sometimes referred to as the *outside diameter,* is the largest diameter of the thread of the screw or nut.

The *minor* or *root diameter* is the smallest diameter of the thread of the screw or nut; that is, it is the diameter at the roots or bottoms of the threads. It is evident from the sketch that the root diameter of a thread is equal to the major diameter minus the double depth.

20-5   Sharp V-thread

The sharp V-thread is so named because the sides of the thread form the letter V. It is now used for special work where additional thread contact is desired, and it is also used with brass pipe work. The study of the sharp V-thread will make clear the dimensions of the other thread forms discussed in this chapter.

In Fig. 20-4 the horizontal line *AB* (joining two successive roots) forms, with the sides of the thread, an equilateral triangle *ABC.* If the pitch of the thread is 1 in., the sides of the triangle will be 1 in. each. The depth of the

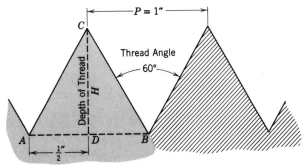

Fig. 20-4

V-thread H, which is a perpendicular from the crest C to the side AB, divides the triangle into two equal right triangles. In the right triangle ACD,

$$H = \sqrt{\overline{AC}^2 - \overline{AD}^2} = \sqrt{1^2 - (\tfrac{1}{2})^2} = \sqrt{1 - \tfrac{1}{4}} = \sqrt{\tfrac{3}{4}} = 0.866 \text{ in.}$$

That is, the depth of a sharp V-thread of 1-in. pitch is 0.866 in. The depth of a $\tfrac{1}{2}$-in. pitch V-thread is equal to one-half of 0.866, or 0.433 in.

RULE
 The depth of a sharp V-thread is found by multiplying the pitch by 0.866.

Example
 Find the depth of a V-thread having 16 threads per inch.

SOLUTION   Depth $=$ pitch $\times$ 0.866 $= \tfrac{1}{16} \times 0.866 = 0.0541$ in.

Problems 20-5

Find the depth of V-threads having the following numbers of threads per inch.

1. 20	2. 18	3. 14	4. 12	5. 11
6. 9	7. 8	8. 7	9. 6	10. 5

20-6   Double Depth of Sharp V-thread

Twice the depth of a thread is called the *double depth*. The double depth of a 1-in. pitch sharp V-thread is 2 × 0.866 = 1.732 in.; the double depth of a $\tfrac{1}{4}$-in. pitch V-thread is $\tfrac{1}{4}$ × 1.732 = 0.433 in.

Problems 20-6

Find the double depth of V-threads with the following numbers of threads per inch.

1. 3   2. $3\frac{1}{4}$ 3. $3\frac{1}{2}$ 4. 4   5. $4\frac{1}{2}$
6. 5   7. 8   8. 12   9. 16   10. 20

20-7 Minor Diameter

Figure 20-3 shows that the minor diameter of a screw thread is equal to the major diameter minus the double depth.

Example

Find the minor diameter of a $\frac{3}{8}$-in.–16 V-thread. ($\frac{3}{8}$ in. = major diameter; 16= number of threads per inch.)

SOLUTION   Major diameter = $\frac{3}{8}$ = 0.3750 in.
                  Double depth = $\frac{1}{16}$ × 1.732 = 0.1082 in.
                  Minor diameter   = 0.2668 in.

Problems 20-7

Find the minor diameters of the following V-thread screws.

1. $\frac{1}{4}''$–20   2. $\frac{7}{16}''$–14   3. $\frac{9}{16}''$–12   4. $\frac{13}{16}''$–10   5. $\frac{15}{16}''$–9
6. $1\frac{1}{4}''$–7   7. $1\frac{1}{2}''$–6   8. $1\frac{3}{4}''$–5   9. $2\frac{1}{4}''$–$4\frac{1}{2}$ 10. $2\frac{7}{8}''$–4

20-8 Tap Drill Sizes

To tap a nut or hole is to cut a thread in it. A tap drill is used for drilling holes before tapping. Although theoretically the diameter of the drill is equal to the minor diameter of the bolt for which the nut is to be used, in practice the tap drills are somewhat larger than the minor diameter of the thread (see Fig. 20-5). The tap drill diameter for a sharp V-thread is equal to the minor diameter plus one-quarter of the pitch.

Example

Find the proper size tap drill for a $\frac{7}{16}$-in.–14 V-thread.

SOLUTION   Major diameter = $\frac{7}{16}$ in. = 0.4375 in.
                  Double depth = $\frac{1}{14}$ × 1.732 = 0.1237 in.
                  Minor diameter   = 0.3138 in.
                  Clearance = $\frac{1}{4}$ × pitch
                        = $\frac{1}{4}$ × $\frac{1}{14}$ = $\frac{1}{56}$ = 0.0179 in.
                  Drill size   = 0.3317 in.

According to the table of decimal equivalents, the nearest $\frac{1}{64}$ in. to 0.3317 is 0.3281, which is the decimal equivalent of $\frac{21}{64}$ in. Hence a $\frac{21}{64}$-in. drill is to be used.

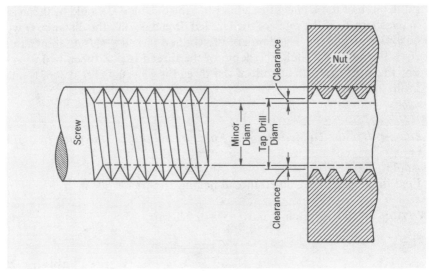

Fig. 20-5

Find the proper tap drill sizes for the following V-threads.

1. $\frac{5}{16}''{-}18$　　2. $\frac{3}{8}''{-}16$　　3. $\frac{1}{2}''{-}12$　　4. $\frac{5}{8}''{-}11$　　5. $\frac{11}{16}''{-}11$
6. $\frac{3}{4}''{-}10$　　7. $\frac{7}{8}''{-}9$　　8. $1''{-}8$　　9. $1\frac{3}{8}''{-}6$　　10. $2''{-}4\frac{1}{2}$

20-9　The Unified Thread

The Unified thread, the standard thread agreed upon by the United States, Britain, and Canada, follows the shape of the older American Sharp V-thread. The top of the thread is flattened and the bottom of the thread is rounded. The sides of the thread, if extended, would form the Sharp V-thread (see Fig. 20-6). The dimensions of the Unified thread are based on the depth of the sharp V-thread, H.

The crest of the thread is flattened (it can be rounded) a distance of $H/8$. The root of the thread is $\frac{17}{24}H$ from the crest. The root may be flat or rounded.

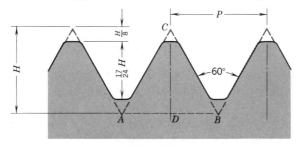

Fig. 20-6

If the pitch of the Unified thread is 1 in., the distance H would be 0.866 in. as in the sharp V. If the pitch of the Unified thread is $\frac{1}{2}$ in., the distance H would be 0.433, and so on. The depth of the Unified thread is $\frac{17}{24}H$ as shown in Fig. 20-6. For the 1-in. pitch, the depth of the thread is $\frac{17}{24} \times 0.866 = 0.61343$ in. For the $\frac{1}{2}$-in. pitch, the depth of the thread is $\frac{17}{24} \times 0.433 = 0.30671$, and so forth.

RULE
___

The depth of the Unified thread is found by multiplying the pitch by 0.61343.

Example
Find the depth of a Unified thread having 16 threads per inch.

SOLUTION   Depth = pitch $\times$ 0.61343 = 0.0383

Problems 20-9

Find the depth of the Unified threads having the following number of threads per inch.

1. 20	2. 18	3. 16	4. 14	5. 13
6. 12	7. 11	8. 10	9. 9	10. 8

20-10   Double Depth of the Unified Thread

The double depth of a 1-in. pitch Unified thread is $1 \times 2 \times 0.61343 = 1.22686$ in.; the double depth of a $\frac{1}{4}$-in. pitch Unified thread is $\frac{1}{4} \times 2 \times 0.61343 = 0.30671$ in.

Problems 20-10

Find the double depth of the Unified threads with the following number of threads per inch.

1. 8	2. 9	3. 10	4. 11	5. 12
6. 13	7. 14	8. 16	9. 18	10. 20

20-11   Minor Diameter of Unified Threads

The minor diameter of the Unified thread is equal to the major diameter of the screw thread minus the double depth.

Example
Find the minor diameter of a $\frac{1}{4}$-20 Unified thread ($\frac{1}{4}$ means major diameter = $\frac{1}{4}$ inch; 20 means 20 threads per inch).

SOLUTION   Major diameter $= \frac{1}{4}$ $= 0.25000$ in.
           Double depth $= \frac{1}{20} \times 2 \times 0.61343 = \underline{0.06134}$ in.
           Minor diameter $= 0.18866$ in.

Problems 20-11

Find the minor diameter of the following Unified threads.

1. $\frac{5}{16} - 18$ 2. $\frac{3}{8} - 16$ 3. $\frac{7}{16} - 14$ 4. $\frac{1}{2} - 13$
5. $\frac{9}{16} - 12$ 6. $\frac{5}{8} - 11$ 7. $\frac{3}{4} - 10$ 8. $\frac{7}{8} - 9$

20-12   American National Thread

There are two series of American National threads: the American National coarse-thread series and the American National fine-thread series.

The American National course-thread series comprises the former U.S. Standard threads supplemented in the sizes below $\frac{1}{4}$ in. by part of the A.S.M.E. Standards (see Table 8 at the back of the text).

The American National fine-thread series comprises the regular screw thread series of the S.A.E. supplemented in sizes below $\frac{1}{4}$ in. by the A.S.M.E. fine-thread series (see Table 9).

The American National thread is a 60-degree thread with flattened tops and bottoms. If we start with a V-thread and divide the height or depth into eight equal parts, the American National is formed by eliminating one-eighth of the depth at the top and one-eighth of the depth at the bottom. The depth of the resulting thread is therefore equal to three-fourths of the depth of the V-thread of the same pitch.

If, as in Fig. 20-7, the pitch is 1 in., the depth of the V-thread is 0.866 in. and the depth of the corresponding American National thread is $\frac{3}{4}$ of 0.866, or 0.6495 in. The width of the thread at top and bottom is $\frac{1}{8}$ of the pitch.

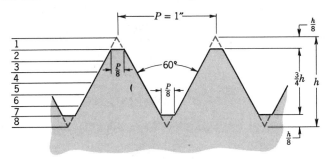

Fig. 20-7

To enable you to compute the proportions of an American National thread of any pitch, memorize the constant 0.6495.

Likewise, you should memorize the constant for the American Standard Unified thread, which is 0.61343.

Example

Find the depth of an American National thread having 16 threads to the inch.

SOLUTION   Depth = pitch × 0.6495 = $\frac{1}{16}$ × 0.6495
                                          = 0.0406 in.

Find the single depth of the following American National threads.

1. Seven threads per inch
2. Eight threads per inch
3. $\frac{1}{10}$″ pitch
4. $\frac{1}{12}$″ pitch
5. $5\frac{1}{2}$ threads per inch
6. Thirteen threads per inch

20-13   Double Depth of American National Thread

The double depth of an American National thread of 1-in. pitch is obviously equal to 2 × 0.6495 in. or 1.299 in. The double depth of an American National thread of any pitch is equal to *the pitch multiplied by* 1.299.

Example

Find the double depth of an American National thread having 16 threads per inch.

SOLUTION   Double depth = pitch × 1.299
                                  = $\frac{1}{16}$ × 1.299 = 0.0812 in.

Find the double depth of the following American National threads.

1. Pitch = $\frac{1}{20}$″
2. $4\frac{1}{2}$ threads per inch
3. Pitch = $\frac{1}{11}$″
4. Nine threads per inch
5. Six threads per inch
6. Eleven threads per inch

20-14   Minor Diameter of American National Thread

The minor diameter is equal to the major diameter minus the double depth.

Example

Find the minor diameter of a $\frac{5}{8}$-in.–11 American National thread.

SOLUTION   Major diameter = $\frac{5}{8}$ in.   = 0.6250 in.
           Double depth = $\frac{1}{11}$ × 1.299 = 0.1181 in.
           Minor diameter   = 0.5069 in.

Problems 20-14

Find the minor diameter of the following American National threads.

1. $\frac{1}{4}''-20$ 2. $\frac{3}{8}''-16$ 3. $\frac{1}{2}''-13$ 4. $\frac{3}{4}''-10$ 5. $1''-8$

6. $1\frac{1}{2}''-6$ 7. $1\frac{3}{4}''-5$ 8. $2''-4\frac{1}{2}$ 9. $2\frac{3}{4}''-4$ 10. $\frac{1}{8}''-7$

20-15 Size of Tap Drill for American National Threads

Commercial tap drills are designed to allow an engagement of three-fourths of the full depth of the screw thread. For a nut to engage the screw three-fourths of the depth of the thread, the tap drill diameter must be equal to the minor diameter of the screw plus one-fourth the double depth of the thread.

Example 1

Find the commercial size tap drill (that is, for $\frac{3}{4}$ engagement) for a $\frac{7}{8}$-in.–9 American National coarse-thread.

SOLUTION   Major diameter $= \frac{7}{8}$ in. $= 0.8750$ in.

Double depth $= \frac{1}{9} \times 1.299$ $= \underline{0.1443}$ in.

Minor diameter $= 0.7307$ in.

$$\text{Clearance} = \frac{\text{double depth of thread}}{4}$$

$$= \frac{0.1443}{4} \qquad\qquad = \underline{0.0361}\ \text{in.}$$

Drill size $= 0.7668$ in.

The nearest $\frac{1}{64}$ in. is $\frac{49}{64}$ in. Use a drill $\frac{49}{64}$ in. in diameter.

The *minimum* standard tap drill adopted by the American Standard Association is for a nut that will engage the screw thread five-sixths of the depth of the thread. The tap drill size for such a nut must be equal to the minor diameter of the screw thread plus one-sixth of the double depth of the thread.

Example 2

Find the tap drill size for a $\frac{7}{8}$-in.–9 American National coarse-thread for $\frac{5}{6}$ engagement.

SOLUTION   Major diameter $= \frac{7}{8}$ in. $= 0.8750$ in.

Double depth $= \frac{1}{9} \times 1.299$ $= \underline{0.1443}$ in.

Minor diameter $= 0.7307$ in.

$$\text{Clearance} = \frac{\text{double depth of thread}}{6}$$

$$= \frac{0.1443}{6} \qquad\qquad = \underline{0.0241}\ \text{in.}$$

Drill size $= 0.7548$ in.

The nearest $\frac{1}{64}$ in. is $\frac{3}{4}$ in. Use a drill $\frac{3}{4}$ in. in diameter.

Find the commercial tap drill size and also the minimum tap drill size for each of the following American National threads.

1. $\frac{5}{16}''$–18   2. $\frac{7}{16}''$–14   3. $\frac{7}{8}''$–9   4. $1''$–8   5. $1\frac{3}{8}''$–6
6. $1\frac{3}{4}''$–5   7. $4''$–4   8. $1\frac{1}{2}''$–6   9. $2\frac{1}{4}''$–$4\frac{1}{4}$ 10. $2\frac{1}{2}''$–4

20-16 Width of Point of Tool

The shape of the point of the tool for cutting an American National thread must be the same as the bottom of the thread. The point of the tool therefore is flattened an amount equal to one-eighth of the pitch of the thread to be cut.

Example
Find the width of tool point for cutting a $\frac{1}{4}$-in. pitch American National thread.

SOLUTION Width of tool point $= \frac{1}{8} \times$ pitch $= \frac{1}{8} \times \frac{1}{4} = \frac{1}{32}$ in.

Find the width of tool point required to cut the following American National threads.

1. Pitch $= \frac{2}{9}''$ 2. Pitch $= \frac{1}{4}''$ 3. Pitch $= \frac{1}{16}''$
4. Pitch $= \frac{1}{7}''$ 5. Pitch $= \frac{1}{12}''$ 6. Pitch $= \frac{2}{7}''$

20-17 Square Thread

The square thread (Fig. 20-8) is so called because a cross section of the thread is theoretically a square. The sketch is self-explanatory. The thickness of the thread and the width of the space are each equal to one-half the pitch. In the nut, however, the space is slightly more than half the pitch, to allow for a sliding fit. The depth of the thread, although theoretically equal to half the pitch, is actually cut 0.005 in. deeper than half the pitch.

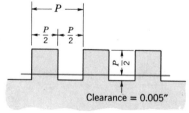

Fig. 20-8

The width of the tool point for cutting a square thread is equal to the width of the space, which is half the pitch.

20-18   Tap Drill for Square Thread

The relation between a square-threaded screw and its nut is shown in Fig. 20-9. It will be seen that there is a clearance of 0.005 in. between the screw threads and the nut threads, both top and bottom. The tap drill therefore has to be 2 × 0.005 in., or 0.01 in. larger than the root diameter of the screw, and the tap diameter will be 2 × 0.005 in., or 0.01 in. larger than the outside diameter of the screw.

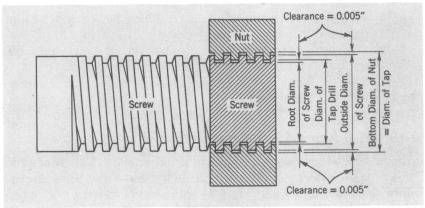

Fig. 20-9

Example 1
Find the double depth of a $\frac{5}{8}$-in.–8 square thread.

SOLUTION   Double depth = pitch + 2 × clearance
$= \frac{1}{8} + 2 \times 0.005$ in. = 0.125 in. + 0.01 in.
= 0.135 in.

Example 2
Find the root diameter of a $\frac{3}{8}$-in.–16 square-thread screw.

SOLUTION   Screw size = $\frac{3}{8}$ in.   = 0.3750 in.
Double depth = $\frac{1}{16}$ + 0.01 = 0.0625 + 0.01 = 0.0725 in.
Root diameter   = 0.3025 in.

Example 3
Find the size of tap drill for a $\frac{5}{8}$-in.–8 square-thread screw.

SOLUTION   Figure 20-9 shows that the diameter of the tap drill is 2 × 0.005 in. larger than the root diameter of the thread.

Screw size = $\frac{5}{8}$ in.   = 0.6250 in.
Pitch = $\frac{1}{8}$ in.   = 0.1250 in.
Diameter of tap drill = 0.5000 in. = $\frac{1}{2}$ in.

Solve the following problems involving square-threaded screws:

1. Find the double depth of a $\frac{3}{4}''$–6 square thread.
2. Find the double depth of a $\frac{7}{8}''$–$4\frac{1}{2}$ square thread.
3. Find the double depth of a square-thread screw 1″ in diameter with a double thread of $\frac{2}{5}''$ lead.
4. Find the double depth of a square-thread screw 4″ in diameter with three threads per inch.
5. Find the double depth of a $2\frac{1}{2}''$ square-thread screw with four threads per inch.
6–10. Find the root diameters of the square-thread screws in Problems 1 to 5.
11–15. Find the tap drill sizes for the square-thread screws in Problems 1 to 5.

20-19   The Acme 29-Degree Screw Thread

The Acme 29-degree screw thread is best understood by a study of its derivation from the square thread. In Fig. 20-10 we start with a square thread and draw lines making an angle of $14\frac{1}{2}°$ with the vertical sides of the square thread. The sides of the resulting Acme thread will therefore make an angle of 29° with each other. The depth of the Acme thread is 0.01 in. greater than half the pitch of the thread.

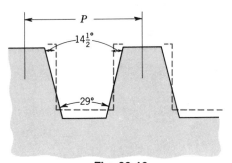

Fig. 20-10

The dimensions of the Acme thread for the screw (Fig. 20-11) are as follows:

F = flat on top = 0.3707 × pitch

S = width of space = 0.6293 × pitch

T = bottom flat = width of tool point for cutting thread

= 0.3707 × pitch − 0.0052 in.

D = depth = $P/2$ + 0.010 in.

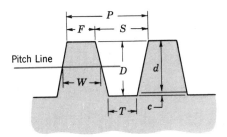

Fig. 20-11

Example

Find the root diameter, top flat, and bottom flat of an $\frac{3}{4}$-in.–8 Acme 29-degree screw thread.

SOLUTION　　　　Screw size = $\frac{3}{4}$ in.　　　　　　　　　　　　　　　　= 0.7500 in.
　　　　　　　　Depth = $P/2$ + 0.010 in. = $\frac{1}{2} \times \frac{1}{8}$ + 0.010 in.
　　　　　　　　　　= 0.0625 + 0.010 = 0.0725 in.
　　　　　　　　Double depth = 2 × 0.0725　　　　　　　　= 0.1450 in.
　　　　　　　　　Root diameter　　　　　　　　　　　　= 0.6050 in.
　　　　Top flat = 0.3707 × pitch = 0.3707 × $\frac{1}{8}$ = 0.0463 in.
　　Bottom flat = 0.3707 × pitch − 0.0052
　　　　　　　　= 0.3707 × $\frac{1}{8}$ − 0.0052 = 0.0463 − 0.0052
　　　　　　　　= 0.0411 in.

20-20　Tap Drill and Tap, Acme Thread

The relative sizes of the Acme screw and the tap drill and tap are shown in Fig. 20-12. The clearances between the screw threads and the nut threads are all 0.010 in. The tap drill for the nut is therefore 2 × 0.010, or 0.020 in. larger

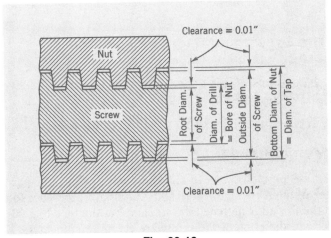

Fig. 20-12

than the root diameter of the screw, and the diameter of the tap is 0.020 in. larger than the outside diameter of the screw. The dimensions of the threads on the tap are as follows:

$$\text{Top flat} = 0.3707 \times \text{pitch} - 0.0052 \text{ in.}$$

$$\text{Bottom flat} = 0.3707 \times \text{pitch} - 0.0052 \text{ in.}$$

$$\text{Depth} = P/2 + 0.020 \text{ in.}$$

Example

A $\frac{3}{4}$-in. bolt has an Acme 29-degree thread of $\frac{1}{8}$-in. pitch. Find tap drill size, bottom and top flats of tap thread, and outside diameter of tap.

SOLUTION

$$\text{Tap drill size} = \text{root diameter} + 2 \times 0.01 \text{ in.}$$
$$= 0.605 + 0.02$$
$$= 0.625 \text{ in.}$$
$$\text{Bottom flat} = 0.3707 \times \text{pitch} - 0.0052$$
$$= 0.3707 \times \tfrac{1}{8} - 0.0052$$
$$= 0.0463 - 0.0052 = 0.0411 \text{ in.}$$
$$\text{Top flat} = \text{same as bottom flat} = 0.0411 \text{ in.}$$
$$\text{Outside diameter of tap} = \text{outside diameter of screw} + 2 \times 0.01$$
$$= 0.750 + 0.020 = 0.7700 \text{ in.}$$

Problems 20-20

In each of the following Acme 29-degree screw threads, find: (a) root diameter; (b) top flat of screw; (c) bottom flat of screw; (d) tap drill size; (e) bottom flat of tap; (f) top flat of tap; (g) outside diameter of tap.

1. 2″–4   2. $1\frac{1}{2}$″–$5\frac{1}{2}$ 3. $2\frac{1}{8}$″–$3\frac{1}{2}$ 4. $\frac{1}{2}$″–10   5. 1″–7

20-21   The Brown and Sharpe 29-Degree Worm Thread

For the dimensions of this thread, which is similar to the 29-degree Acme thread, see Section 20-19.

20-22   Metric Standard Screw Threads

The form of the metric thread and the dimensions of the parts are exactly like those of the American National thread; that is, the sides form an angle of 60° and the thread is flattened $h/8$ top and bottom (see Fig. 20-13). Both the diameter and the pitch of the screw are given in millimeters. See Appendix Table 3, Converting Millimeters to Inches and Decimals.

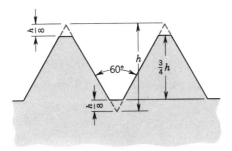

Fig. 20-13

Example 1

Find the double depth of a 6.5-mm-pitch thread.

SOLUTION 6.5 mm = 6.5 × 0.0394 in.　　= 0.2559 in.
　　　　　Double depth = pitch × 1.299 = 0.2559 × 1.299
　　　　　　　　　　　　　　　　　　　= 0.3324 in.

Example 2

Find the root diameter of a metric screw having an outside diameter of 10 mm and a pitch of 1.5 mm.

SOLUTION 10 mm = 0.3937 in.; 1.5 mm　　= 0.0591 in.
　　　　　Outside diameter　　　　　　　= 0.3937 in.
　　　　　Double depth = 0.0591 × 1.299 = 0.0767 in.

　　　　　Root diameter　　　　　　　　= 0.3170 in.

Example 3

What size drill should be used for a metric thread 26 mm in diameter and 3-mm pitch?

SOLUTION Diameter = 26 mm = 1.0236 in.
　　　　　Pitch = 3 mm = 0.1181 in.
　　　　　Diameter　　　　　　　　　　　　= 1.0236 in.
　　　　　Double depth = 0.1181 × 1.299　= 0.1534 in.

　　　　　Root diameter　　　　　　　　　= 0.8702 in.
　　　　　Clearance = $\frac{1}{8}$ pitch = $\frac{1}{8}$ × 0.1181 = 0.0148 in.

　　　　　Drill size　　　　　　　　　　　= 0.8850 in.

The nearest 64th of an inch above this is $\frac{57}{64}$ in. Hence a $\frac{57}{64}$-in. drill should be used.

Example 4

Find the width of tool point for a 7-mm-pitch thread.

SOLUTION   Pitch = 7 mm = 0.2756 in.
            Width of tool point = $\frac{1}{8}$ pitch = $\frac{1}{8}$ × 0.2756
                                 = 0.0345 in.

NOTE. To compute the various parts of a metric thread, first convert the given metric dimensions into inches and then proceed exactly as with American National threads.

Problems 20-22

Compute the parts of the metric threads in the following chart.

	Diameter (mm)	Pitch (mm)	Double Depth (in.)	Root Diameter (in.)	Drill Size (in.)	Width of Tool Point (in.)
1.	3	0.5	____	____	____	____
2.	5	0.75	____	____	____	____
3.	6	1.0	____	____	____	____
4.	9	1.0	____	____	____	____
5.	12	1.5	____	____	____	____
6.	20	2.5	____	____	____	____
7.	22	2.5	____	____	____	____
8.	24	3.0	____	____	____	____
9.	30	3.5	____	____	____	____
10.	40	4.0	____	____	____	____

20-23   Whitworth Standard Threads

The Whitworth or British standard thread is a 55-degree V-thread with rounded top and bottom. If we start with a 55-degree V-thread, the Whitworth is obtained by eliminating one-sixth of the height of the V both top and bottom and rounding off the thread at the point and at the root as shown in Fig. 20-14. The depth of the resulting thread is therefore two-thirds of the depth of the corresponding V-thread.

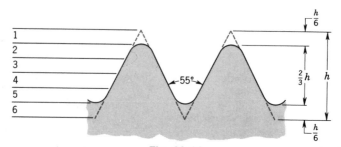

Fig. 20-14

The dimensions of the Whitworth standard thread follow:

$$\text{Depth} = \text{pitch} \times 0.6403$$

$$\text{Radius} = \text{pitch} \times 0.1373$$

Example 1

Find the double depth of a Whitworth standard thread having 12 threads per inch.

SOLUTION Double depth $= 2 \times \frac{1}{12} \times 0.6403 = 0.1067$ in.

Example 2

Find the root diameter of a $\frac{1}{4}$-in.–20 Whitworth standard thread.

SOLUTION Outside diameter $= \frac{1}{4}$ in. $= 0.2500$ in.

Double depth $= 2 \times \frac{1}{20} \times 0.6403 = \underline{0.0640}$ in.

Root diameter $= 0.1860$ in.

Problems 20-23

Find the double depth of the following Whitworth threads.

1. Pitch $= \frac{1}{24}''$
2. Pitch $= \frac{1}{10}''$
3. 9 threads per inch
4. 6 threads per inch
5. $3\frac{1}{2}$ threads per inch

Find the root diameter of the following Whitworth threads.

6. $\frac{3}{8}''$–16
7. $\frac{5}{8}''$–11
8. $\frac{3}{4}''$–10
9. $1\frac{7}{8}''$–$4\frac{1}{2}$
10. $3\frac{5}{8}''$–$3\frac{1}{4}$
11. $\frac{11}{16}''$–11

20-24 Tap Drill Size, Whitworth Thread

The diameter of the tap drill used in boring a nut blank for a Whitworth thread is larger by one-eighth of the pitch than the root diameter of the thread.

Example

Find the proper tap drill size for a $\frac{7}{8}$-in.–9 Whitworth thread.

SOLUTION Outside diameter $= \frac{7}{8}$ in. $= 0.8750$ in.

Double depth $= 2 \times \frac{1}{9} \times 0.6403 = \underline{0.1423}$ in.

Root diameter $= 0.7327$ in.

Clearance $= \frac{1}{8} \times$ pitch $= \frac{1}{8} \times \frac{1}{9} = \underline{0.0139}$ in.

Drill size $= 0.7466$ in.

The nearest $\frac{1}{64}$ in. above this decimal is $\frac{3}{4}$ in. Hence a $\frac{3}{4}$-in. drill should be used.

Problems 20-24

Find the proper size drill to be used for each of the following Whitworth threads.

1. $\frac{5}{16}''$–18   2. $\frac{1}{2}''$–12   3. $\frac{13}{16}''$–10   4. $1''$–8   5. $1\frac{1}{4}''$–7
6. $1\frac{1}{2}''$–6   7. $2''$–$4\frac{1}{2}$ 8. $2\frac{1}{2}''$–4   9. $3''$–$3\frac{1}{2}$ 10. $5\frac{1}{4}''$–$2\frac{5}{8}$

20-25   Radius of Tool Point, Whitworth Thread

Example
Find the radius of tool point for cutting a Whitworth standard thread having $4\frac{1}{2}$ threads per inch.

SOLUTION   Pitch $= \dfrac{1}{4\frac{1}{2}} = \dfrac{2}{9}$ in.

Radius $=$ pitch $\times 0.1373 = \frac{2}{9} \times 0.1373$

$= 0.0305$ in.

Problems 20-25

1–10. Find the radius of tool point for cutting the threads in each of the problems of Problem Set 20-24.

20-26   American Standard Taper Pipe Threads

A study of Fig. 20-15 will show the following characteristics of the American Standard taper pipe threads.

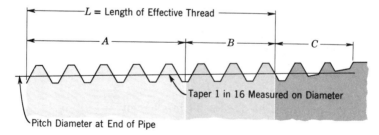

Fig. 20-15

There are three varieties of thread on the pipe. At the end of the pipe, in the section marked A, the threads are perfect both at top and bottom. The length of section A is found by Briggs' formula:

$$A = (0.8D + 4.8) \times P$$

where D is the actual external diameter and P is the pitch of the thread. The taper of section A is $\frac{1}{16}$ in. per inch or $\frac{3}{4}$ in. per foot measured on the diameter.

The threads in this section are 60-degree V-threads with top and bottom slightly flattened. The depth of the threads, therefore, instead of being 0.866 × P, is only 0.8 × P.

In the section marked B in the diagram, there are *two* threads having the same taper at the bottom as the threads in section A but with flat tops.

In the section marked C, there are three to four imperfect threads with flat tops and bottoms; the imperfections are caused by the chamfer of the threading die.

The value of section L, the length of effective thread, which consists of A + B, is found by the formula:

$$L = (0.8D + 6.8)P$$

Example
Find the depth of a perfect American Standard pipe thread of $\frac{1}{8}$-in. pitch.

SOLUTION Depth = $0.8 \times P = 0.8 \times \frac{1}{8} = 0.1$ in.

Problems 20-26

Find the depth of the following perfect American Standard pipe threads.

1. $\frac{1}{27}''$ pitch   2. $\frac{1}{18}''$ pitch   3. $\frac{1}{14}''$ pitch   4. $11\frac{1}{2}$ threads per inch

20-27 Length of Part Having Perfect Threads and Length of Effective Thread

Example
A pipe whose actual external diameter is 2.875 in. is threaded with American Standard pipe threads of $\frac{1}{8}$-in. pitch. What is the length of the part with the perfect threads and what is the length of effective thread?

SOLUTION $A = (0.8D + 4.8) \times P$
$= (0.8 \times 2.875 + 4.8) \times \frac{1}{8}$
$= (2.3 + 4.8) \times \frac{1}{8}$
$= 7.1 \times \frac{1}{8} = 0.89$ in.
$L = (0.8D + 6.8)P$
$= (0.8 \times 2.875 + 6.8)P$
$= (2.3 + 6.8)P$
$= 9.1 \times \frac{1}{8} = 1.14$ in.

Problems 20-27

Find the length of the part in which the threads are perfect and the effective length of thread in the following pipes having American Standard threads.

	Actual Outside Diameter	Number of Threads per Inch	Length of Part with Perfect Threads	Effective Length of Thread
1.	0.405″	27	_____	_____
2.	0.675″	18	_____	_____
3.	1.050″	14	_____	_____
4.	1.9″	11½	_____	_____
5.	4.0″	8	_____	_____

20-28 Thickness of Metal between Bottom of Thread and Inside of Pipe in Straight Pipe Thread

The thickness of the metal between the bottom of the thread and the inside of the pipe is equal to half the difference between the root diameter of the thread and the actual inside diameter of the pipe. See Fig. 20-16.

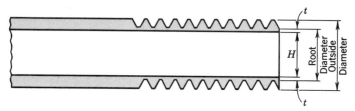

Fig. 20-16

Example

Find the amount of metal left between the bottom of a ⅛-in. thread and the inside of the pipe on a pipe having an actual inside diameter of 3.548 in. and an outside diameter of 4 in.

SOLUTION   Outside diameter   = 4.000 in.
                 Double depth = 2 × ⅛ × 0.8   = 0.200 in.
                 Root diameter   = 3.800 in.
                 Inside diameter of pipe   = 3.548 in.
                                                 0.252 in.
           Thickness of metal = ½ × 0.252 = 0.126 in.

Problems 20-28

Find the thickness of the metal between the bottom of the thread and the inside of the pipe in each of the following pipes.

	Actual Inside Diameter	Actual Outside Diameter	Number of Threads per Inch	Thickness of Metal
1.	0.364″	0.540″	18	_____
2.	0.623″	0.840″	14	_____

3.	1.048″	1.315″	11½	————
4.	3.067″	3.5″	8	————
5.	9″	9.688″	8	————

20-29 Lathe Gearing for Cutting Screw Threads

One of the most important uses to which a lathe is put is cutting screw threads. This is accomplished by closing the split nut attached to the carriage so as to engage the lead screw. In cutting a desired number of threads per inch on a screw, it is evident that while the tool travels a distance of 1 in., the piece of work must make a number of revolutions equal to the required number of threads per inch.

The speed of the tool depends on the speed of the lead screw, which in turn depends upon the pitch of the lead screws; the speed of rotation of the work depends upon the speed of the spindle. Hence, if the lathe is so geared that the spindle and the lead screw revolve at the same rate of speed, the number of threads per inch cut on the work will be the same as the number of threads per inch on the lead screw. If, for example, the lead screw has 6 threads per inch, the piece of work will revolve 6 times while the tool travels 1 in., and the resulting screw will have 6 threads per inch.

If we want to cut 12 threads per inch, the piece of work will have to revolve 12 times while the tool travels 1 in., that is, while the lead screw revolves 6 times. But for the work to revolve twice as fast as the lead screw, the spindle must turn twice as fast as the lead screw. This is accomplished by the change gears D and F in Fig. 20-17, which shows the arrangement of the gears in a simple geared lathe.

$$A = \text{Spindle gear that turns the work}$$
B and B' = Reverse gears to give direction to the lead screw as right or left hand screw is required.
$$C = \text{Fixed stud gear}$$
$$D = \text{Change stud gear}$$
$$E = \text{Intermediate or idler gear}$$
$$F = \text{Lead screw gear}$$

Case 1

If the ratio of spindle to stud is 1 to 1, that is, if gear A and gear C have the same number of teeth, then the pitch of the work will be the same as the lead screw pitch if gears D and F are of equal size. If a screw is required having three times as many threads per inch as the lead screw, the spindle will have to revolve three times as fast as the lead screw, and therefore a gear with one-third as many teeth must be placed on the stud as on the lead screw.

The preceding may be expressed by the following simple formula:

$$\frac{\text{Lead screw}}{\text{Screw to cut}} = \frac{\text{Stud gear}}{\text{Lead screw gear}}$$

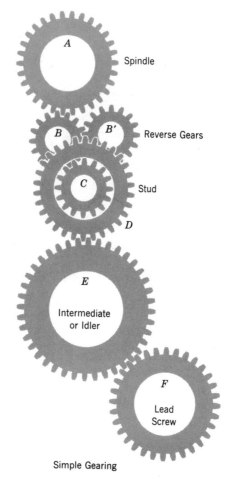

Simple Gearing

Fig. 20-17

which means that the ratio of the stud gear to the lead screw gear must be the same as the ratio of the lead screw to the screw we wish to cut.

Example

What change gears are required to cut a screw with ten threads per inch if the lead screw has six threads per inch?

SOLUTION $\dfrac{\text{Lead screw}}{\text{Screw to cut}} = \dfrac{6}{10} = \dfrac{6 \times 4}{10 \times 4} = \dfrac{24}{40}$

EXPLANATION Write a fraction with the number of threads per inch on the lead screw as the numerator and the number of threads per inch on the screw to be cut as the denominator. Since the teeth of the stud gear and the lead screw gear must be in the same ratio as the terms of this fraction, we multiply both terms of the fraction by any multiplier that will give us an equivalent fraction whose terms will represent suitable gears. In this instance 4 was chosen as the multiplier, giving gears 24 and 40. If the 24-tooth gear is placed on the stud D and the 40-tooth gear on the lead screw F, the ratio of the speed of the stud (and therefore of the spindle) to the speed of the lead screw will be as 40 is to 24, or 10 to 6. Therefore, the piece of work will revolve 10 times to 6 turns of the lead screw, and the result will be a screw having 10 threads per inch.

Any other multiplier might be selected that would give other gears that would produce the same result as 24 and 40. For example,

$$\frac{\text{Lead screw}}{\text{Screw to cut}} = \frac{6}{10} = \frac{6 \times 6}{10 \times 6} = \frac{36}{60}$$

Multiplying both terms of the fraction by 6 gives the gears 36 and 60. Placing the 36 on the stud and the 60 on the lead screw, the speed ratio between the spindle and the lead screw is the same as before, namely, 10 to 6, and we get a screw with 10 threads per inch.

NOTE. Since idler E does not affect the relative speeds of D and F, it does not enter into the computations.

Case 2

If spindle gear A and stud gear C do not have the same number of teeth, that is, if the ratio of spindle to stud is not 1 to 1, we proceed as follows in computing the change gears:

1. As in Case 1, write a fraction expressing the ratio of the number of threads on the lead screw to the number of threads to be cut.
2. Multiply this ratio by the ratio of the number of teeth on the stud gear to the number of teeth on the spindle gear.
3. Find suitable gears by multiplying the result of step 2 by any convenient multiplier.

Example

What change gears are required to cut 16 threads per inch with a lead screw having 6 threads per inch when the spindle gear has 20 teeth and the stud gear has 40 teeth?

SOLUTION $\quad \dfrac{\text{Lead screw}}{\text{Screw to cut}} = \dfrac{6}{16}$

$$\frac{\text{Fixed stud gear}}{\text{Spindle gear}} = \frac{40}{20}$$

$$\text{Change gears} = \frac{6}{16} \times \frac{40}{20} = \frac{3}{4}$$

$$\frac{3}{4} \times \frac{12}{12} = \frac{36}{48}$$

$$\text{Stud gear} = 36 \text{ teeth}$$

$$\text{Lead screw gear} = 48 \text{ teeth}$$

Problems 20-29

Find suitable change gears to be used in the cases given. Select gears from the following set, which is furnished with a certain lathe.

24 28 32 36 40 44 48 52 56 60 64 68 72 76
80 84 88 92 96 100 104 108 112 116 120

1. To cut 8 threads per inch with a 6-pitch lead screw
2. To cut 14 threads per inch with an 8-pitch lead screw
3. To cut 12 threads per inch with an 8-pitch lead screw
4. To cut a thread of $\frac{1}{13}''$ pitch with a lead screw whose pitch is $\frac{1}{6}''$
5. To cut 10 threads per inch with a 4-pitch lead screw
6. Determine the gears for stud and lead screw when 18 threads per inch are to be cut with a lead screw whose pitch is $\frac{1}{8}''$.
7. A lathe having an 8-pitch lead screw is found to have a 32 gear on the stud and an 88 gear on the lead screw. How many threads per inch will it cut on a piece of work?
8. A lathe whose lead screw has 6 threads per inch has a lead screw gear of 36 teeth. What gear must be placed on the spindle to cut a thread of $\frac{1}{4}''$ pitch?
9. It is desired to cut 20 threads per inch on a screw. If the pitch of the lead screw is $\frac{1}{8}''$ and the stud gear has 32 teeth, what gear must be placed on the lead screw?
10. It is found that a lathe having a stud gear of 36 teeth and a lead screw gear of 84 teeth cuts a screw with 7 threads per inch. What is the pitch of the lead screw?

Solve the following problems concerning change gears.

11. If the ratio of spindle to stud is 1 to 3 and the lead screw has 8 threads per inch, what change gears are required to cut 12 threads per inch?
12. Compute the necessary change gears for cutting 20 threads per inch with a lead screw having 8 threads per inch when the spindle gear has 18 teeth and the stud gear has 36 teeth.
13. Find the change gears for cutting a $\frac{1}{8}''$-pitch screw with a $\frac{1}{4}''$-pitch lead screw when the spindle gear has 16 teeth and the stud gear has 32 teeth.
14. If the spindle gear has 16 teeth and the stud gear 48 teeth, what change gears are required for cutting a screw with 8 threads to the inch if the lead screw has 12 threads to the inch?
15. What gear must be placed on the stud of a lathe if the ratio of spindle to stud is 1 to 2, the lead screw pitch is $\frac{1}{6}''$, an 80-tooth gear is on the lead screw, and a pitch of $\frac{1}{10}''$ is wanted on a screw?

20-30 Fractional Threads

To cut a fractional number of threads per inch, the same method is followed as for whole threads.

Example

Find the change gears necessary for cutting $11\frac{1}{2}$ threads per inch with a lead screw having 6 threads per inch.

SOLUTION $\quad \dfrac{\text{Lead screw}}{\text{Screw to cut}} = \dfrac{6}{11\frac{1}{2}} = \dfrac{6 \times 8}{11\frac{1}{2} \times 8} = \dfrac{48}{92}$

$$\text{Stud gear} = 48 \text{ teeth}$$

$$\text{Lead screw gear} = 92 \text{ teeth}$$

Problems 20-30

Find the change gears for cutting the following fractional number of threads per inch.

1. Find the change gears for cutting $4\frac{1}{2}$ threads per inch with an 8-pitch lead screw and a spindle to stud ratio of 1 to 1.
2. With a spindle to stud ratio of 1 to 2, find the change gears for cutting a screw with $3\frac{1}{2}$ threads per inch, the lead screw having 6 threads per inch.
3. It is desired to cut $5\frac{1}{2}$ threads per inch on a bolt. With a spindle to stud ratio of 1 to 1 and a lead screw having 6 threads per inch, what change gears are necessary?
4. What change gears would be required in Problem 3 if the spindle to stud ratio were 1 to 2?
5. What change gears would be required in Problem 3 if the spindle to stud ratio were 1 to 3?

In Problems 6, 7, 8, 9, and 10, assume that the spindle to stud ratio is 1 to 1 and compute the change gears required.

6. To cut $2\frac{3}{8}$ threads per inch with a 6-pitch lead screw.
7. To cut $2\frac{3}{4}$ threads per inch with a 6-pitch lead screw.
8. To cut $2\frac{1}{4}$ threads per inch with an 8-pitch lead screw.
9. To cut $2\frac{5}{8}$ threads per inch with an 8-pitch lead screw.
10. To cut $4\frac{1}{2}$ threads per inch with a 6-pitch lead screw.

20-31 Use of Compound Gearing

Sometimes it is impossible to obtain the required speed ratio between spindle and lead screw for cutting a desired thread by simple gearing with the gears available. And certain cases may require gears with an unusually large number of teeth. To obviate these possibilities, recourse is had to **compound gearing**. In this system the idler E of Fig. 20-17 is replaced by two gears G and H as shown in Fig. 20-18. One of these compound gears meshes with the change gear D and the other meshes with the lead screw gear F. The following examples will illustrate the method of computing compound gearing.

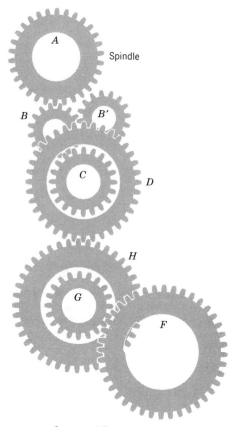

Compound Gearing

Fig. 20-18

Example 1

Find the compound gearing required to cut a screw having 35 threads per inch with a lead screw having 6 threads per inch; the ratio of spindle to stud is 1 to 1.

SOLUTION $\dfrac{\text{Lead screw}}{\text{Screw to cut}} = \dfrac{6}{35} = \dfrac{2}{5} \times \dfrac{3}{7} = \dfrac{2 \times 16}{5 \times 16} \times \dfrac{3 \times 8}{7 \times 8} = \dfrac{32}{80} \times \dfrac{24}{56}$

EXPLANATION

1. Express in the form of a fraction the ratio of the number of threads per inch in the lead screw to the number of threads per inch in the screw to be cut. This gives the fraction $\frac{6}{35}$.

2. Separate each term of this fraction into two factors: $6 = 2 \times 3$ and $35 = 5 \times 7$. This gives the two fractions $\frac{2}{5}$ and $\frac{3}{7}$.

3. Treat each of the fractions exactly as in simple gearing; that is, multiply both numerator and denominator by any convenient number that

will give an equivalent fraction whose terms will represent suitable gears. In this case the terms of the fraction $\frac{2}{5}$ are multiplied by 16, giving the fraction $\frac{32}{80}$; and the terms of the fraction $\frac{3}{7}$ are multiplied by 8, giving the fraction $\frac{24}{56}$.

4. The numerators 32 and 24 represent driving gears and the denominators 80 and 56 represent driven gears. Either of the drivers may be placed on the stud and either of the driven may be placed on the lead screw.

    If any of these four gears are not among those furnished with the lathe, use, instead of 16 and 8, such other multipliers as will give available gears.

    If the ratio of fixed stud gear to spindle is not 1 to 1, multiply the ratio of lead screw to be cut by the ratio of stud to spindle and then proceed as in Example 1 of compound gearing.

Example 2

Find the compound gearing required to cut 5 threads per inch on a lathe having a lead screw with 6 threads per inch; the ratio of fixed stud gear to spindle gear is 2 to 1.

SOLUTION $\quad \dfrac{\text{Lead screw}}{\text{Screw to cut}} = \dfrac{6}{5}$

$$\frac{\text{Fixed stud gear}}{\text{spindle gear}} = \frac{2}{1}$$

$$\frac{6}{5} \times \frac{2}{1} = \frac{12}{5}$$

$$\frac{12}{5} = \frac{3}{2} \times \frac{4}{2\frac{1}{2}} = \frac{3 \times 12}{2 \times 12} \times \frac{4 \times 12}{2\frac{1}{2} \times 12} = \frac{36}{24} \times \frac{48}{30}$$

The driving gears are 36 and 48.
The driven gears are 24 and 30.

NOTE. In giving the sizes of the compounding gears, the first mentioned is the driver; the second, the driven gear.

Problems 20-31

Find the compound gearing necessary to cut the following threads per inch on the lathe described.

NOTE. Unless otherwise stated, the ratio of spindle to stud is 1 to 1.

1. Calculate by compounding, the change gears necessary to cut 20 threads per inch in a lathe with a lead of 6.
2. Find by compounding, the change gears required for cutting $3\frac{1}{2}$ threads per inch in a lathe with a screw constant of 6.

3. Find by compounding, the change gears for cutting a $\frac{1}{24}''$-pitch thread on a lathe with a lead of 6.

4. On a lathe having 8 threads per inch on the lead screw, a 72 gear is used on the lead screw and a 36 gear on the stud. If the compound gears are 48 and 24, how many threads per inch will be cut?

5. Find by compounding, suitable gears to be used in cutting 23 threads per inch with a lead screw having 6 threads per inch.

6. What gears should be used to cut 16 threads per inch on a lathe having 8 threads per inch on the lead screw if the compounding gears are 30 and 60?

7. With a lead screw having 6 threads per inch, a 20-tooth stud gear, and an 80-tooth lead screw gear, what compounding gears could be used to cut 48 threads per inch?

8. It is desired to cut a screw with $5\frac{1}{2}$ threads per inch. If the lead screw has 6 threads per inch and the ratio of spindle to stud is 1 to 2, find the necessary change gears.

9. What change gears are required to cut $3\frac{3}{4}$ threads per inch on a lathe having a lead screw of $\frac{1}{8}''$ pitch if the compound gears are 24 and 48?

10. What change gears are required to cut a double-threaded screw of $\frac{3}{4}''$ lead if the lead screw has 6 threads per inch?

11. What change gears are required to cut $2\frac{2}{3}$ threads per inch on a lathe with a lead of 6, and 24 and 48 compound gears?

12. What change gears are required to cut $2\frac{7}{8}$ threads per inch on a lathe with a lead of 6?

13. We are required to cut a triple-threaded screw of $\frac{3}{8}''$ pitch. What change gears are required if the lead screw has 8 threads per inch?

14. We are required to cut a quadruple-threaded screw with a lead of $1\frac{1}{4}''$. Compute the compound gearing required if the lead screw has 6 threads per inch.

20-32   Cutting Metric Threads

To cut a metric thread on a lathe whose lead screw pitch is given in inches, we proceed as follows.

Example
Find the change gears for cutting a thread of 2.5-mm pitch on a lathe whose lead screw pitch is $\frac{1}{6}$ in.

SOLUTION   First find the number of threads per inch in the screw to be cut. (1 in. = 25.4 mm)

$$\text{Number of threads per inch} = \frac{25.4}{2.5} = 10\frac{4}{25}$$

$$\text{Number of threads per inch in the lead screw} = \frac{1}{\frac{1}{6}} = 6$$

$$\frac{\text{Lead screw}}{\text{Screw to cut}} = \frac{6}{10\frac{4}{25}} = \frac{6}{\frac{254}{25}} \quad 6 \times \frac{25}{254} = \frac{150}{254}$$

$$\frac{150}{254} = \frac{2}{2} \times \frac{75}{127} = \frac{2 \times 24}{2 \times 24} \times \frac{75 \times 1}{127 \times 1} = \frac{48}{48} \times \frac{75}{127}$$

Gears 48 and 75 are drivers.
Gears 48 and 127 are driven.

NOTE. A special 127-tooth gear is generally used in cutting metric threads on a lathe whose lead screw pitch is given in inches.

Problems 20-32

Find the compound gearing necessary to cut the following metric threads.

1. Compute the gears required to cut a thread of 3-mm pitch with a lead screw of 6.
2. What change gears are necessary by compounding to cut a 3.5-mm-pitch thread with a lead screw of 6?
3. It is required to cut a thread of 4-mm pitch with a lead screw of 8. Find the compound gearing necessary.
4. Calculate, by compounding, the change gears necessary for cutting a thread of 7.5-mm pitch with a lead screw of 6.
5. What change gears are required for cutting a thread of 10-mm pitch with a lead screw of 6?

Self-Test

Solve the following problems involving screw threads. Check your answer with the answers in the back of the book.

1. What is the pitch of a screw having 13 threads per inch?
2. A single threaded screw has 20 threads per inch. How far does it advance if it is turned into a nut for five complete turns?
3. A double threaded screw has 20 threads per inch. What is its lead?
4. What is the double depth of a Sharp V-thread having 18 threads per inch?
5. What is the double depth of a Unified thread having 18 threads per inch?
6. What is the minor diameter of a $\frac{1}{4}$–20 American National threaded bolt?
7. What is the tap drill size for the bolt in Problem 6?
8. What is the root diameter of a $\frac{3}{4}$–6 square thread screw?
9. Find the depth of a perfect American Standard pipe thread of $\frac{1}{14}$ in. pitch.
10. What is the ratio of the change gears required to cut a screw with 9 threads per inch if the lead screw has 6 threads per inch?

Solve the following problems involving screw threads. Use the procedures explained in this chapter.

1. What is the pitch of a screw having 20 threads per inch?
2. A single-threaded screw has 13 threads per inch. What is its pitch? What is its lead?
3. Find the depth of a V-thread having 20 threads per inch.
4. Find the double depth of a V-thread having 12 threads per inch.
5. Find the minor diameter of a $\frac{1}{4}''$–20 V-threaded bolt.
6. Find the proper tap drill size for a $\frac{5}{16}''$–18 V-thread.
7. Find the depth of an American Standard thread having 13 threads per inch.
8. Find the minor diameter of an $\frac{1}{4}''$–20 American Standard thread.
9. Find the double depth of a $\frac{3}{4}''$–6 square thread.
10. Find the depth of a perfect American Standard pipe thread of $\frac{1}{18}''$ pitch.
11. What is the ratio of the change gears required to cut a screw with 20 threads per inch if the lead screw has 6 threads per inch?
12. Find the change gear ratio for cutting $4\frac{1}{2}$ threads per inch with an 8-pitch lead screw and a spindle to stud ratio of 1 to 1.

21

Cutting Speed
and Feed

21-1 Cutting Speed and Surface, or Rim, Speed

In the interest of efficiency and economy, the machinist should run machines at various speeds, depending on the nature of the metal that is being worked, the kind of cut (that is, whether it is a roughing or a finishing cut), the feed and depth of cut, the material and shape of the cutting tool, and the strength and design of the machine. Machinery handbooks contain tables giving the proper speeds at which various metals should be worked on the different machines. Likewise, in the operation of grindstones, emery wheels, and flywheels, it is necessary to know the speed at which they may be run to obtain the best results. The proper speed for grindstones and emery wheels is determined by the strength of the stone and the nature of the grinding to be done.

Problems involving cutting speed or rim speed may be grouped in the following three classes:

1. Given the diameter of the piece of work or wheel and the revolutions per minute, to find the cutting speed or rim speed in feet per minute

469

2. Given the diameter and the speed in feet per minute, to find the number of revolutions per minute
3. Given the speed in feet per minute and the revolutions per minute, to find the diameter

21-2   Cutting Speed on the Lathe

The *cutting speed* is the rate at which a tool passes over a piece of work. It is always expressed in feet per minute. For the lathe, cutting speed may be defined as the number of linear feet on the surface of the work that passes the point of the tool in one minute. In every complete turn of a piece of work in the lathe, a distance equal to the circumference of the work passes the point of the tool. The cutting speed in feet per minute is found by multiplying the circumference of the work expressed in feet by the number of revolutions per minute.

Example
Find the cutting speed of a piece of cast iron $\frac{7}{8}$ in. in diameter turned in a lathe at 180 rpm.

SOLUTION   Cutting speed = circumference × rpm
$$= \tfrac{7}{8} \times \tfrac{22}{7} \times \tfrac{1}{12} \times 180$$
$$= 41\tfrac{1}{4} \text{ ft per minute}$$

EXPLANATION   Multiplying the diameter $\frac{7}{8}$ by $\frac{22}{7}$ gives the circumference in inches. Dividing by 12 gives the circumference in feet. That is, in each turn of the piece of work, $\frac{7}{8} \times \frac{22}{7} \times \frac{1}{12}$ feet will pass the point of the tool. And, since the piece revolves 180 times in 1 minute, multiplying $\frac{7}{8} \times \frac{22}{7} \times \frac{1}{12}$ by 180 gives the number of feet on the surface of the work that will pass the tool in 1 minute.

21-3   Surface Speed or Rim Speed

By the *surface speed* or *rim speed* of an emery wheel or flywheel is meant the number of feet that a point on the rim or circumference of the wheel travels in 1 minute. Every time that a wheel makes one complete turn, a point on its rim travels a distance equal to its circumference. The number of feet that a point on the rim will travel in a minute is equal to the number of turns that the wheel makes in 1 minute multiplied by its circumference expressed in feet.

Example
Find the surface speed of an emery wheel 8 in. in diameter making 240 rpm.

SOLUTION $8 \times \frac{22}{7} \times \frac{1}{12} \times 240 = 503$ ft per minute

EXPLANATION   In each turn of the wheel, a point on its rim travels a distance
equal to its circumference, that is, $8 \times \frac{22}{7}$. Divide this by 12 to express the
circumference in feet. Since the wheel turns 240 times in 1 minute, a point
on the rim will travel in 1 minute, 240 times the length of one circumference; therefore, multiply the circumference by 240.

Problems 21-3

Find the cutting speed in the following problems. Use $\frac{22}{7}$ as the approximation for π when finding the circumference.

1. A piece $1\frac{1}{4}''$ in diameter is turned in the lathe at 165 rpm. Find the cutting
   speed in feet per minute.
2. Calculate the cutting speed in turning a brass rod $1\frac{3}{8}''$ in diameter at 240
   rpm.
3. The diameter of a piece that is being turned is $2\frac{1}{4}''$. If the piece makes 110
   rpm, what is the cutting speed?
4. Find the cutting speed when a piece $1\frac{1}{8}''$ in diameter is turned at 120 rpm.
5. A $3''$ rod is turned in a lathe at 75 rpm. What is the cutting speed?
6. Find the surface speed of an emery wheel $12''$ in diameter making 210
   rpm.
7. An $8'$ flywheel makes 200 rpm. Find the rim speed.
8. A $12''$ emery wheel is driven by a motor that makes 2,000 rpm. A $3''$
   pulley on the motor drives an $8''$ pulley on the emery wheel spindle. Find
   the surface speed of the emery wheel in feet per minute.
9. A bar of cold-rolled machine steel $3\frac{1}{2}''$ in diameter is turned in the lathe at
   32 rpm. What is the cutting speed?
10. A brass rod $\frac{7}{8}''$ in diameter is turned in the lathe at 320 rpm. What is the
    cutting speed?

21-4   Finding the Revolutions per Minute, Given the Diameter and the Cutting Speed

Example 1

At how many revolutions per minute should a lathe be run to give a cutting
speed of 36 ft per minute when turning a rod $2\frac{1}{2}$ in. in diameter?

SOLUTION   Circumference $= \frac{5}{2} \times \frac{22}{7} \times \frac{1}{12} = \frac{55}{84}$ ft

rpm = cutting speed $\div$ circumference

$= 36 \div \frac{55}{84} = 36 \times \frac{84}{55} = 55$

EXPLANATION   Since in every turn of the piece of work a distance equal to its
circumference passes the cutting tool, for 36 ft of the surface of the work
to pass the tool, the piece must make as many times one turn as the length

of the circumference is contained in 36 ft. The rule is:

$$\text{rpm} = \text{cutting speed} \div \text{circumference}$$

Example 2

A 14-in. emery wheel (Fig. 21-1) is required to have a surface speed of 5,500 ft per minute. The wheel is driven by a 4-in. pulley on the same spindle belted to a 10-in. pulley. What must be the revolutions per minute of the 10-in. pulley?

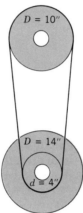

Fig. 21-1

SOLUTION

$$\text{rpm of emery wheel} = \frac{\text{rim speed}}{\text{circumference}}$$

$$\text{Circumference of wheel} = 14 \times \tfrac{22}{7} \times \tfrac{1}{12} = \tfrac{22}{6} \text{ ft}$$
$$\text{rpm of wheel} = 5500 \div \tfrac{22}{6}$$
$$= 5500 \times \tfrac{6}{22} = 1500$$
$$\text{rpm of 4-in. pulley} = 1500$$
$$\text{rpm of 10-in. pulley} = \frac{4 \times 1500}{10} = 600$$

EXPLANATION   The circumference of the emery wheel is found by multiplying its diameter by $\tfrac{22}{7}$. Divide by 12 to express the circumference in feet. This gives $\tfrac{22}{6}$ ft. Next find the revolutions per minute of the wheel by dividing the rim speed by the circumference, which gives 1,500 rpm. But since the 4-in. pulley is on the same spindle as the emery wheel, it will make the same number of revolutions per minute. Since the 10-in. pulley is belted to the 4-in. pulley, the 10-in. pulley will make $\tfrac{4}{10}$ as many revolutions per minute as the 4-in. pulley, or $\tfrac{4}{10} \times 1,500 = 600$ rpm.

Problems 21-4

Find the revolutions per minute required in the following problems. Use $\tfrac{22}{7}$ as an approximation for π when finding the circumference.

1. What should be the maximum number of revolutions per minute of a 6′ flywheel if any point on its rim should travel not more than a mile per minute?
2. At how many revolutions per minute should a 6″ gear blank be turned if the cutting speed is to be 42 ft per minute?
3. A certain grindstone may run with a rim speed of 1,200 ft per minute. What should be the revolutions per minute if the diameter of the stone is 30″?
4. At how many revolutions per minute may a 4′ grindstone be run if it can stand a surface speed of 700 ft per minute?
5. The cutting speed of a piece of work in a lathe is 250 ft per minute. If the diameter is 2½″, what must be its revolutions per minute?
6. How many revolutions per minute must a 24″ circular saw make to give a cutting speed of 3,600 ft per minute?
7. How many revolutions per minute should a steel shaft 1¾″ in diameter make if it is to be turned in the lathe with a cutting speed of 30 ft per minute?
8. A crank pin of annealed tool steel 3¾″ in diameter is turned in the lathe with a cutting speed of 20 ft per minute. How many revolutions per minute must it make?
9. The surface of a cast iron pulley 12″ in diameter is turned in the lathe with a cutting speed fo 40 ft per minute. How many revolutions per minute must it make?
10. The cutting speed on a cast iron cylinder 4½″ in diameter that is bored in a lathe is 42 ft per minute. How many revolutions per minute does it make?

21-5 Finding the Diameter, Given the Rim Speed and the Revolutions per Minute

Example
An emery wheel that is required to have a rim speed of a mile a minute is to be placed on a spindle that makes 1,600 rpm. What must be the diameter of the emery wheel?

SOLUTION $\text{Circumference} = \dfrac{\text{rim speed}}{\text{rpm}} = \dfrac{5,280}{1,600} \times 12 = 39.6 \text{ in.}$

$$\begin{aligned} \text{Diameter} &= \text{circumference} \div \tfrac{22}{7} \\ &= 39.6 \div \tfrac{22}{7} = 39.6 \times \tfrac{7}{22} \\ &= 12.6 \text{ in.} \end{aligned}$$

EXPLANATION   In a mile there are 5,280 ft. Since in 1,600 turns a point on the rim travels 5,280 ft, in one turn it will travel $\frac{1}{1600}$ of 5,280 ft. But in one turn a point on the rim travels a distance equal to the circumference; therefore, dividing the rim speed by the turns made in one minute gives the length of the circumference in feet. Multiply by 12 to change it to inches. To find the diameter, divide the circumference by $\frac{22}{7}$, giving 12.6 in. A 12-in. wheel would be used.

Problems 21-5

Solve the following problems involving rim speed and revolutions per minute. Use $\frac{22}{7}$ as an approximation for π.

1. An emery wheel having an allowable surface speed of 4,500 ft per minute makes 1,500 rpm. What is its diameter?
2. A flywheel making 330 rpm has a rim speed of a mile per minute. What is its diameter?
3. What must be the diameter of a circular saw to give a cutting speed of 3,000 ft per minute if it revolves at the rate of 400 rpm?
4. An emery wheel which is required to have a surface speed of 5,000 ft per minute is driven by a 2,000-rpm motor. What must be the diameter of the emery wheel if the motor pulley is 3″ in diameter and the pulley on the emery wheel spindle is 8″ in diameter?
5. If the motor in Problem 4 is used to drive a 12″ emery wheel, what must be the size of the pulley on the emery wheel spindle if a speed of 4,500 ft per minute is required?
6. What is the diameter of a rod that makes 48 rpm and has a cutting speed of 36 ft per minute?
7. What is the diameter of a crank that makes 24 rpm and has a cutting speed of 20 ft per minute?
8. A brass rod is turned in the lathe at 400 rpm with a cutting speed of 75 ft per minute. What is the diameter?
9. The bearings at the ends of a crank shaft are turned in a lathe at 18 rpm with a cutting speed of 30 ft per minute. What is the diameter?
10. The rim speed of a pulley making 50 rpm is 150 ft per minute. What is the diameter of the pulley?

21-6   Drill Press and Milling Machine

The cutting speed of a drill is the number of feet traveled in 1 minute by the outer corners of the cutting edges.

Example 1

A 2-in. hole is drilled in a piece of work with the drill revolving at the rate of 240 rpm. What is the cutting speed?

SOLUTION   Cutting speed = circumference of drill × rpm
$$= 2 \times \tfrac{22}{7} \times \tfrac{1}{12} \times 240$$
$$= 126 \text{ ft per minute}$$

EXPLANATION　In each complete revolution of the drill, the outer corners of the cutting edges travel a distance equal to the circumference of the drill. Since the drill makes 240 rpm, the outer corners of the cutting edges will travel 240 circumferences in 1 minute. Multiplying the diameter by $\frac{22}{7}$ gives $2 \times \frac{22}{7}$. Dividing by 12 gives the circumference in feet. Multiplying by 240 gives the speed in feet per minute.

RULE _____

For milling machines, the cutting speed is found by multiplying the circumference of the cutter by the revolutions per minute.

Example 2
Find the cutting speed of a 5-in. milling cutter making 30 rpm.

SOLUTION　Cutting speed = circumference × rpm
$$= 5 \times \tfrac{22}{7} \times \tfrac{1}{12} \times 30$$
$$= 39\tfrac{2}{7} \text{ ft per minute}$$

EXPLANATION　Multiply the diameter by $\frac{22}{7}$ to get the circumference in inches. Dividing by 12 gives the circumference in feet. Multiplying by 30 (the revolutions per minute) gives the cutting speed in feet per minute.

To find the revolutions per minute of a drill when the diameter and the cutting speed are known, proceed as in the following example.

Example
How many revolutions per minute must a $1\frac{1}{2}$-in. drill make in order to have a cutting speed of 30 ft per minute?

SOLUTION　　　　　rpm = cutting speed ÷ circumference
$$\text{Circumference} = 1\tfrac{1}{2} \times \tfrac{22}{7} \times \tfrac{1}{12} = \tfrac{11}{28} \text{ ft}$$
$$\text{rpm} = 30 \div \tfrac{11}{28} = 30 \times \tfrac{28}{11} = 76\tfrac{4}{11}$$

EXPLANATION　In each turn of the drill, the outer corners of the cutting edges travel a distance equal to one circumference. Find the circumference in feet by multiplying the diameter by $\frac{22}{7}$ and dividing by 12. This gives $\frac{3}{2} \times \frac{22}{7} \times \frac{1}{12} = \frac{11}{28}$ ft. That is, while the drill makes one revolution, the outer corners of the cutting edges travel $\frac{11}{28}$ ft. If the corners are to travel 30 ft in a minute, the drill must revolve as many times as $\frac{11}{28}$ is contained in 30. Dividing 30 by $\frac{11}{28}$, we get $30 \times \frac{28}{11} = 76\frac{4}{11}$ rpm.

To find the revolutions per minute of the milling cutter for a desired cutting speed when the diameter of the cutter is known, proceed exactly as in finding the revolutions per minute of a drill when the diameter and the cutting speed are given; that is, divide the cutting speed by the circumference of the cutter.

To find the size of cutter to use with a given number of revolutions per minute and cutting speed, proceed as in the following example.

Example

What must be the diameter of a milling cutter to have a cutting speed of 24 ft per minute when it is run at 15 rpm?

SOLUTION   Circumference = cutting speed ÷ rpm

$$= \tfrac{24}{15} \times 12 = 19.2 \text{ in.}$$

$$\text{Diameter} = \text{circumference} \div \tfrac{22}{7}$$

$$= 19.2 \times \tfrac{7}{22} = 6.1 \text{ in.}$$

Use a 6-in. cutter.

Problems 21-6

Solve the following cutting speed problems. Use $\tfrac{22}{7}$ as an approximation for π.

1. A $1\tfrac{3}{4}''$ hole is bored with a drill running at 80 rpm. Find the cutting speed.
2. What is the cutting speed of a drill boring a $\tfrac{5}{8}''$ hole at 220 rpm?
3. Find the cutting speed of a $2\tfrac{3}{4}''$ milling cutter making 75 rpm.
4. Calculate the cutting speed of a $3\tfrac{1}{2}''$ milling cutter making 120 rpm.
5. A milling cutter $2\tfrac{1}{4}''$ in diameter makes 50 rpm. What is the cutting speed?
6. What is the cutting speed of a $1\tfrac{7}{8}''$ drill making 60 rpm?
7. Compute the cutting speed of a $1\tfrac{1}{2}''$ drill making 72 rpm.
8. Find the cutting speed of a $5\tfrac{1}{2}''$ milling cutter making 40 rpm.
9. A $2''$ milling cutter makes 100 rpm. Calculate the cutting speed.
10. What is the cutting speed of a $\tfrac{3}{4}''$ drill making 128 rpm?

Find the revolutions per minute in the following problems. Use $\tfrac{22}{7}$ as an approximation for π.

11. A $\tfrac{7}{8}''$ hole is bored in the drill press with a cutting speed of 32 ft per minute. Find the revolutions per minute of the drill.
12. A $1\tfrac{1}{4}''$ drill is boring a hole in a piece of iron with a cutting speed of 35 ft per minute. How many revolutions per minute must it make?
13. How many revolutions per minute must a $7\tfrac{1}{2}''$ milling cutter make to have a cutting speed of 50 ft per minute?
14. An $8''$ milling cutter has a cutting speed of 38 ft per minute. Find the revolutions per minute.
15. At what speed must a $\tfrac{3}{4}''$ drill be run to have a cutting speed of 25 ft per minute?
16. How fast must a $9''$ milling cutter be run to have a cutting speed of 48 ft per minute?
17. What must be the revolutions per minute of a $2\tfrac{3}{4}''$ milling cutter to have a speed of 120 ft per minute?

18. How many revolutions per minute does a $1\frac{3}{16}''$ drill make if its cutting speed is 33 ft per minute?

19. A $2\frac{1}{4}''$ drill has a cutting speed of 50 ft per minute. At how many revolutions per minute is the drill running?

20. A $6\frac{1}{2}''$ milling cutter has a cutting speed of 44 ft per minute. Compute the revolutions per minute.

Find the diameters of the following milling cutters. Use $\frac{22}{7}$ as an approximation for π.

21. What size milling cutter should be used at 40 rpm to give a cutting speed of 40 ft per minute?

22. A milling cutter having a cutting speed of 60 ft per minute is making 120 rpm. What is its diameter?

23. Find the size of the milling cutter running at 46 rpm with a cutting speed of 46 ft per minute.

24. What is the diameter of a milling cutter having a cutting speed of 28 ft per minute and making 52 rpm?

25. What must be the diameter of a milling cutter if it is to have a cutting speed of 90 ft per minute and run at 45 rpm?

21-7   Cutting Speed on the Planer

In figuring the *actual* or *net cutting speed* on the planer, take into account the time occupied by the idle return stroke. If a planer has a forward stroke of 16 ft per minute and a return stroke of 32 ft per minute, the return stroke will take half as much time as the forward stroke. Assuming the cut to be 16 ft long, the forward stroke will take 1 minute and the return $\frac{1}{2}$ minute. The complete stroke, cut and return, will take $1\frac{1}{2}$ minutes. But since it takes $1\frac{1}{2}$ minutes to cut 16 ft, it will cut in 1 minute a distance equal to $16 \div 1\frac{1}{2} = 16 \times \frac{2}{3} = 10\frac{2}{3}$ ft, which is called the net or actual cutting speed of the planer. That is, the tool in the forward stroke moves at the rate of 16 ft per minute while in contact with the work, but because of the time lost by the idle return stroke, the actual or net result is only an average of $10\frac{2}{3}$ ft cut for every minute that the planer is operating.

Example
What is the net or actual cutting speed of a planer making 10 strokes per minute, each 2 ft 6 in. long, if the reverse is 3 to 1? Find the theoretical cutting speed.

SOLUTION   2 ft 6 in. $= 2\frac{1}{2}$ ft

Distance traveled in one
direction in one minute $= 10 \times 2\frac{1}{2} = 25$ ft $=$ net cutting speed

Since the reverse is 3 to 1, the forward stroke will take up $\frac{3}{4}$ of each minute, whereas the return takes $\frac{1}{4}$. That is, it takes the planer $\frac{3}{4}$ of a minute to travel the 25 ft. In 1 minute the theoretical cutting speed would be

$$25 \div \tfrac{3}{4} = 25 \times \tfrac{4}{3} = 33\tfrac{1}{3} \text{ ft}$$

Problems 21-7

Find the actual and the theoretical cutting speeds of the following planers.

1. A planer has a forward stroke of 18 ft per minute and a return twice as fast. Find the actual and the theoretical cutting speeds.
2. A planer with a 4 to 1 return makes 12 strokes a minute, each stroke 27" long. Find the net and the theoretical cutting speeds.
3. Find the actual and the theoretical cutting speeds of a planer making 14 strokes per minute; the length of stroke is 3′6″ and the return stroke 3 to 1.
4. What are the net and the theoretical cutting speeds of a planer having a forward stroke of 24 ft per minute and a return of 2 to 1?
5. With a return of 2 to 1 and a stroke of 4′3″, what are the net and the theoretical cutting speeds of a planer making 8 strokes a minute?
6. A planer with a 3 to 1 return makes 15 strokes per minute. Find the actual and the theoretical cutting speeds if the stroke is 28″ long.

21-8   Time Required for a Job on the Planer

Example

Find the time required for one complete cut on a casting 1 ft 3 in. × 3 ft 6 in. if the cutting speed is 20 ft per minute, the feed $\tfrac{1}{4}$ in. per stroke, and the return 2 to 1. Allow a clearance of 3 in. at each end.

NOTE.   A feed of $\tfrac{1}{4}$ in. per stroke means that for each forward stroke there is a sidewise motion of the tool of $\tfrac{1}{4}$ in.

SOLUTION   Total length of forward stroke = 3 ft 6 in. + clearance
$$= 3 \text{ ft } 6 \text{ in.} + 6 \text{ in.} = 4 \text{ ft } 0 \text{ in.}$$

Since the cutting speed is 20 ft per minute, whereas the length of the forward stroke is only 4 ft, the forward stroke will take only $\tfrac{4}{20}$, or $\tfrac{1}{5}$ of a minute. The return stroke, being twice as fast, will take $\tfrac{1}{2}$ of $\tfrac{1}{5}$, or $\tfrac{1}{10}$ of a minute.

The time required for one complete stroke = $\tfrac{1}{5} + \tfrac{1}{10} = \tfrac{3}{10}$ of a minute. Since the feed is $\tfrac{1}{4}$ in., four strokes will be required for every inch of width of the casting.

Total number of strokes = 15 × 4 = 60.

Since every complete stroke takes $\tfrac{3}{10}$ of a minute, 60 strokes will take 60 × $\tfrac{3}{10}$ = 18 minutes.

Problems 21-8

Find the time required for one complete cut in each of the following planer problems.

1. Find the time required for taking a complete cut on a plate $2' \times 3'$ if the cutting speed is 30 ft per minute, the return is 4 to 1, and the feed $\frac{1}{8}''$. The clearance at each end is $3''$.
2. A planer has a forward movement of 24 ft a minute and a return of 3 to 1. How long will it take for a complete cut on a casting $1'8'' \times 5'6''$ with a feed of $\frac{3}{16}''$? Allow $3''$ at each end for tool clearance.
3. How long will it take to finish one face of a casting $2'6'' \times 4'0''$ if the planer has a forward stroke of 20 ft per minute, a return of 3 to 1, a roughing feed of $\frac{1}{16}''$, and a finishing feed of $\frac{1}{8}''$? Allow $3''$ at each end for the tool to clear.
4. With a forward stroke of 36 ft per minute, a return of 2 to 1, and a feed of $\frac{3}{32}''$, how long will it take for one complete cut on a face $15'' \times 3'4''$? Allow $3''$ at each end for tool clearance.
5. Find the time required for finishing a casting $2'6'' \times 6'0''$ on a planer having a forward movement of 40 ft per minute and a return of 2 to 1, if a feed of $\frac{1}{8}''$ is used for roughing and $\frac{1}{4}''$ for finishing. Clearance for the tool is $3''$ at each end.

21-9   Number of Strokes per Minute

To find the number of strokes per minute for a given length of stroke and given cutting and return speeds, proceed as in the following example.

Example

Find the number of strokes per minute a planer must make if the forward or cutting speed is 28 ft per minute, the return 2 to 1, and the length of stroke 8 ft including clearance.

SOLUTION   Time required for forward stroke $= \frac{8}{28} = \frac{2}{7}$ minute
Time required for return stroke $= \frac{1}{2}$ of $\frac{2}{7} = \frac{1}{7}$ minute
Time required for complete stroke $= \frac{1}{7} + \frac{2}{7} = \frac{3}{7}$ minute

Since each complete stroke takes $\frac{3}{7}$ of a minute, in 1 minute the planer will make $1 \div \frac{3}{7}$, or $1 \times \frac{7}{3} = 2\frac{1}{3}$ strokes.

Problems 21-9

Find the number of strokes per minute in the following planer problems.

NOTE.   In each of the following problems, the given length of stroke includes the clearance for the tool at both ends.

1. Find the number of strokes per minute of a planer having a forward stroke of 24 ft per minute, a return of 2 to 1, and a length of stroke of $6'$.

2. How many strokes per minute must a planer make if its forward movement is 30 ft per minute, its return 2 to 1, and length of stroke 4′?
3. A planer has a cutting stroke of 20 ft per minute, a return of 3 to 1, and a length of stroke of 5′6″. How many strokes will it make in a minute?
4. Compute the number of strokes per minute for a cut 6′6″ long if the cutting speed of the planer is 24 ft per minute and the return 2 to 1.
5. The forward movement of a planer is 32 ft per minute and the return is 2 to 1. Compute the number of strokes per minute required for a stroke 5′ long.

21-10 Cutting Speed of Shapers

The cutting speed of a **geared shaper** is obtained in exactly the same way as the cutting speed of a planer.

The method of computing the cutting speed of a **crank shaper** is illustrated in the following example.

Example
Find the cutting speed of a shaper making 30 rpm with a reverse of 2 to 1 if the length of stroke is 14 in.

SOLUTION

$$\text{Length of stroke} = 14 \text{ in.} = 1\tfrac{1}{6} \text{ ft}$$
$$\text{Total length of metal cut in one minute} = 30 \times 1\tfrac{1}{6} = 35 \text{ ft}$$
$$\text{Cutting speed} = 35 \times \tfrac{3}{2} = 52.5 \text{ ft per minute}$$

EXPLANATION Since there is a stroke for each revolution, there will be 30 strokes per minute. The actual total length of metal cut per minute is 30 times the length of stroke, or $30 \times 1\tfrac{1}{6} = 35$ ft. That is, 35 ft is the actual net cutting speed. But since the return is twice as fast as the forward stroke, the return takes $\tfrac{1}{3}$ of the minute, whereas the forward stroke takes $\tfrac{2}{3}$ of the minute. That is, 35 ft of metal are cut in $\tfrac{2}{3}$ of a minute. For a whole minute the cutting speed of the shaper is $35 \div \tfrac{2}{3} = 35 \times \tfrac{3}{2} = 52.5$ ft per minute.

Problems 21-10

Find the cutting speed in the following shaper problems.

1. A crank shaper with a 2 to 1 return makes 28 strokes of 1′ length in a minute. Find the cutting speed.
2. Find the cutting speed of a crank shaper making 25 rpm with a reverse of 3 to 1 if the length of stroke is 10″.
3. With a return of 2 to 1 and a speed of 27 rpm, what is the cutting speed of a crank shaper if the length of stroke is 15″?
4. What is the cutting speed of a crank shaper making 32 rpm with a 2 to 1 return if the length of stroke is 14″?

5. A crank shaper having a 3 to 1 return makes 28 rpm and the length of cut is 9″. What is the cutting speed?

21-11   Cutting Feed of a Lathe

The *cutting feed* of a lathe is the distance the tool travels sideways along the work while the piece of work makes one revolution. If, for example, the tool moves $\frac{1}{25}$ in. while the piece turns around once, the feed is $\frac{1}{25}$ in. Sometimes the feed is stated as so many revolutions of the piece for every inch of travel of the tool. A feed of 25 means that the work makes 25 revolutions while the tool travels 1 in.

Example 1
A piece of work revolves 60 times while the tool travels $\frac{3}{4}$ in. Find the feed.

SOLUTION 1   Feed = $60 \div \frac{3}{4} = 60 \times \frac{4}{3} = 80$

EXPLANATION   Since the piece revolves 60 times while the tool travels $\frac{3}{4}$ in., for the tool to move a whole inch, the piece will revolve $60 \times \frac{4}{3}$, or 80 times.

SOLUTION 2   Feed = $\frac{1}{60}$ of $\frac{3}{4} = \frac{1}{80}$ in.

EXPLANATION   Since in 60 turns of the piece of work the tool moves $\frac{3}{4}$ in., in one turn of the work, the tool will travel $\frac{1}{60}$ of $\frac{3}{4}$, or $\frac{1}{80}$ in.

Example 2
With a feed of 28, how many revolutions are required to take a cut over a piece of work 8 in. long?

SOLUTION   Total revolutions required = $8 \times 28 = 224$

EXPLANATION   Since the piece must revolve 28 times for each inch of travel of the tool, for 8 in. it must revolve 8×28, or 224 times.

Example 3
Find the time required to take one cut over a piece of work 14 in. long and $1\frac{1}{2}$ in. in diameter if the cutting speed is 32 ft per minute and the feed 20.

SOLUTION
$$\text{Circumference of work} = \tfrac{22}{7} \times 1\tfrac{1}{2} = \tfrac{33}{7} \text{ in.}$$
$$\text{Cutting speed in inches} = 32 \times 12 = 384 \text{ in.}$$
$$\text{rpm} = \text{cutting speed} \div \text{circumference}$$
$$= 384 \div \tfrac{33}{7} = 384 \times \tfrac{7}{33} = \tfrac{896}{11}$$
$$\text{Total revolutions required for one cut} = 14 \times 20 = 280$$
$$\text{Time required for one cut} = \text{total revolutions} \div \text{rpm}$$
$$= 280 \div \tfrac{896}{11} = 280 \times \tfrac{11}{896}$$
$$= 3.4 \text{ minutes}$$

Find the cutting feed of the lathes in the following problems.

1. Find the feed of a lathe if the tool travels $3\frac{1}{4}''$ while the work makes 120 rpm.
2. With a feed of 30, how many revolutions are required to take a cut over a piece of work $6\frac{1}{2}''$ long?
3. A shaft 28″ long is being turned in a lathe at a speed of 48 rpm and a feed of 25. Find the time required for one cut.
4. Find the time required for one cut over a piece of work $1\frac{3}{4}''$ in diameter and 20″ long that is being turned in a lathe with a cutting speed of 36 and a feed of 18.
5. With a feed of 32 and a cutting speed of 24 ft per minute, how long will it take for a cut over a piece $2\frac{1}{2}''$ in diameter and 8″ long?
6. A piece of tool steel $1\frac{7}{8}''$ in diameter and 15″ long is being turned in a lathe. If the cutting speed is 24 ft per minute and the feed $\frac{1}{10}''$, find the time required for one cut.
7. With a feed of 22 and a cutting speed of 27 ft per minute, how many inches of a piece of 2″ steel will be turned in 10 minutes?
8. How long will it take for one cut over a $3\frac{1}{8}''$ rod 18″ long with a cutting speed of 35 ft per minute and a feed of $\frac{1}{25}''$?
9. A lathe tool moves $3\frac{1}{4}''$ along the work in 1 minute while the lathe makes 250 rpm. What is the feed?

21-12 Drill Press Feed

The feed of a drill press is the distance the drill penetrates the work in one revolution. If the drill revolves 100 times while it penetrates a distance of 1 in., the feed is $\frac{1}{100}$ in. per revolution.

Example

Find the time required to drill through a piece of work 3 in. thick with a drill making 240 rpm if the feed per revolution is 0.007 in.

SOLUTION   Penetration in 1 minute = 240 × 0.007
                                        = 1.680 in.
        Time to drill through 3 in. = 3 ÷ 1.68
                                    = 1.786 minutes
                                    = 1 minute 47 seconds

EXPLANATION   Since the drill penetrates 0.007 in. for each turn, in a minute it will advance 240 times 0.007 in., or 1.68 in. To drill through 3 in. it will take as many times 1 minute as 1.68 in. is contained in 3 in., or 3 ÷ 1.68 = 1.786 minutes.

Problems 21-12

Find the time required to do the following drill press operations.

1. With a speed of 250 rpm and a feed of 0.018″ per revolution, how long will it take to drill through a piece of cast iron $4\frac{1}{4}$″ thick?

2. A $1\frac{1}{8}$″ drill making 280 rpm is fed 0.016″ per revolution. Find the time required to drill through a piece $3\frac{7}{16}$″ thick.

3. How long will it take to drill 24 holes in a plate $1\frac{1}{8}$″ thick with a drill making 225 rpm and a feed of 0.015″ per revolution? Allow $\frac{1}{2}$ minute for adjusting the plate after each hole is drilled.

4. With a feed of 0.025″ and a speed of 280 rpm how long will it take to drill through a piece $3\frac{7}{8}$″ thick?

5. At a speed of 300 rpm and a feed of 0.02″, how long will it take to drill through a piece $4\frac{1}{2}$″ thick?

21-13 Milling Machine Feed

The feed of the milling machine is the distance traveled by the table while the cutter makes one revolution. If a milling cutter revolves 80 times while the table moves 1 in., the feed is $\frac{1}{80}$ in.

Example 1

Find the feed per minute of a milling cutter if the feed is 0.02 in. per revolution and the cutter makes 120 rpm.

SOLUTION   Feed per minute = $120 \times 0.02 = 2.4$ in.

EXPLANATION   Since the cutter makes 120 revolutions in a minute, it will cut 120×0.02 in., or 2.4 in. in a minute.

Example 2

How long will it take for a cutter to traverse a piece of work 10 in. long if it makes 32 rpm and the feed is $\frac{1}{40}$ in.?

SOLUTION   Total revolutions required = $10 \times 40 = 400$
Time required = $400 \div 32$
$= 12.5$ minutes

EXPLANATION   With a feed of $\frac{1}{40}$ in., 40 revolutions of the cutter are required for each inch traversed. Since the cutter makes 32 revolutions in 1 minute, to make 400 revolutions it will take $400 \div 32$, or 12.5 minutes.

Example 3

A 2-in. milling cutter makes 210 rpm. If the table feed is 7 in. per minute, find the cutting speed. Find the feed per tooth if the cutter has 14 teeth.

SOLUTION   Cutting speed = circumference × rpm
$$= 2 \times \tfrac{22}{7} \times \tfrac{1}{12} \times 210 = 110 \text{ ft per minute}$$
$$\text{Feed per revolution} = 7 \div 210 = \tfrac{7}{210} = 0.0333 \text{ in.}$$
$$\text{Feed per tooth} = 0.0333 \div 14 = 0.0024 \text{ in.}$$

EXPLANATION   The cutting speed is found by multiplying the circumference by the revolutions per minute. Since the table moves 7 in. in the time that the cutter makes 210 revolutions, in one revolution of the cutter the table will move $\tfrac{7}{210}$ of 7 in., or $\tfrac{1}{30}$ in. Since there are 14 teeth in the cutter, the feed per tooth will be $\tfrac{1}{14}$ of the feed per revolution, or $\tfrac{1}{14}$ of 0.0333 in. = 0.0024 in.

Problems 21-13

Solve the following problems involving the milling machine.

1. A milling cutter revolves 75 times while the table moves 1″. Find the feed per revolution.
2. A milling cutter making 150 rpm has a feed of 0.02″. What is the feed per minute?
3. What is the feed per revolution of a milling cutter which makes 80 rpm with a table feed of $6\tfrac{1}{2}$″? What is the feed per tooth if the cutter has 10 teeth?
4. A $\tfrac{3}{4}$″ end mill makes 210 rpm and the table feed is $8\tfrac{1}{2}$″ per minute. Find the cutting speed. Find the feed per tooth if the end mill has 8 teeth.
5. A milling cutter makes 100 rpm with a table feed of $5\tfrac{3}{8}$″. What is the feed per revolution? What is the feed per tooth if the cutter has 12 teeth?
6. A $\tfrac{3}{4}$″ cutter with 10 teeth makes 200 rpm with a table feed of $8\tfrac{3}{4}$″. Find the cutting speed and the feed per tooth.
7. A milling cutter $2\tfrac{1}{4}$″ in diameter has 12 teeth. If the cutting speed is 75 ft per minute, find the feed per minute when every tooth is allowed to cut a chip 0.005″ thick.
8. A milling cutter 3″ in diameter has 15 teeth. If it makes 48 rpm and the table feed is $2\tfrac{3}{4}$″ per minute, find the thickness of the chip cut by each tooth.
9. A $\tfrac{7}{8}$″ end mill with 8 teeth has a surface speed of 35 ft per minute. What is the feed per minute if each tooth cuts a chip 0.004″ thick?

Self-Test

Solve the following problems involving cutting speed and feed. Check your answers with the answers in the back of the book. Use $\tfrac{22}{7}$ for π.

1. Find the cutting speed of a piece of steel $1\tfrac{1}{4}$ in. in diameter turned in a lathe at 200 rpm.

2. An 18-in. flywheel makes 400 rpm. What is its rim speed in feet per minute?

3. A 12-in. emery wheel is to have a rim speed of 6,000 ft per minute. What is its rpm?

4. A cut-off silicon carbide blade, 8 in. in diameter, is not to exceed 7,000 rpm. What is its maximum rim speed in feet per minute?

5. What is the cutting speed of a 4-in. milling cutter making 40 rpm?

6. What must be the diameter of a milling cutter to have a cutting speed of 30 ft per minute when it is run at 12 rpm?

7. What is the actual cutting speed of a planer making 12 strokes per minute, each 2 ft long, if the reverse is 3 to 1?

8. What is the theoretical cutting speed of the planer in Problem 7?

9. With a speed of 400 rpm and a feed of 0.012 in. per revolution, how long will it take to drill through 6 in. of steel?

10. A milling cutter revolves 100 times while the table moves 1 in. What is the feed per revolution?

Chapter Test

Solve the following problems involving cutting speed and feed.

1. Find the cutting speed of a piece of brass $1\frac{1}{2}''$ in diameter turned in a lathe at 220 rpm.

2. Find the surface speed of an emery wheel $6''$ in diameter making 1,750 rpm.

3. At how many revolutions per minute should a lathe be run to give a cutting speed of 40 ft per minute when turning a shaft $3\frac{1}{8}''$ in diameter?

4. A $10''$ emery wheel is required to have a surface speed of 6,000 ft per minute. How fast must the emery wheel turn in rpm?

5. If the rim speed of an emery wheel should be 5,600 ft per minute and the spindle speed is 1,750 rpm, what should be the diameter of the emery wheel?

6. A brass rod is turned in a lathe at 450 rpm with a cutting speed of 80 ft per minute. What is the diameter of the brass rod?

7. A $1\frac{1}{2}''$ hole is drilled in a piece of work with the drill revolving at 280 rpm. What is the cutting speed?

8. Find the cutting speed of a $10''$ milling cutter making 40 rpm.

9. How many revolutions per minute must a $2''$ drill make in order to have a cutting speed of 25 ft per minute?

10. What must be the diameter of a milling cutter to have a cutting speed of 30 ft per minute when it is run at 18 rpm?

11. Find the actual cutting speed of a planer making 20 strokes per minute, each $3'6''$ long, if the reverse is 3 to 1?

12. Find the theoretical cutting speed of the planer in Problem 11.

13. Find the time required to complete one cut on a casting $3'6''$ long and

1′8″ wide, if the cutting speed is 20 ft per minute, the feed is $\frac{1}{4}$″ per stroke and the return is 2 to 1. Allow 3″ clearance at each end.

14. Find the cutting speed of a shaper making 30 rpm with a reverse of 2 to 1 if the length of the stroke is 16″.

15. A piece of work revolves 80 times while the tool travels $\frac{1}{2}$″. Find the feed.

16. A shaft 16″ long is being turned in a lathe at a speed of 60 rpm and a feed of 20. Find the time required for one cut.

17. With a feed of 24 and a cutting speed of 30 ft per minute, how many inches of a piece of 2″ steel will be turned in 10 minutes?

18. With a feed of 0.025″ and a speed of 300 rpm, how long will it take to drill through a piece $4\frac{1}{2}$″ thick?

19. If the feed were 0.0125″ in Problem 18, how much faster will the drill have to turn in order to drill the piece in the same amount of time?

20. A milling cutter revolves 100 times while the table moves 1″. Find the feed per revolution.

22

Gears

22-1 Spur Gears

Two round steel disks like those shown in Fig. 22-1 will drive each other if pressed close together. This is called a *friction drive* and may be used where little power is to be transmitted. It is quite evident, however, that if any considerable power is required, the two cylindrical disks will slip on each other. To prevent this slippage and make a positive drive, teeth are provided on each disk and corresponding grooves for the teeth to enter. Such cylindrical disks or wheels are called *gear wheels* or *gears*. A pair of gears is shown in Fig. 22-2.

Pitch Circle

The circle representing the disk on which the teeth are built is called the *pitch circle*. It is not visible at all on the actual gear, yet it remains the most important element in the design of the gear (see Figs. 22-3 and 22-4).

Pitch Diameter

The diameter of the pitch circle is called the *pitch diameter* of the gear.

Fig. 22-1   Fig. 22-2

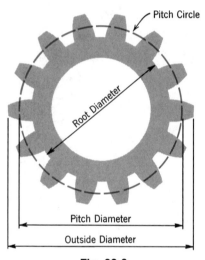

Fig. 22-3

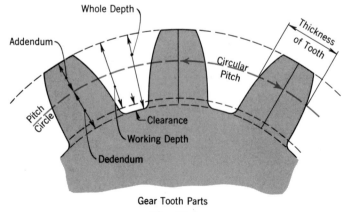

Gear Tooth Parts

Fig. 22-4

Gear Tooth Parts
The parts of a gear tooth are illustrated in Fig. 22-4.

Addendum, Dedendum, and Clearance
The projection of the gear tooth above the pitch circle is called the **addendum**. The part of the tooth below the pitch circle is called the **dedendum**. Its depth is equal to that of the addendum plus an additional depth called the **clearance**.

Whole Depth and Working Depth
The sum of the addendum and the dedendum constitutes the **whole depth**. This is also called the **cutting depth** because it is the depth of cut required in cutting the gear from the blank. The whole depth minus the clearance is called the **working depth**, that is, the depth to which the teeth of the mating gear enter. The part of the working depth below the pitch circle is equal to the depth of the addendum. The working depth is equal to twice the depth of the addendum. The whole depth of a gear tooth is therefore equal to two addenda plus the clearance.

Outside Diameter
The **outside diameter**, or the diameter of the blank, is equal to the pitch diameter plus twice the addendum (see Fig. 22-3).

Root Diameter
The **root diameter**, or the diameter between the bottoms of the teeth, is equal to the outside diameter minus twice the whole depth.

Circular Pitch
The distance P_c between the centers of two adjoining teeth is called the **circular pitch**. It is measured on the circumference of the pitch circle and is equal to that circumference divided by the number of teeth. If we call the diameter of the pitch circle D and the number of teeth N, then the circumference of the pitch circle $= \pi \times D$, and the circular pitch $= \dfrac{\pi \times D}{N}$, or, $P_c = \dfrac{\pi \times D}{N}$.

Also, since $N \times P_c =$ the pitch circumference, we have

$$\text{Pitch diameter} = \frac{\text{pitch circumference}}{\pi} \quad \text{or} \quad D = \frac{N \times P_c}{\pi}$$

Also, since $\dfrac{\text{pitch circumference}}{\text{circular pitch}} =$ number of teeth, we have

$$\text{Number of teeth}, \ N = \frac{\pi \times D}{P_c}$$

The **thickness of tooth** is half the circular pitch.

Example 1

Find the circular pitch of a gear which has 48 teeth and a pitch diameter of 6 in.

SOLUTION $P_c = \dfrac{\text{pitch circumference}}{\text{number of teeth}} = \dfrac{\pi \times D}{N}$

$$= \frac{3.1416 \times 6}{48} = 0.3927 \text{ in.}$$

Example 2

Find the pitch diameter of a gear that has 45 teeth and a circular pitch of 0.3142 in.

SOLUTION   Pitch diameter $= \dfrac{\text{circumference of pitch circle}}{\pi}$

$$D = \frac{N \times P_c}{\pi} = \frac{45 \times 0.3142}{3.1416} = 4.5 \text{ in.}$$

Example 3

Find the number of teeth in a gear which has a pitch diameter of $12\frac{1}{2}$ in. and a circular pitch of 0.7854 in.

SOLUTION   Number of teeth $= \dfrac{\text{circumference of pitch circle}}{P_c}$

$$N = \frac{\pi \times D}{P_c} = \frac{3.1416 \times 12.5}{0.7854} = 50 \text{ teeth}$$

Example 4

Find the thickness of tooth of the gear in Example 3.

SOLUTION   Thickness of tooth $= \dfrac{P_c}{2} = \dfrac{0.7854}{2} = 0.3927 \text{ in.}$

Problems 22-1

Solve the following problems involving spur gears.

1. Find the circular pitch of a gear that has 60 teeth and a pitch diameter of $7\frac{1}{2}''$.
2. What is the circular pitch of a 70-tooth gear if the pitch diameter is $3\frac{1}{2}''$?
3. A gear that has a pitch diameter of $22''$ has 55 teeth. Find the circular pitch.
4. Find the circular pitch of a 70-tooth gear with a $40''$ pitch diameter.
5. Find the circular pitch of a 36-tooth gear with a $4''$ pitch diameter.

6. A gear has 65 teeth and 0.6283″ circular pitch. Find the pitch diameter.
7. A gear has 31 teeth and a circular pitch of 1.5708″. Find the pitch diameter.
8. Find the pitch diameter of an 82-tooth gear whose circular pitch is 0.1963″.
9. A gear has 30 teeth and 0.3927″ circular pitch. Find the pitch diameter.
10. Find the pitch diameter of a 45-tooth gear whose circular pitch is 2.0944″.
11. How many teeth has a gear whose circular pitch is 3.1416″ and whose pitch diameter is 82″?
12. Find the number of teeth on a gear with a pitch diameter of 48″ and a circular pitch of 2.5133″.
13. Find the number of teeth on a gear with a pitch diameter of 4″ and a circular pitch of 0.6283″.
14. Find the number of teeth on a gear with a pitch diameter of $5\frac{1}{2}$″ and a circular pitch of 0.2244″.
15. Find the number of teeth on a gear with a pitch diameter of $2\frac{1}{4}$″ and a circular pitch of 0.2618″.
16. Find the thickness of tooth in the gears in Problems 11 to 15.

22-2 Diametral Pitch

In modern gear design, the circular pitch is used very little. Another pitch, called the diametral pitch, is commonly used because of its convenience in computation. When we speak of the pitch of a gear, we mean diametral pitch unless circular pitch is specified. The *diametral pitch* is the number of teeth per inch of pitch diameter.

Diametral pitch is not a distance, as pitch is commonly understood. For instance, if a gear has a pitch diameter of 3 in. and 24 teeth, it has $\frac{24}{3}$ or 8 teeth for every inch of its pitch diameter; that is, its diametral pitch is 8. It is called an 8-pitch gear. It will mesh only with other 8-pitch gears. A 60-tooth, 8-pitch gear has a pitch diameter of $\frac{60}{8}$ or $7\frac{1}{2}$ in.; a 40-tooth, 8-pitch gear has a pitch diameter of $\frac{40}{8}$ or 5 in.

Since the diametral pitch means the number of teeth per inch of pitch diameter, the number of teeth on a gear is equal to the pitch diameter times the pitch. A 6-pitch gear with a pitch diameter of $5\frac{1}{2}$ in. will have $5\frac{1}{2} \times 6$, or 33 teeth. This relation between pitch, number of teeth, and pitch diameter simplifies gear computation to a great extent.

Using the following letters: D for pitch diameter, P for diametral pitch, and N for number of teeth, we have, by definition:

$$\text{Diametral pitch} = \frac{\text{number of teeth}}{\text{pitch diameter}} \quad \text{or} \quad P = \frac{N}{D}$$

Also,

$$\text{Number of teeth} = \text{pitch diameter} \times \text{pitch} \quad \text{or} \quad N = D \times P$$

Also,

$$\text{Pitch diameter} = \frac{\text{number of teeth}}{\text{diametral pitch}} \quad \text{or} \quad D = \frac{N}{P}$$

Example 1
Find the diametral pitch of a gear that has 72 teeth and a pitch diameter of $4\frac{1}{2}$ in.

SOLUTION $P = \dfrac{N}{D} = \dfrac{72}{4\frac{1}{2}} = 16$. The gear is a 16-pitch gear.

Example 2
What is the pitch diameter of a 12-pitch gear with 45 teeth?

SOLUTION $D = \dfrac{N}{P} = \dfrac{45}{12} = 3\frac{3}{4}$. The pitch diameter is $3\frac{3}{4}$ in.

Example 3
How many teeth must a 5-pitch gear have if its pitch diameter is 6.2 in.?

SOLUTION $N = D \times P = 6.2 \times 5 = 31$ teeth. The gear has 31 teeth.

Problems 22-2

Solve the following problems involving diametral pitch and circular pitch.

1. Find the diametral pitch of a 68-tooth gear whose pitch diameter is 17″.
2. Find the diametral pitch of a 40-tooth gear whose pitch diameter is 16″.
3. Find the diametral pitch of a 30-tooth gear whose pitch diameter is $2\frac{1}{2}$″.
4. Find the diametral pitch of a 65-tooth gear whose pitch diameter is $3\frac{1}{4}$″.
5. Find the diametral pitch of a 124-tooth gear whose pitch diameter is $7\frac{3}{4}$″.
6. How many teeth are there on a 12-pitch gear whose pitch diameter is $5\frac{1}{2}$″?
7. How many teeth are there on a 9-pitch gear whose pitch diameter is 11″?
8. How many teeth are there on a 20-pitch gear whose pitch diameter is 9.1″?
9. How many teeth are there on a 4-pitch gear whose pitch diameter is $16\frac{3}{4}$″?
10. How many teeth are there on a 16-pitch gear whose pitch diameter is $3\frac{7}{8}$″?
11. Find the pitch diameter of a 70-tooth, 5-pitch gear.
12. Find the pitch diameter of a 48-tooth, 9-pitch gear.
13. Find the pitch diameter of an 18-tooth, 8-pitch gear.
14. Find the pitch diameter of a 45-tooth, $2\frac{1}{2}$-pitch gear.
15. Find the pitch diameter of a 27-tooth, $1\frac{1}{2}$-pitch gear.

22-3 Proportions of Gear Teeth

In standard gears, *the addendum is the reciprocal of the pitch,* that is, addendum = 1/pitch. In a 10-pitch gear the addendum is $\frac{1}{10}$ in.; in a 4-pitch gear the addendum is $\frac{1}{4}$ in. The dedendum is equal in depth to the addendum plus the clearance. The *clearance* is usually stated as $\frac{1}{20}$ of the circular pitch.

22-4 Relation between Circular Pitch and Diametral Pitch

It is often necessary to convert circular pitch into diametral pitch, and vice versa. The circular pitch, being part of the circumference of the pitch circle, is found by dividing that circumference by the number of teeth in the gear.

Let D = pitch diameter
N = number of teeth
P = pitch
P_c = circular pitch

$$\text{Pitch circumference} = \pi \times \text{pitch diameter} = \pi \times D$$

$$\text{Circular pitch} = \frac{\text{pitch circumference}}{\text{number of teeth}}$$

that is,

$$P_c = \frac{\pi \times D}{N}$$

But

$$N = D \times P$$

Writing this in place of N, we get

$$P_c = \frac{\pi \times D}{P \times D}$$

from which, by cancellation, we get the formula for converting diametral pitch into circular pitch:

$$P_c = \pi/P$$

To convert circular pitch into diametral pitch we proceed as follows:

$$\text{Diametral pitch} = \frac{\text{number of teeth}}{\text{pitch diameter}}$$

that is,

$$P = N/D$$

But

$$N = \frac{\pi \times D}{P_c}$$

Putting this value in place of N in the formula, we get

$$P = \frac{\pi \times D}{P_c \times D}$$

which, by cancellation of D, becomes the formula for converting circular pitch into diametral pitch:

$$P = \pi / P_c$$

Example 1
Find the circular pitch of a 12-pitch gear.

SOLUTION $P_c = \dfrac{\pi}{P} = \dfrac{3.1416}{12} = 0.2618$ in.

Example 2
Find the diametral pitch of a gear whose circular pitch is 0.6283 in.

SOLUTION $P = \dfrac{\pi}{P_c} = \dfrac{3.1416}{0.6283} = 5$

Problems 22-4

Find the circular pitch of the following gears.

1. 6-pitch gear
2. $2\frac{1}{2}$-pitch gear
3. 10-pitch gear
4. $1\frac{1}{4}$-pitch gear
5. 18-pitch gear
6. 20-pitch gear
7. $1\frac{1}{2}$-pitch gear
8. 3-pitch gear
9. 8-pitch gear
10. 14-pitch gear

Find the diametral pitch of the following gears.

11. A gear whose circular pitch is 1.2566″
12. A gear whose circular pitch is 0.1571″
13. A gear whose circular pitch is 0.3491″
14. A gear whose circular pitch is 1.7952″
15. A gear whose circular pitch is 1.5708″
16. A gear whose circular pitch is $\frac{3}{4}$″
17. A gear whose circular pitch is $1\frac{1}{2}$″
18. A gear whose circular pitch is 0.3927″
19. A gear whose circular pitch is 0.1963″
20. A gear whose circular pitch is 2.5133″

22-5   Clearance

The *clearance,* which is usually stated as $\frac{1}{20}$ of the circular pitch, can also be expressed in terms of the diametral pitch.

$$\text{Clearance} = \tfrac{1}{20} \text{ of } P_c = \tfrac{1}{20} \text{ of } \pi/P$$

$$= \frac{1}{20} \times \frac{3.1416}{P} = \frac{0.157}{P}$$

Example 1
Find the clearance of a gear whose circular pitch is 0.3491 in.

SOLUTION   Clearance $= \dfrac{P_c}{20} = \dfrac{0.3491}{20} = 0.0175$ in.

Example 2
Find the clearance of a 12-pitch gear.

SOLUTION   Clearance $= \dfrac{0.157}{P} = \dfrac{0.157}{12} = 0.0131$ in.

Problems 22-5

Find the clearance in each of the following gears.

1. 16-pitch gear   2. 12-pitch gear
3. $2\tfrac{1}{2}$-pitch gear   4. 8-pitch gear
5. $1\tfrac{3}{4}$-pitch gear
6. A gear whose circular pitch is 1.5708"
7. A gear whose circular pitch is 0.2244"
8. A gear whose circular pitch is 0.6283"
9. A gear whose circular pitch is 0.2618"
10. A gear whose circular pitch is 0.7500"

22-6   Depth of Tooth

The whole depth of tooth = addendum + dedendum

$$= \frac{1}{P} + \frac{1}{P} + \frac{0.157}{P} = \frac{2.157}{P}$$

Since

$$\text{Dedendum} = \text{addendum} + \text{clearance}$$

then

$$\text{Whole depth} = \text{addendum} + \text{addendum} + \text{clearance}$$

Example 1
Find the whole depth of tooth of an 8-pitch gear.

SOLUTION   Whole depth $= \dfrac{2.157}{P} = \dfrac{2.157}{8} = 0.2696$ in.

Example 2
Find the whole depth of tooth of a gear whose circular pitch is 0.6283 in.

SOLUTION
$$P = \frac{\pi}{P_c} = \frac{3.1416}{0.6283} = 5$$

$$\text{Whole depth} = \frac{2.157}{P} = \frac{2.157}{5} = 0.4314 \text{ in.}$$

Problems 22-6

Find the whole depth of tooth in each of the following gears.

1. 16-pitch gear
2. 12-pitch gear
3. $2\frac{1}{2}$-pitch gear
4. 8-pitch gear
5. $1\frac{3}{4}$-pitch gear
6. A gear whose circular pitch is 1.5708″
7. A gear whose circular pitch is 0.2244″
8. A gear whose circular pitch is 0.6283″
9. A gear whose circular pitch is 0.2618″
10. A gear whose circular pitch is 0.7500″

22-7 Outside Diameter

The outside diameter of a gear = pitch diameter + 2 × addendum

$$= D + \frac{1}{P} + \frac{1}{P}$$

$$= D + \frac{2}{P}$$

Example
Find the outside diameter of an 8-pitch gear whose pitch diameter is $5\frac{1}{2}$ in.

SOLUTION   Outside diameter $= D + 2/P = 5\frac{1}{2} + \frac{2}{8} = 5\frac{1}{2} + \frac{1}{4} = 5\frac{3}{4}$ in.

Problems 22-7

Find the outside diameters of the following gears.

1. A 16-pitch gear with a pitch diameter of $8\frac{1}{4}$″
2. A $2\frac{1}{2}$-pitch gear whose pitch diameter is 24″
3. A 16-pitch gear whose pitch diameter is 1.875″
4. A 10-pitch gear whose pitch diameter is 3.1″
5. A 20-pitch gear whose pitch diameter is $9\frac{1}{4}$″
6. A 4-pitch gear whose pitch diameter is 12″

7. A 12-pitch gear with a pitch diameter of $3\frac{1}{4}''$
8. A 9-pitch gear whose pitch diameter is 5.333"
9. A 14-pitch gear whose pitch diameter is $11\frac{1}{2}''$
10. A 6-pitch gear whose pitch diameter is 3.667"

22-8 Pitch Diameter

To find the pitch diameter when the outside diameter and the pitch are given, use the following method:

Pitch diameter = outside diameter − 2 × addendum

Using symbol OD for outside diameter, we have the formula

$$D = OD - 2/P$$

The formula for finding the pitch diameter when the outside diameter and the number of teeth are given is as follows:

$$D = \frac{N \times OD}{N + 2}$$

This formula states that we must multiply the outside diameter by the number of teeth and divide that product by the number of teeth plus 2.

Example 1
Find the pitch diameter of an 8-pitch gear whose outside diameter is $6\frac{1}{4}$ in.

SOLUTION $D = OD - 2/P = 6\frac{1}{4} - \frac{2}{8} = 6\frac{1}{4} - \frac{1}{4} = 6$ in.

Example 2
Find the pitch diameter of a 66-tooth gear whose outside diameter is $8\frac{1}{2}$ in.

SOLUTION $D = \frac{N \times OD}{N + 2} = \frac{66 \times 8\frac{1}{2}}{66 + 2} = \frac{561}{68} = 8\frac{1}{4}$ in.

Problems 22-8

Find the pitch diameter of the following gears.

1. A gear that has 58 teeth and an outside diameter of $7\frac{1}{2}''$
2. A 40-tooth gear whose outside diameter is 16.8"
3. A gear which has 114 teeth and an outside diameter of 9.667"
4. A 31-tooth gear whose outside diameter is 3.3"
5. A 57-tooth gear whose outside diameter is $14\frac{3}{4}''$
6. An 18-pitch gear whose outside diameter is $5\frac{1}{2}''$
7. A 14-pitch gear whose outside diameter is 3.286"
8. A $2\frac{1}{2}$-pitch gear whose outside diameter is 40.8"

9. A 6-pitch gear whose outside diameter is 15″
10. A 12-pitch gear whose outside diameter is 7.417″

22-9   Use of Formulas

You will notice that frequently there are several formulas for finding a required item. You must select the formula that contains the information given. For instance, to find the pitch diameter of a gear whose outside diameter is $7\frac{3}{4}$ in. and that has 60 teeth, we select the formula that gives D in terms of OD and N, namely,

$$D = \frac{OD \times N}{N + 2}$$

When the circular pitch is given, it will often be convenient to convert it into diametral pitch and continue the calculation in terms of diametral pitch.

Example
Find the depth of cut and size of blank for a 6-pitch gear of 80 teeth.

SOLUTION   Depth of cut is the same as the whole depth.

$$\text{Depth of cut} = \frac{2.157}{P} = \frac{2.157}{6} = 0.3595 \text{ in.}$$

Size of blank is the same as the outside diameter.

$$\text{Size of blank} = \frac{N + 2}{P} = \frac{80 + 2}{6} = \frac{82}{6} = 13.667 \text{ in.}$$

Problems 22-9

Compute the missing dimensions of the gears in the following chart.

	Pitch	Pitch Diameter	Number of Teeth	Circular Pitch	Addendum	Clearance	Dedendum	Working Depth	Whole Depth	Outside Diameter
1.	10	$2\frac{1}{2}$″								
2.	16		25							
3.	12									$4\frac{2}{3}$″
4.			32							$8\frac{1}{2}$″
5.		3.417″	41							
6.		3.8″		0.1571″						
7.			55	1.2566″						
8.				0.3927″						$9\frac{1}{2}$″
9.	$1\frac{1}{2}$		90							
10.	14		75							

FORMULAS FOR GEAR COMPUTATION

To find	Symbol	Having	Formula
Pitch	P	Circular pitch	$P = \pi/P_c$
	P_c	Pitch	$P_c = \pi/P$
Circular pitch	P_c	Pitch diameter and number of teeth	$P_c = \dfrac{\pi \times D}{N}$
	P	Pitch diameter and number of teeth	$P = N/D$
Pitch	P	Outside diameter and number of teeth	$P = \dfrac{N + 2}{OD}$
	D	Number of teeth and pitch	$D = N/P$
	D	Number of teeth and circular pitch	$D = \dfrac{N \times P_c}{\pi}$
Pitch diameter	D	Outside diameter and number of teeth	$D = \dfrac{OD \times N}{N + 2}$
	D	Outside diameter and pitch	$D = OD - 2/P$
	OD	Number of teeth and pitch	$OD = \dfrac{N + 2}{P}$
	OD	Pitch diameter and pitch	$OD = D + 2/P$
Outside diameter	OD	Number of teeth and circular pitch	$OD = \dfrac{(N + 2) \times P_c}{\pi}$
	OD	Pitch diameter and circular pitch	$OD = D + 2P_c/\pi$
Number of teeth	N	Pitch diameter and pitch	$N = D \times P$
	N	Pitch diameter and circular pitch	$N = \dfrac{D \times \pi}{P_c}$
Thickness of tooth	T	Circular pitch	$T = P_c/2$
Addendum	Add	Pitch	$Add = 1/P$
	Add	Circular pitch	$Add = P_c/\pi$
Dedendum	Ded	Pitch	$Ded = 1.157/P$
	Ded	Circular pitch	$Ded = 0.3683P_c$
Clearance	Cl	Circular pitch	$Cl = P_c/20$
	Cl	Pitch	$Cl = 0.157/P$
Whole depth	WD	Pitch	$WD = 2.157/P$
	WD	Circular pitch	$WD = 0.6866 \times P_c$
Working depth	wd	Pitch	$wd = 2/P$
	wd	Circular pitch	$wd = \dfrac{2 \times P_c}{\pi}$

22-10 Racks

The proportions of the teeth in a rack are the same as in a gear. The linear pitch of the rack is the same as the circular pitch of the gears that mesh with the rack. (See Fig. 22-5, which shows the tooth proportions for a rack to mesh with a 6-pitch gear.)

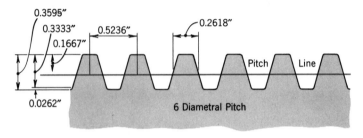

Fig. 22-5

Example 1
Find the linear pitch of a 5-pitch rack.

SOLUTION   Linear pitch = circular pitch

$$P_c = \frac{\pi}{P} = \frac{3.1416}{5} = 0.6283 \text{ in.}$$

Example 2
Find the depth of cut for a 12-pitch rack.

SOLUTION   Depth of cut = whole depth $= \dfrac{2.157}{P} = \dfrac{2.157}{12} = 0.1798$ in.

<hr>

Problems 22-10

1. Find the whole depth of tooth of a 4-pitch rack.
2. Find the linear pitch of a 3-pitch rack.
3. A 6-pitch rack is driven by a 27-tooth pinion. Find the distance traveled for each turn of the pinion.
4. A rack is required to have a linear pitch of ¾″. What must be the pitch diameter of a 20-tooth pinion to mesh with the rack?
5. Find the pitch diameter of a 30-tooth pinion to mesh with a rack whose linear pitch is 1⅛″.

22-11 Center-to-Center Distance of Gears

Since the pitch circles of gears in mesh are supposed to be in contact, the distance from the center of one gear to the center of the other gear is equal to the sum of the pitch radii. In other words, the center-to-center distance of a pair of gears in mesh is equal to half the sum of the pitch diameters (Fig. 22-6).

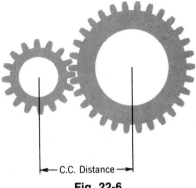

←— C.C. Distance —→

Fig. 22-6

In a rack and pinion, the distance from the center of the pinion to the pitch line of the rack is equal to half the pitch diameter of the pinion (Fig. 22-7).

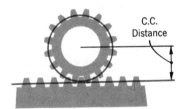

C.C. Distance

Fig. 22-7

Example 1

Two gears in mesh have pitch diameters of 8 in. and $3\frac{1}{4}$ in., respectively. Find their center-to-center distance.

SOLUTION   Center-to-center distance $= \dfrac{8 + 3\frac{1}{4}}{2} = \dfrac{11\frac{1}{4}}{2} = 5\frac{5}{8}$ in.

Example 2

Find the center-to-center distance of a pair of 8-pitch gears having 42 teeth and 75 teeth, respectively.

SOLUTION

$$75\text{-tooth gear, } D = \frac{N}{P} = \frac{75}{8} = 9\frac{3}{8} \text{ in.}$$

$$42\text{-tooth gear, } D = \frac{N}{P} = \frac{42}{8} = 5\frac{1}{4} \text{ in.}$$

$$\text{Center-to-center distance} = \frac{9\frac{3}{8} + 5\frac{1}{4}}{2} = \frac{14\frac{5}{8}}{2} = 7\frac{5}{16} \text{ in.}$$

Problems 22-11

Find the center-to-center distance of the following gears.

1. Find the center-to-center distance of a pair of gears whose pitch diameters are $7\frac{1}{2}''$ and $5\frac{1}{4}''$.
2. A pair of 5-pitch gears have 45 teeth and 62 teeth, respectively. Find the distance center to center.
3. Find the distance center to center of a 32-tooth gear and a 100-tooth gear of 12-pitch.
4. A pair of gears of $2\frac{1}{2}$-pitch have 30 teeth and 45 teeth, respectively. Find the distance center to center.
5. A 35-tooth, 14-pitch pinion meshes with a rack. Find the distance from center of pinion to pitch line of rack.
6. Find the distance from the center of an 8-pitch, 21-tooth pinion to the pitch line of a rack which is driven by this pinion.
7. A pair of gears of 45 teeth and 75 teeth, respectively, have a center-to-center distance of $12''$. Find their pitch diameters.
8. Find the pitch diameters of a pair of gears of 49 teeth and 72 teeth if the distance center to center is $4.321''$.
9. Gear A has 40 teeth and an outside diameter of $7''$; gear B has 18 teeth and an outside diameter of $3.333''$. What should be the distance center to center when properly adjusted?
10. Find the center-to-center distance of the following pair of gears: gear A has 90 teeth and an outside diameter of $7\frac{2}{3}''$; gear B has 30 teeth and an outside diameter of $2\frac{2}{3}''$.

22-12 Selection of Cutters

Because the shape of the tooth changes with the number of teeth on a gear, cutters are made in sets, usually eight to the set for each pitch. The following table gives the cutter number to select for different numbers of teeth. Theoretically a different-shaped cutter is required for each number of teeth, but for ordinary work the approximation is close enough if we use the cutters within the limits listed in the table.

TABLE OF CUTTERS
Use No. 8 gear cutter for cutting gears with 12 or 13 teeth.
Use No. 7 gear cutter for cutting gears with 14 to 16 teeth.
Use No. 6 gear cutter for cutting gears with 17 to 20 teeth.
Use No. 5 gear cutter for cutting gears with 21 to 25 teeth.
Use No. 4 gear cutter for cutting gears with 26 to 34 teeth.
Use No. 3 gear cutter for cutting gears with 35 to 54 teeth.
Use No. 2 gear cutter for cutting gears with 55 to 134 teeth.
Use No. 1 gear cutter for cutting gears with 135 teeth to a rack.

Problems 22-12

Select cutters for all the gears in the chart of Problem Set 22-9.

22-13 Bevel Gears

Bevel gears differ from spur gears in that the teeth are cut on the surface of a
cone instead of a cylinder. Figure 22-8 shows two frustums of cones in

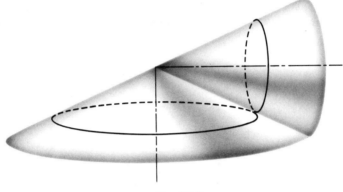

Fig. 22-8

contact. Such frustums of cones, if pressed firmly together, will drive each
other. To secure a positive drive, teeth are cut on the surface of these cones,
producing a pair of bevel gears, as shown in Fig. 22-9.

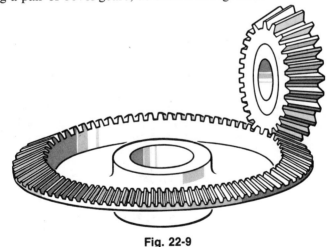

Fig. 22-9

22-14 Definitions Applying to Bevel Gears—Pitch Cones

The cones that form the basis of the bevel gears are called the *pitch cones,*
and their diameters are the *pitch diameters* of the bevel gears. The pitch

cones are not visible at all on the finished gears, but they are the most important elements in the design.

The *addendum* is the projection of the tooth above the pitch cone surface.

The depth of the tooth below the pitch cone surface is called the *dedendum*. The depth of the dedendum is equal to the depth of the addendum plus the *clearance*.

The terms *addendum, dedendum, clearance, pitch diameter, diametral pitch, number of teeth, circular pitch, thickness of tooth,* and *width of space at pitch line* have the same meaning as in spur gears and are computed in exactly the same way. For further definitions in reference to bevel gears it will be necessary to refer to Fig. 22-10.

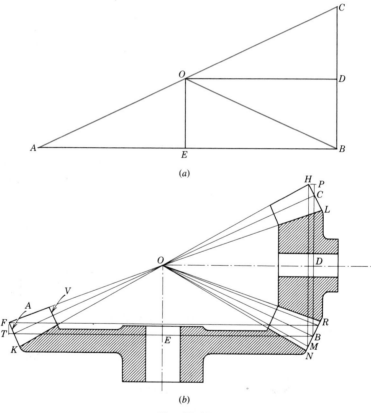

Fig. 22-10

Figure 22-10*a* shows the pitch cones of a pair of bevel gears in mesh; the axes are at right angles to each other. Figure 22-10*b* shows a cross section through the bevel gears built on these cones. We will call the large one the *gear* and the small one the *pinion*.

$$AB = \text{pitch diameter of the gear}$$
$$AC = \text{pitch diameter of the pinion}$$

FR = outside diameter of the gear
HM = outside diameter of the pinion
$OA = OB = OC$ = pitch cone radius
Angle EOB = pitch cone angle of the gear
Angle DOB = pitch cone angle of the pinion
Angle AOF = angle COH = addendum angle
(same for both gear and pinion)
Angle AOK = angle LOC = dedendum angle
(same for both gear and pinion)
AT = angular addendum of gear
CP = angular addendum of pinion
Angle EOF = turning angle of gear
Angle DOH = turning angle of pinion
Angle EOK = cutting angle of gear
Angle DOL = cutting angle of pinion

The following relations can be seen from the construction of the figure:

Turning angle = pitch cone angle + addendum angle
Cutting angle = pitch cone angle − dedendum angle
Outside diameter = pitch diameter + twice the angular addendum

The following example will serve to illustrate the principles of bevel gear design.

Example
Compute the dimensions of a pair of 4-pitch bevel gears with shafts at right angles, the gear to have 60 teeth and the pinion 18 teeth.

SOLUTION   To find pitch diameters:

$$\text{Pitch diameter of gear} = \frac{N}{P} = \frac{60}{4} = 15 \text{ in.}$$

$$\text{Pitch diameter of pinion} = \frac{N}{P} = \frac{18}{4} = 4\tfrac{1}{2} \text{ in.}$$

To find pitch cone angles, construct the pitch cones by drawing $AB =$ 15 in. at the right angles to $BC = 4\tfrac{1}{2}$ in. Connect A and C, and from the center O of AC, draw OB and the perpendiculars OE and OD (Fig. 22-11).

Angle BOE is the pitch cone angle of the gear; angle BOD is the pitch cone angle of the pinion.

$$EB = \tfrac{1}{2} \text{ of } AB = \tfrac{1}{2} \text{ of } 15 \text{ in.} = 7\tfrac{1}{2} \text{ in.}$$
$$BD = \tfrac{1}{2} \text{ of } BC = \tfrac{1}{2} \text{ of } 4\tfrac{1}{2} \text{ in.} = 2\tfrac{1}{4} \text{ in.}$$

$$\text{Tangent } BOE = \frac{BE}{OE} = \frac{7\tfrac{1}{2}}{2\tfrac{1}{4}} = 3.33333$$

Therefore angle $BOE = 73°18'$ = pitch cone angle of gear.

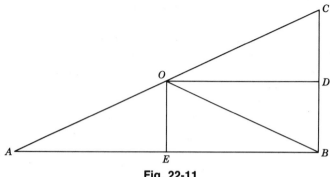

Fig. 22-11

$$\text{Tangent } BOD = \frac{BD}{OD} = \frac{2\frac{1}{4}}{7\frac{1}{2}} = 0.30000$$

Therefore angle $BOD = 16°42' =$ pitch cone angle of pinion.

Proof
Sum of pitch cone angle $= 73°18' + 16°42' = 90°$

Problems 22-14

Compute the pitch diameters and the pitch cone angles of the following pairs of bevel gears at right angles. Draw the pitch cone diagram in each instance.

1. Gear 60 teeth, pinion 25 teeth, pitch 4.
2. Gear 44 teeth, pinion 24 teeth, pitch 8.
3. Gear 72 teeth, pinion 36 teeth, pitch 6.
4. Gear 35 teeth, pinion 30 teeth, pitch 5.
5. Gear 50 teeth, pinion 25 teeth, pitch 10.
6. Gear 54 teeth, pinion 30 teeth, pitch 3.
7. Gear 100 teeth, pinion 27 teeth, pitch 5.
8. Gear 62 teeth, pinion 40 teeth, pitch 8.
9. Gear 45 teeth, pinion 36 teeth, pitch 4.
10. Gear 75 teeth, pinion 24 teeth, pitch 6.

22-15   Tooth Parts

Continuing the example of the preceding section and referring to Fig. 22-10:

$$\text{Addendum} = \frac{1}{P} = \frac{1}{4} = 0.25 \text{ in.} = AF \text{ in Fig. 22-10}b$$

$$\text{Clearance} = \frac{0.157}{P} = \frac{0.157}{4} = 0.0393 \text{ in.}$$

$$\text{Dedundum} = \text{depth under pitch cone}$$
$$= \text{addendum} + \text{clearance}$$
$$= \frac{1}{P} + \frac{0.157}{P} = \frac{1.157}{P} = \frac{1.157}{4} = 0.2893 \text{ in.}$$

Whole depth of tooth at pitch circle
$$= 0.25 + 0.25 + 0.0393 = 0.5393 \text{ in.}$$
$$= FK \text{ in Fig. } 22\text{-}10b$$

$$\text{Circular pitch at pitch line} = \frac{\pi}{P} = \frac{3.1416}{4} = 0.7854 \text{ in.}$$

$$\text{Thickness of tooth} = \text{space between teeth}$$
$$= \frac{0.7854}{2} = 0.3927 \text{ in.}$$

Problems 22-15

1–10. Compute the addendum, clearance, dedendum, whole depth, depth under pitch cone, circular pitch, thickness of tooth, and space between teeth in each of the ten pairs of bevel gears in Problem Set 22-14.

22-16   Pitch Cone Radius

In Fig. 22-11, the pitch cone radius
$$\sin \text{ angle } EOA = AE/OA$$
$$OA = \frac{AE}{\sin \text{ angle } EOA}$$
$$OA = \frac{7.5}{\sin 73°18'} = \frac{7.5}{0.95782} = 7.83 \text{ in.}$$

22-17   Addendum Angle and Turning Angle

Since the line FK in Fig. 22-10b is perpendicular to the pitch cone radius, then FA/FO = the tangent of the addendum angle AOF; that is,

$$\text{Tangent of addendum angle} = \frac{\text{addendum}}{\text{pitch cone radius}}$$
$$= \frac{0.25}{7.83} = 0.03193$$

$$\text{Addendum angle } AOF = 1°50'$$

The turning angle is equal to the pitch cone angle plus the addendum angle.

Turning angle of the gear = 73°18′ + 1°50′ = 75°08′
Turning angle of the pinion = 16°42′ + 1°50′ = 18°32′

22-18   Dedendum Angle and Cutting Angle

$$\text{Tangent of dedendum angle } AOK = \frac{KA}{AO} = \frac{\text{depth under pitch cone}}{\text{pitch cone radius}}$$

$$= \frac{0.2893}{7.83} = 0.03695$$

Dedendum angle AOK = 2°07′
Cutting angle = pitch cone angle minus
dedendum angle
Cutting angle of gear = 73°18′ − 2°07′ = 71°11′
Cutting angle of pinion = 16°42′ − 2°07′ = 14°35′

Problems 22-18

1–10. Compute the pitch cone radius, addendum angle, turning angle, dedendum angle, and cutting angle for the ten pairs of bevel gears in Problem Set 22-14.

22-19   Outside Diameters

The outside diameter = pitch diameter + twice the angular addendum.

Outside Diameter of Gear
In triangle ATF (Fig. 22-10b), AF is the addendum, 0.25 in.; angle TAF is equal to the pitch cone angle of the gear, EOA = 73°18′.

Angular addendum $AT = AF \times \cos TAF = 0.25 \times \cos 73°18′$
$= 0.25 \times 0.28736 = 0.072$ in.
Outside diameter of gear = 15 + 2 × 0.072 = 15 + 0.144
= 15.144 in.

Outside Diameter of Pinion
In triangle PCH, CH is the addendum, 0.25 in.; angle PCH = pitch cone angle of the pinion DOC, which is 16°42′.

Angular addendum $PC = CH \times \cos PCH = 0.25 \times \cos 16°42′$
$= 0.25 \times 0.95782 = 0.239$ in.
Outside diameter of pinion = 4½ + 2 × 0.239 = 4.5 + 0.478
= 4.978 in.

1–10. Compute the outside diameters of the ten pairs of bevel gears in Problem Set 22-14.

22-20 Selecting Cutters for Bevel Gears

Bevel gears may be cut on the milling machine with spur gear cutters of the same pitch. The cutters, however, are not selected for the number of teeth on the bevel gear but for an imaginary number of teeth computed as follows. In Fig. 22-12, showing the pitch cones of the bevel gears under consideration, the line XY is drawn perpendicular to the pitch cone radius OB; that is, the line XY is a prolongation of the edge of tooth RM in 22-10b. The center lines of the gear and the pinion are prolonged to meet this line.

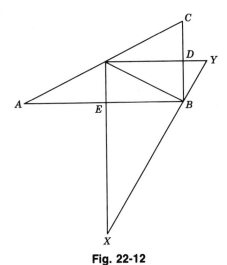

Fig. 22-12

$BY = BO \times$ tangent angle BOY
$\quad\quad =$ pitch cone radius $\times$ tangent of pitch cone angle of pinion
$\quad\quad = 7.83 \times \tan 16°42'$
$\quad\quad = 7.83 \times 0.30001 = 2.349$ in.
$BX = BO \times$ tangent angle BOX
$\quad\quad =$ pitch cone radius $\times$ tangent of pitch cone angle of gear
$\quad\quad = 7.83 \times \tan 73°18'$
$\quad\quad = 7.83 \times 3.33317 = 26.099$ in.

BY is the pitch radius of the imaginary pinion for which to select the cutter.

$\quad\quad$ Diameter $= 2 \times 2.349 = 4.698$ in.
Number of teeth $=$ pitch diameter $\times$ pitch $= 4.698 \times 4$
$\quad\quad\quad\quad\quad\quad\quad\quad = 18.792$ teeth

The pinion is to be cut with a 19-tooth cutter, that is, a No. 6 cutter.

BX is the pitch radius of the imaginary gear for which to select the cutter.

Diameter = 2 × 26.1 = 52.2 in.
Number of teeth = pitch diameter × pitch = 52.2 × 4
= 208.8 teeth

The gear is to be cut with a 209-tooth cutter, that is, a No. 1 cutter.

Problems 22-20

1–10. Select the cutters for the ten pairs of bevel gears in Problem Set 22-14.

22-21 Dimensions of Teeth at Small End

The *width* or *face* of the teeth, *AV* in 22-10*b*, is usually made about one-quarter of the pitch cone radius. The dimensions of the teeth at the small end are proportional to their dimensions at the large end. The ratio between the small end dimensions and the large end dimensions

$$= \frac{\text{pitch cone radius at small end}}{\text{pitch cone radius at large end}}$$

The pitch cone radius is 7.83 in.; therefore, let us assume a width of face of 2 in., making the pitch cone radius at the small end of the teeth *OV* = 7.83 − 2.00 = 5.83 in. Then the ratio between the small pitch cone radius and the large pitch cone radius = 5.83/7.83 = 0.745. We can now find the tooth proportions at the small end by multiplying the large end dimensions by the factor 0.745.

Addendum at small end = 0.745 × 0.25 = 0.1862 in.
Clearance at small end = 0.745 × 0.0393 = 0.0293 in.
Dedendum at small end = 0.745 × 0.2893 = 0.2155 in.
Whole depth at small end = 0.745 × 0.5393 = 0.4018 in.
Circular pitch at small end = 0.745 × 0.7854 = 0.5851 in.
Thickness of tooth at small end = space at small end
= 0.745 × 0.3927 = 0.2926 in.

Problems 22-21

1–10. Compute the addendum, clearance, dedendum, whole depth, circular pitch, thickness of tooth, and space at the small ends of the teeth of the ten pairs of bevel gears in Problem Set 22-14. In each instance assume a tooth face of approximately ¼ of the pitch cone radius but not more than 0.3 of the pitch cone radius.

22-22 Miter Gears

When the gear and the pinion are both the same size and their axes are at right angles, they are called *miter gears.* The computation in this case is greatly simplified; all elements are the same for both gears.

The pitch cone angle is 45°.

$$\text{Pitch cone radius} = \frac{\text{pitch radius}}{\sin 45°} = \frac{\text{pitch diameter}}{2 \times \sin 45°}$$

$$= \frac{\text{pitch diameter}}{2 \times 0.70711} = \frac{\text{pitch diameter}}{1.41422}$$

$$= 0.707 \times \text{pitch diameter}$$

$$\text{Face angle} = 45° + \text{addendum angle}$$

$$\text{Cutting angle} = 45° - \text{dedendum angle}$$

$$\text{Angular addendum} = \text{addendum} \times \cos 45°$$

$$= 0.70711 \times \text{addendum}$$

Self-Test

Solve the following problems involving gears. Check your answers with the answers in the back of the book.

1. Find the circular pitch of a gear that has 64 teeth and a pitch diameter of 8 in.
2. Find the pitch diameter of a gear that has 60 teeth and a circular pitch of 0.2356 in.
3. Find the number of teeth on a gear with a pitch diameter of 8 in. and a circular pitch of 0.5236 in.
4. What is the diametral pitch of a gear with 64 teeth and a pitch diameter of 4 in.?
5. What is the pitch diameter of a 16-pitch gear with 48 teeth?
6. What is the circular pitch of an 8-pitch gear?
7. Find the clearance of an 8-pitch gear.
8. Find the whole depth of tooth of a 12-pitch gear.
9. Find the outside diameter of a 12-pitch gear with a pitch diameter of 4 in.
10. Two gears in mesh have pitch diameters of 8 in. and 6 in., respectively. Find their center-to-center distance.

Chapter Test

Solve the following problems involving gears.

1. Find the circular pitch of a gear that has 64 teeth and a pitch diameter of 6".
2. Find the pitch diameter of a gear that has 60 teeth and a circular pitch of 0.3927".

TABLE OF RULES FOR BEVEL GEARS

To Find	Rule
Pitch diameter	Pitch diameter $= N/P$
Pitch cone angles	$\text{Tan pitch cone angle, gear} = \dfrac{\text{pitch diameter, gear}}{\text{pitch diameter, pinion}}$ $\text{Tan pitch cone angle, pinion} = \dfrac{\text{pitch diameter, pinion}}{\text{pitch diameter, gear}}$
Pitch cone radius	$\text{Pitch cone radius} = \dfrac{\frac{1}{2}\text{-pitch diameter of gear}}{\text{sine pitch cone angle, gear}}$
Tooth parts, large end. Addendum	Addendum $= 1/P$
Clearance	Clearance $= 0.157/P$
Dedendum or depth below pitch cone	Depth below pitch cone $= 1.157/P$
Whole depth	Whole depth $= 2.157/P$
Circular pitch at pitch line	Circular pitch $= \pi/P$
Thickness of tooth or space at pitch line	Thickness $=$ space $= P_c/2$
Tooth dimensions at small end	Multiply large end dimensions by following ratio: $\dfrac{\text{pitch cone radius at small end}}{\text{pitch cone radius at large end}}$
Addendum angle	$\text{Tan addendum angle} = \dfrac{\text{addendum}}{\text{pitch cone radius}}$
Dedendum angle	$\text{Tan dedendum angle} = \dfrac{\text{dedendum}}{\text{pitch cone radius}}$
Turning angle	Face angle $=$ pitch cone angle $+$ addendum angle
Cutting angle	Cutting angle $=$ pitch cone angle $-$ dedendum angle
Angular addendum	Gear, angular addendum $=$ addendum $\times$ cosine of pitch cone angle of gear Pinion, angular addendum $=$ addendum $\times$ cosine of pitch cone angle of pinion
Outside diameter	Outside diameter $=$ pitch diameter $+$ two times the angular addendum
Number of teeth for which to select cutter	Number of teeth for which to select the cutter $= \dfrac{\text{actual number of teeth}}{\text{cosine of pitch cone angle}}$

$$\text{Number of teeth for which to select cutter} = \frac{\text{number of teeth}}{\cos 45°}$$

$$= \frac{\text{number of teeth}}{0.70711}$$

$$= 1.414 \times \text{number of teeth}$$

512

3. Find the number of teeth in a gear that has a pitch diameter of 10″ and a circular pitch of 0.3927″.
4. Find the thickness of tooth of the gear in Problem 3.
5. Find the diametral pitch of a gear that has 64 teeth and a pitch diameter of 4″.
6. What is the pitch diameter of a 16-pitch gear with 128 teeth?
7. How many teeth must an 8-pitch gear have if its pitch diameter is $3\frac{1}{2}$″?
8. What is the circular pitch of a 10-pitch gear?
9. Find the diametral pitch of a gear whose circular pitch is 0.31416″.
10. Find the clearance of a gear whose circular pitch is 0.2836″.
11. Find the clearance of an 8-pitch gear.
12. Find the whole depth of a 16-pitch gear.
13. Find the outside diameter of an 8-pitch gear whose pitch diameter is $6\frac{3}{4}$″.
14. Find the whole depth of tooth of a 6-pitch rack.
15. Find the center-to-center distance of two gears whose pitch diameters are $6\frac{1}{2}$″ and $4\frac{3}{4}$″.
16. Compute the pitch diameters of a pair of bevel gears at right angles if the gear has 64 teeth, the pinion has 20 teeth, and the pitch is 4.

Appendix

TABLE 1
POWERS, ROOTS, CIRCUMFERENCES, AND AREAS

No.	Square	Cube	Square Root	Cube Root	No. = Diam.		No.
					Circum.	Area	
1	1	1	1.0000	1.0000	3.142	0.7854	1
2	4	8	1.4142	1.2599	6.283	3.1416	2
3	9	27	1.7321	1.4422	9.425	7.0686	3
4	16	64	2.0000	1.5874	12.566	12.5664	4
5	25	125	2.2361	1.7100	15.708	19.6350	5
6	36	216	2.4495	1.8171	18.850	28.2743	6
7	49	343	2.6458	1.9129	21.991	38.4845	7
8	64	512	2.8284	2.0000	25.133	50.2655	8
9	81	729	3.0000	2.0801	28.274	63.6173	9
10	100	1000	3.1623	2.1544	31.416	78.5398	10
11	121	1331	3.3166	2.2240	34.558	95.0332	11
12	144	1728	3.4641	2.2894	37.699	113.097	12
13	169	2197	3.6056	2.3513	40.841	132.732	13
14	196	2744	3.7417	2.4101	43.982	153.938	14
15	225	3375	3.8730	2.4662	47.124	176.715	15
16	256	4096	4.0000	2.5198	50.265	201.062	16
17	289	4913	4.1231	2.5713	53.407	226.980	17
18	324	5832	4.2426	2.6207	56.549	254.469	18
19	361	6859	4.3589	2.6684	59.690	283.529	19
20	400	8000	4.4721	2.7144	62.832	314.159	20
21	441	9261	4.5826	2.7589	65.973	346.361	21
22	484	10648	4.6904	2.8020	69.115	380.133	22
23	529	12167	4.7958	2.8439	72.257	415.476	23
24	576	13824	4.8990	2.8845	75.398	452.389	24
25	625	15625	5.0000	2.9240	78.540	490.874	25
26	676	17576	5.0990	2.9625	81.681	530.929	26
27	729	19683	5.1962	3.0000	84.823	572.555	27
28	784	21952	5.2915	3.0366	87.965	615.752	28
29	841	24389	5.3852	3.0723	91.106	660.520	29
30	900	27000	5.4772	3.1072	94.248	706.858	30
31	961	29791	5.5678	3.1414	97.389	754.768	31
32	1024	32768	5.6569	3.1748	100.531	804.248	32
33	1089	35937	5.7446	3.2075	103.673	855.299	33
34	1156	39304	5.8310	3.2396	106.814	907.920	34
35	1225	42875	5.9161	3.2711	109.956	962.113	35
36	1296	46656	6.0000	3.3019	113.097	1017.88	36
37	1369	50653	6.0828	3.3322	116.239	1075.21	37
38	1444	54872	6.1644	3.3620	119.381	1134.11	38
39	1521	59319	6.2450	3.3912	122.522	1194.59	39

No.	Square	Cube	Square Root	Cube Root	No. = Diam. Circum.	No. = Diam. Area	No.
40	1600	64000	6.3246	3.4200	125.66	1256.64	40
41	1681	68921	6.4031	3.4482	128.81	1320.25	41
42	1764	74088	6.4807	3.4760	131.95	1385.44	42
43	1849	79507	6.5574	3.5034	135.09	1452.20	43
44	1936	85184	6.6332	3.5303	138.23	1520.53	44
45	2025	91125	6.7082	3.5569	141.37	1590.43	45
46	2116	97336	6.7823	3.5830	144.51	1661.90	46
47	2209	103823	6.8557	3.6088	147.65	1734.94	47
48	2304	110592	6.9282	3.6342	150.80	1809.56	48
49	2401	117649	7.0000	3.6593	153.94	1885.74	49
50	2500	125000	7.0711	3.6840	157.08	1963.50	50
51	2601	132651	7.1414	3.7084	160.22	2042.82	51
52	2704	140608	7.2111	3.7325	163.36	2123.72	52
53	2809	148877	7.2801	3.7563	166.50	2206.18	53
54	2916	157464	7.3485	3.7798	169.65	2290.22	54
55	3025	166375	7.4162	3.8030	172.79	2375.83	55
56	3136	175616	7.4833	3.8259	175.93	2463.01	56
57	3249	185193	7.5498	3.8485	179.07	2551.76	57
58	3364	195112	7.6158	3.8709	182.21	2642.08	58
59	3481	205379	7.6811	3.8930	185.35	2733.97	59
60	3600	216000	7.7460	3.9149	188.50	2827.43	60
61	3721	226981	7.8102	3.9365	191.64	2922.47	61
62	3844	238328	7.8740	3.9579	194.78	3019.07	62
63	3969	250047	7.9373	3.9791	197.92	3117.25	63
64	4096	262144	8.0000	4.0000	201.06	3216.99	64
65	4225	274625	8.0623	4.0207	204.20	3318.31	65
66	4356	287496	8.1240	4.0412	207.35	3421.19	66
67	4489	300763	8.1854	4.0615	210.49	3525.65	67
68	4624	314432	8.2462	4.0817	213.63	3631.68	68
69	4761	328509	8.3066	4.1016	216.77	3739.28	69
70	4900	343000	8.3666	4.1213	219.91	3848.45	70
71	5041	357911	8.4261	4.1408	223.05	3959.19	71
72	5184	373248	8.4853	4.1602	226.19	4071.50	72
73	5329	389017	8.5440	4.1793	229.34	4185.39	73
74	5476	405224	8.6023	4.1983	232.48	4300.84	74
75	5625	421875	8.6603	4.2172	235.62	4417.86	75
76	5776	438976	8.7178	4.2358	238.76	4536.46	76
77	5929	456533	8.7750	4.2543	241.90	4656.63	77
78	6084	474552	8.8318	4.2727	245.04	4778.36	78
79	6241	493039	8.8882	4.2908	248.19	4901.67	79

No.	Square	Cube	Square Root	Cube Root	No. = Diam. Circum.	No. = Diam. Area	No.
80	6400	512000	8.9443	4.3089	251.33	5026.55	80
81	6561	531441	9.0000	4.3267	254.47	5153.00	81
82	6724	551368	9.0554	4.3445	257.61	5281.02	82
83	6889	571787	9.1104	4.3621	260.75	5410.61	83
84	7056	592704	9.1652	4.3795	263.89	5541.77	84
85	7225	614125	9.2195	4.3968	267.04	5674.50	85
86	7396	636056	9.2736	4.4140	270.18	5808.80	86
87	7569	658503	9.3274	4.4310	273.32	5944.68	87
88	7744	681472	9.3808	4.4480	276.46	6082.12	88
89	7921	704969	9.4340	4.4647	279.60	6221.14	89
90	8100	729000	9.4868	4.4814	282.74	6361.73	90
91	8281	753571	9.5394	4.4979	285.88	6503.88	91
92	8464	778688	9.5917	4.5144	289.03	6647.61	92
93	8649	804357	9.6437	4.5307	292.17	6792.91	93
94	8836	830584	9.6954	4.5468	295.31	6939.78	94
95	9025	857375	9.7468	4.5629	298.45	7088.22	95
96	9216	884736	9.7980	4.5789	301.59	7238.23	96
97	9409	912673	9.8489	4.5947	304.73	7389.81	97
98	9604	941192	9.8995	4.6104	307.88	7542.96	98
99	9801	970299	9.9499	4.6261	311.02	7697.69	99
100	10000	1000000	10.0000	4.6416	314.16	7853.98	100
101	10201	1030301	10.0499	4.6570	317.30	8011.85	101
102	10404	1061208	10.0995	4.6723	320.44	8171.28	102
103	10609	1092727	10.1489	4.6875	323.58	8332.29	103
104	10816	1124864	10.1980	4.7027	326.73	8494.87	104
105	11025	1157625	10.2470	4.7177	329.87	8659.01	105
106	11236	1191016	10.2956	4.7326	333.01	8824.73	106
107	11449	1225043	10.3441	4.7475	336.15	8992.02	107
108	11664	1259712	10.3923	4.7622	339.29	9160.88	108
109	11881	1295029	10.4403	4.7769	342.43	9331.32	109
110	12100	1331000	10.4881	4.7914	345.58	9503.32	110
111	12321	1367631	10.5357	4.8059	348.72	9676.89	111
112	12544	1404928	10.5830	4.8203	351.86	9852.03	112
113	12769	1442897	10.6301	4.8346	355.00	10028.7	113
114	12996	1481544	10.6771	4.8488	358.14	10207.0	114
115	13225	1520875	10.7238	4.8629	361.28	10386.9	115
116	13456	1560896	10.7703	4.8770	364.42	10568.3	116
117	13689	1601613	10.8167	4.8910	367.57	10751.3	117
118	13924	1643032	10.8682	4.9049	370.71	10935.9	118
119	14161	1685159	10.9087	4.9187	373.85	11122.0	119

No.	Square	Cube	Square Root	Cube Root	No. = Diam.		No.
					Circum.	Area	
120	14400	1728000	10.9545	4.9324	376.99	11309.7	120
121	14641	1771561	11.0000	4.9461	380.13	11499.0	121
122	14884	1815848	11.0454	4.9597	383.27	11689.9	122
123	15129	1860867	11.0905	4.9732	386.42	11882.3	123
124	15376	1906624	11.1355	4.9866	389.56	12076.3	124
125	15625	1953125	11.1803	5.0000	392.70	12271.8	125
126	15876	2000376	11.2250	5.0133	395.84	12469.0	126
127	16129	2048383	11.2694	5.0265	398.98	12667.7	127
128	16384	2097152	11.3137	5.0397	402.12	12868.0	128
129	16641	2146689	11.3578	5.0528	405.27	13069.8	129
130	16900	2197000	11.4018	5.0658	408.41	13273.2	130
131	17161	2248091	11.4455	5.0788	411.55	13478.2	131
132	17424	2299968	11.4891	5.0916	414.69	13684.8	132
133	17689	2352637	11.5326	5.1045	417.83	13892.9	133
134	17956	2406104	11.5758	5.1172	420.97	14102.6	134
135	18225	2460375	11.6190	5.1299	424.12	14313.9	135
136	18496	2515456	11.6619	5.1426	427.26	14526.7	136
137	18769	2571353	11.7047	5.1551	430.40	14741.1	137
138	19044	2628072	11.7473	5.1676	433.54	14957.1	138
139	19321	2685619	11.7898	5.1801	436.68	15174.7	139
140	19600	2744000	11.8322	5.1925	439.82	15393.8	140
141	19881	2803221	11.8743	5.2048	442.96	15614.5	141
142	20164	2863288	11.9164	5.2171	446.11	15836.8	142
143	20449	2924207	11.9583	5.2293	449.25	16060.6	143
144	20736	2985984	12.0000	2.2415	452.39	16286.0	144
145	21025	3048625	12.0416	5.2536	455.53	16513.0	145
146	21316	3112136	12.0830	5.2656	458.67	16741.5	146
147	21609	3176523	12.1244	5.2776	461.81	16971.7	147
148	21904	3241792	12.1655	5.2896	464.96	17203.4	148
149	22201	3307949	12.2066	5.3015	468.10	17436.6	149
150	22500	3375000	12.2474	5.3133	471.24	17671.5	150
151	22801	3442951	12.2882	5.3251	474.38	17907.9	151
152	23104	3511808	12.3288	5.3368	477.52	18145.8	152
153	23409	3581577	12.3693	5.3485	480.66	18385.4	153
154	23716	3652264	12.4097	5.3601	483.81	18626.5	154
155	24025	3723875	12.4499	5.3717	486.95	18869.2	155
156	24336	3796416	12.4900	5.3832	490.09	19113.4	156
157	24649	3869893	12.5300	5.3947	493.23	19359.3	157
158	24964	3944312	12.5698	5.4061	496.37	19606.7	158
159	25281	4019679	12.6095	5.4175	499.51	19855.7	159

No.	Square	Cube	Square Root	Cube Root	No. = Diam. Circum.	Area	No.
160	25600	4096000	12.6491	5.4288	502.65	20106.2	160
161	25921	4173281	12.6886	5.4401	505.80	20358.3	161
162	26244	4251528	12.7279	5.4514	508.94	20612.0	162
163	26569	4330747	12.7671	5.4626	512.08	20867.2	163
164	26896	4410944	12.8062	5.4737	515.22	21124.1	164
165	27225	4492125	12.8452	5.4848	518.36	21382.5	165
166	27556	4574296	12.8841	5.4959	521.50	21642.4	166
167	27889	4657463	12.9228	5.5069	524.65	21904.0	167
168	28224	4741632	12.9615	5.5178	527.79	22167.1	168
169	28561	4826809	13.0000	5.5288	530.93	22431.8	169
170	28900	4913000	13.0384	5.5397	534.07	22698.0	170
171	29241	5000211	13.0767	5.5505	537.21	22965.8	171
172	29584	5088448	13.1149	5.5613	540.35	23235.2	172
173	29929	5177717	13.1529	5.5721	543.50	23506.2	173
174	30276	5268024	13.1909	5.5828	546.64	23778.7	174
175	30625	5359375	13.2288	5.5934	549.78	24052.8	175
176	30976	5451776	13.2665	5.6041	552.92	24328.5	176
177	31329	5545233	13.3041	5.6147	556.06	24605.7	177
178	31684	5639752	13.3417	5.6252	559.20	24884.6	178
179	32041	5735339	13.3791	5.6357	562.35	25164.9	179
180	32400	5832000	13.4164	5.6462	565.49	25446.9	180
181	32761	5929741	13.4536	5.6567	568.63	25730.4	181
182	33124	6028568	13.4907	5.6671	571.77	26015.5	182
183	33489	6128487	13.5277	5.6774	574.91	26302.2	183
184	33856	6229504	13.5647	5.6877	578.05	26590.4	184
185	34225	6331625	13.6015	5.6980	581.19	26880.3	185
186	34596	6434856	13.6382	5.7083	584.34	27171.6	186
187	34969	6539203	13.6748	5.7185	587.48	27464.6	187
188	35344	6644672	13.7113	5.7287	590.62	27759.1	188
189	35721	6751269	13.7477	5.7388	593.76	28055.2	189
190	36100	6859000	13.7840	5.7489	596.90	28352.9	190
191	36481	6967871	13.8203	5.7590	600.04	28652.1	191
192	36864	7077888	13.8564	5.7690	603.19	28952.9	192
193	37249	7189057	13.8924	5.7790	606.33	29255.3	193
194	37636	7301384	13.9284	5.7890	609.47	29559.2	194
195	38025	7414875	13.9642	5.7989	612.61	29864.8	195
196	38416	7529536	14.0000	5.8088	615.75	30171.9	196
197	38809	7645373	14.0357	5.8186	618.89	30480.5	197
198	39204	7762392	14.0712	5.8285	622.04	30790.7	198
199	39601	7880599	14.1067	5.8383	625.18	31102.6	199

No.	Square	Cube	Square Root	Cube Root	No. = Diam. Circum.	No. = Diam. Area	No.
200	40000	8000000	14.1421	5.8480	628.32	31415.9	200
201	40401	8120601	14.1774	5.8578	631.46	31730.9	201
202	40804	8242408	14.2127	5.8675	634.60	32047.4	202
203	41209	8365427	14.2478	5.8771	637.74	32365.5	203
204	41616	8489664	14.2829	5.8868	640.89	32685.1	204
205	42025	8615125	14.3178	5.8964	644.03	33006.4	205
206	42436	8741816	14.3527	5.9059	647.17	33329.2	206
207	42849	8869743	14.3875	5.9155	650.31	33653.5	207
208	43264	8998912	14.4222	5.9250	653.45	33979.5	208
209	43681	9129329	14.4568	5.9345	656.59	34307.0	209
210	44100	9261000	14.4914	5.9439	659.73	34636.1	210
211	44521	9393931	14.5258	5.9533	662.88	34966.7	211
212	44944	9528128	14.5602	5.9627	666.02	35298.9	212
213	45369	9663597	14.5945	5.9721	669.16	35632.7	213
214	45796	9800344	14.6287	5.9814	672.30	35968.1	214
215	46225	9938375	14.6629	5.9907	675.44	36305.0	215
216	46656	10077696	14.6969	6.0000	678.58	36643.5	216
217	47089	10218313	14.7309	6.0092	681.73	36983.6	217
218	47524	10360232	14.7648	6.0185	684.87	37325.3	218
219	47961	10503459	14.7986	6.0277	688.01	37668.5	219
220	48400	10648000	14.8324	6.0368	691.15	38013.3	220
221	48841	10793861	14.8661	6.0459	694.29	38359.6	221
222	49284	10941048	14.8997	6.0550	697.43	38707.6	222
223	49729	11089567	14.9332	6.0641	700.58	39057.1	223
224	50176	11239424	14.9666	6.0732	703.72	39408.1	224
225	50625	11390625	15.0000	6.0822	706.86	39760.8	225
226	51076	11543176	15.0333	6.0912	710.00	40115.0	226
227	51529	11697083	15.0665	6.1002	713.14	40470.8	227
228	51984	11852352	15.0997	6.1091	716.28	40828.1	228
229	52441	12008989	15.1327	6.1180	719.42	41187.1	229
230	52900	12167000	15.1658	6.1269	722.57	41547.6	230
231	53361	12326391	15.1987	6.1358	725.71	41909.6	231
232	53824	12487168	15.2315	6.1446	728.85	42273.3	232
233	54289	12649337	15.2643	6.1534	731.99	42638.5	233
234	54756	12812904	15.2971	6.1622	735.13	43005.3	234
235	55225	12977875	15.3297	6.1710	738.27	43373.6	235
236	55696	13144256	15.3623	6.1797	741.42	43743.5	236
237	56169	13312053	15.3948	6.1885	744.56	44115.0	237
238	56644	13481272	15.4272	6.1972	747.70	44488.1	238
239	57121	13651919	15.4596	6.2058	750.84	44862.7	239

No.	Square	Cube	Square Root	Cube Root	No. = Diam. Circum.	No. = Diam. Area	No.
240	57600	13824000	15.4919	6.2145	753.98	45238.9	240
241	58081	13997521	15.5242	6.2231	757.12	45616.7	241
242	58564	14172488	15.5563	6.2317	760.27	45996.1	242
243	59049	14348907	15.5885	6.2403	763.41	46377.0	243
244	59536	14526784	15.6205	6.2488	766.55	46759.5	244
245	60025	14706125	15.6525	6.2573	769.69	47143.5	245
246	60516	14886936	15.6844	6.2658	772.83	47529.2	246
247	61009	15069223	15.7162	6.2743	775.97	47916.4	247
248	61504	15252992	15.7480	6.2828	779.12	48305.1	248
249	62001	15438249	15.7797	6.2912	782.26	48695.5	249
250	62500	15625000	15.8114	6.2996	785.40	49087.4	250
251	63001	15813251	15.8430	6.3080	788.54	49480.9	251
252	63504	16003008	15.8745	6.3164	791.68	49875.9	252
253	64009	16194277	15.9060	6.3247	794.82	50272.6	253
254	64516	16387064	15.9374	6.3330	797.96	50670.7	254
255	65025	16581375	15.9687	6.3413	801.11	51070.5	255
256	65536	16777216	16.0000	6.3496	804.25	51471.9	256
257	66049	16974593	16.0312	6.3579	807.39	51874.8	257
258	66564	17173512	16.0624	6.3661	810.53	52279.2	258
259	67081	17373979	16.0935	6.3743	813.67	52685.3	259
260	67600	17576000	16.1245	6.3825	816.81	53092.9	260
261	68121	17779581	16.1555	6.3907	819.96	53502.1	261
262	68644	17984728	16.1864	6.3988	823.10	53912.9	262
263	69169	18191447	16.2173	6.4070	826.24	54325.2	263
264	69696	18399744	16.2481	6.4151	829.38	54739.1	264
265	70225	18609625	16.2788	6.4232	832.52	55154.6	265
266	70756	18821096	16.3095	6.4312	835.66	55571.6	266
267	71289	19034163	16.3401	6.4393	838.81	55990.3	267
268	71824	19248832	16.3707	6.4473	841.95	56410.4	268
269	72361	19465109	16.4012	6.4553	845.09	56832.2	269
270	72900	19683000	16.4317	6.4633	848.23	57255.5	270
271	73441	19902511	16.4621	6.4713	851.37	57680.4	271
272	73984	20123648	16.4924	6.4792	854.51	58106.9	272
273	74529	20346417	16.5227	6.4872	857.66	58534.9	273
274	75076	20570824	16.5529	6.4951	860.80	58964.6	274
275	75625	20796875	16.5831	6.5030	863.94	59395.7	275
276	76176	21024576	16.6132	6.5108	867.08	59828.5	276
277	76729	21253933	16.6433	6.5187	870.22	60262.8	277
278	77284	21484952	16.6733	6.5265	873.36	60698.7	278
279	77841	21717639	16.7033	6.5343	876.50	61136.2	279

No.	Square	Cube	Square Root	Cube Root	No. = Diam.		No.
					Circum.	Area	
280	78400	21952000	16.7332	6.5421	879.65	61575.2	280
281	78961	22188041	16.7631	6.5499	882.79	62015.8	281
282	79524	22425768	16.7929	6.5577	885.93	62458.0	282
283	80089	22665187	16.8226	6.5654	889.07	62901.8	283
284	80656	22906304	16.8523	6.5731	892.21	63347.1	284
285	81225	23149125	16.8819	6.5808	895.35	63794.0	285
286	81796	23393656	16.9115	6.5885	898.50	64242.4	286
287	82369	23639903	16.9411	6.5962	901.64	64692.5	287
288	82944	23887872	16.9706	6.6039	904.78	65144.1	288
289	83521	24137569	17.0000	6.6115	907.92	65597.2	289
290	84100	24389000	17.0294	6.6191	911.06	66052.0	290
291	84681	24642171	17.0587	6.6267	914.20	66508.3	291
292	85264	24897088	17.0880	6.6343	917.35	66966.2	292
293	85849	25153757	17.1172	6.6419	920.49	67425.6	293
294	86436	25412184	17.1464	6.6494	923.63	67886.7	294
295	87025	25672375	17.1756	6.6569	926.77	68349.3	295
296	87616	25934336	17.2047	6.6644	929.91	68813.5	296
297	88209	26198073	17.2337	6.6719	933.05	69279.2	297
298	88804	26463592	17.2627	6.6794	936.19	69746.5	298
299	89401	26730899	17.2916	6.6869	939.34	70215.4	299
300	90000	27000000	17.3205	6.6943	942.48	70685.8	300
301	90601	27270901	17.3494	6.7018	945.62	71157.9	301
302	91204	27543608	17.3781	6.7092	948.76	71631.5	302
303	91809	27818127	17.4069	6.7166	951.90	72106.6	303
304	92416	28094464	17.4356	6.7240	955.04	72583.4	304
305	93025	28372625	17.4642	6.7313	958.19	73061.7	305
306	93636	28652616	17.4929	6.7387	961.33	73541.5	306
307	94249	28934443	17.5214	6.7460	964.47	74023.0	307
308	94864	29218112	17.5499	6.7533	967.61	74506.0	308
309	95481	29503629	17.5784	6.7606	970.75	74990.6	309
310	96100	29791000	17.6068	6.7679	973.89	75476.8	310
311	96721	30080231	17.6352	6.7752	977.04	75964.5	311
312	97344	30371328	17.6635	6.7824	980.18	76453.8	312
313	97969	30664297	17.6918	6.7897	983.32	76944.7	313
314	98596	30959144	17.7200	6.7969	986.46	77437.1	314
315	99225	31255875	17.7482	6.8041	989.60	77931.1	315
316	99856	31554496	17.7764	6.8113	992.74	78426.7	316
317	100489	31855013	17.8045	6.8185	995.88	78923.9	317
318	101124	32157432	17.8326	6.8256	999.03	79422.6	318
319	101761	32461759	17.8606	6.8328	1002.2	79922.9	319

No.	Square	Cube	Square Root	Cube Root	No. = Diam.		No.
					Circum.	Area	
320	102400	32768000	17.8885	6.8399	1005.3	80424.8	320
321	103041	33076161	17.9165	6.8470	1008.5	80928.2	321
322	103684	33386248	17.9444	6.8541	1011.6	81433.2	322
323	104329	33698267	17.9722	6.8612	1014.7	81939.8	323
324	104976	34012224	18.0000	6.8683	1017.9	82448.0	324
325	105625	34328125	18.0278	6.8753	1021.0	82957.7	325
326	106276	34645976	18.0555	6.8824	1024.2	83469.0	326
327	106929	34965783	18.0831	6.8894	1027.3	83981.8	327
328	107584	35287552	18.1108	6.8964	1030.4	84496.3	328
329	108241	35611289	18.1384	6.9034	1033.6	85012.3	329
330	108900	35937000	18.1659	6.9104	1036.7	85529.9	330
331	109561	36264691	18.1934	6.9174	1039.9	86049.0	331
332	110224	36594368	18.2209	6.9244	1043.0	86569.7	332
333	110889	36926037	18.2483	6.9313	1046.2	87092.0	333
334	111556	37259704	18.2757	6.9382	1049.3	87615.9	334
335	112225	37595375	18.3030	6.9451	1052.4	88141.3	335
336	112896	37933056	18.3303	6.9521	1055.6	88668.3	336
337	113569	38272753	18.3576	6.9589	1058.7	89196.9	337
338	114244	38614472	18.3848	6.9658	1061.9	89727.0	338
339	114921	38958219	18.4120	6.9727	1065.0	90258.7	339
340	115600	39304000	18.4391	6.9795	1068.1	90792.0	340
341	116281	39651821	18.4662	6.9864	1071.3	91326.9	341
342	116964	40001688	18.4932	6.9932	1074.4	91863.3	342
343	117649	40353607	18.5203	7.0000	1077.6	92401.3	343
344	118336	40707584	18.5472	7.0068	1080.7	92940.9	344
345	119025	41063625	18.5742	7.0136	1083.8	93482.0	345
346	119716	41421736	18.6011	7.0203	1087.0	94024.7	346
347	120409	41781923	18.6279	7.0271	1090.1	94569.0	347
348	121104	42144192	18.6548	7.0338	1093.3	95114.9	348
349	121801	42508549	18.6815	7.0406	1096.4	95662.3	349
350	122500	42875000	18.7083	7.0473	1099.6	96211.3	350
351	123201	43243551	18.7350	7.0540	1102.7	96761.8	351
352	123904	43614208	18.7617	7.0607	1105.8	97314.0	352
353	124609	43986977	18.7883	7.0674	1109.0	97867.7	353
354	125316	44361864	18.8149	7.0740	1112.1	98423.0	354
355	126025	44738875	18.8414	7.0807	1115.3	98979.8	355
356	126736	45118016	18.8680	7.0873	1118.4	99538.2	356
357	127449	45499293	18.8944	7.0940	1121.5	100098	357
358	128164	45882712	18.9209	7.1006	1124.7	100660	358
359	128881	46268279	18.9473	7.1072	1127.8	101223	359

No.	Square	Cube	Square Root	Cube Root	No. = Diam. Circum.	No. = Diam. Area	No.
360	129600	46656000	18.9737	7.1138	1131.0	101788	360
361	130321	47045881	19.0000	7.1204	1134.1	102354	361
362	131044	47437928	19.0263	7.1269	1137.3	102922	362
363	131769	47832147	19.0526	7.1335	1140.4	103491	363
364	132496	48228544	19.0788	7.1400	1143.5	104062	364
365	133225	48627125	19.1050	7.1466	1146.7	104635	365
366	133956	49027896	19.1311	7.1531	1149.8	105209	366
367	134689	49430863	19.1572	7.1596	1153.0	105785	367
368	135424	49836032	19.1833	7.1661	1156.1	106362	368
369	136161	50243409	19.2094	7.1726	1159.2	106941	369
370	136900	50653000	19.2354	7.1791	1162.4	107521	370
371	137641	51064811	19.2614	7.1855	1165.5	108103	371
372	138384	51478848	19.2873	7.1920	1168.7	108687	372
373	139129	51895117	19.3132	7.1984	1171.8	109272	373
374	139876	52313624	19.3391	7.2048	1175.0	109858	374
375	140625	52734375	19.3649	7.2112	1178.1	110447	375
376	141376	53157376	19.3907	7.2177	1181.2	111036	376
377	142129	53582633	19.4165	7.2240	1184.4	111628	377
378	142884	54010152	19.4422	7.2304	1187.5	112221	378
379	143641	54439939	19.4679	7.2368	1190.7	112815	379
380	144400	54872000	19.4936	7.2432	1193.8	113411	380
381	145161	55306341	19.5192	7.2495	1196.9	114009	381
382	145924	55742968	19.5448	7.2558	1200.1	114608	382
383	146689	56181887	19.5704	7.2622	1203.2	115209	383
384	147456	56623104	19.5959	7.2685	1206.4	115812	384
385	148225	57066625	19.6214	7.2748	1209.5	116416	385
386	148996	57512456	19.6469	7.2811	1212.7	117021	386
387	149769	57960603	19.6723	7.2874	1215.8	117628	387
388	150544	58411072	19.6977	7.2936	1218.9	118237	388
389	151321	58863869	19.7231	7.2999	1221.1	118847	389
390	152100	59319000	19.7484	7.3061	1225.2	119459	390
391	152881	59776471	19.7737	7.3124	1228.4	120072	391
392	153664	60236288	19.7990	7.3186	1231.5	120687	392
393	154449	60698457	19.8242	7.3248	1234.6	121304	393
394	155236	61162984	19.8494	7.3310	1237.8	121922	394
395	156025	61629875	19.8746	7.3372	1240.9	122542	395
396	156816	62099136	19.8997	7.3434	1244.1	123163	396
397	157609	62570773	19.9249	7.3496	1247.2	123786	397
398	158404	63044792	19.9499	7.3558	1250.4	124410	398
399	159201	63521199	19.9750	7.3619	1253.5	125036	399

No.	Square	Cube	Square Root	Cube Root	No. = Diam.		No.
					Circum.	Area	
400	160000	64000000	20.0000	7.3681	1256.6	125664	400
401	160801	64481201	20.0250	7.3742	1259.8	126293	401
402	161604	64964808	20.0499	7.3803	1262.9	126923	402
403	162409	65450827	20.0749	7.3864	1266.1	127556	403
404	163216	65939264	20.0998	7.3925	1269.2	128190	404
405	164025	66430125	20.1246	7.3986	1272.3	128825	405
406	164836	66923416	20.1494	7.4047	1275.5	129462	406
407	165649	67419143	20.1742	7.4108	1278.6	130100	407
408	166464	67917312	20.1990	7.4169	1281.8	130741	408
409	167281	68417929	20.2237	7.4229	1284.9	131382	409
410	168100	68921000	20.2485	7.4290	1288.1	132025	410
411	168921	69426531	20.2731	7.4350	1291.2	132670	411
412	169744	69934528	20.2978	7.4410	1294.3	133317	412
413	170569	70444997	20.3224	7.4470	1297.5	133965	413
414	171396	70957944	20.3470	7.4530	1300.6	134614	414
415	172225	71473375	20.3715	7.4590	1303.8	135265	415
416	173056	71991296	20.3961	7.4650	1306.9	135918	416
417	173889	72511713	20.4206	7.4710	1310.0	136572	417
418	174724	73034632	20.4450	7.4770	1313.2	137228	418
419	175561	73560059	20.4695	7.4829	1316.3	137885	419
420	176400	74088000	20.4939	7.4889	1319.5	138544	420
421	177241	74618461	20.5183	7.4948	1322.6	139205	421
422	178084	75151448	20.5426	7.5007	1325.8	139867	422
423	178929	75686967	20.5670	7.5067	1328.9	140531	423
424	179776	76225024	20.5913	7.5126	1332.0	141196	424
425	180625	76765625	20.6155	7.5185	1335.2	141863	425
426	181476	77308776	20.6398	7.5244	1338.3	142531	426
427	182329	77854483	20.6640	7.5302	1341.5	143201	427
428	183184	78402752	20.6882	7.5361	1344.6	143872	428
429	184041	78953589	20.7123	7.5420	1347.7	144545	429
430	184900	79507000	20.7364	7.5478	1350.9	145220	430
431	185761	80062991	20.7605	7.5537	1354.0	145896	431
432	186624	80621568	20.7846	7.5595	1357.2	146574	432
433	187489	81182737	20.8087	7.5654	1360.3	147254	433
434	188356	81746504	20.8327	7.5712	1363.5	147934	434
435	189225	82312875	20.8567	7.5770	1366.6	148617	435
436	190096	82881856	20.8806	7.5828	1369.7	149301	436
437	190969	83453453	20.9045	7.5886	1372.9	149987	437
438	191844	84027672	20.9284	7.5944	1376.0	150674	438
439	192721	84604519	20.9523	7.6001	1379.2	151363	439

No.	Square	Cube	Square Root	Cube Root	No. = Diam. Circum.	No. = Diam. Area	No.
440	193600	85184000	20.9762	7.6059	1382.3	152053	440
441	194481	85766121	21.0000	7.6117	1385.4	152745	441
442	195364	86350888	21.0238	7.6174	1388.6	153439	442
443	196249	86938307	21.0476	7.6232	1391.7	154134	443
444	197136	87528384	21.0713	7.6289	1394.9	154830	444
445	198025	88121125	21.0950	7.6346	1398.0	155528	445
446	198916	88716536	21.1187	7.6403	1401.2	156228	446
447	199809	89314623	21.1424	7.6460	1404.3	156930	447
448	200704	89915392	21.1660	7.6517	1407.4	157633	448
449	201601	90518849	21.1896	7.6574	1410.6	158337	449
450	202500	91125000	21.2132	7.6631	1413.7	159043	450
451	203401	91733851	21.2368	7.6688	1416.9	159751	451
452	204304	92345408	21.2603	7.6744	1420.0	160460	452
453	205209	92959677	21.2838	7.6801	1423.1	161171	453
454	206116	93576664	21.3073	7.6857	1426.3	161883	454
455	207025	94196375	21.3307	7.6914	1429.4	162597	455
456	207936	94818816	21.3542	7.6970	1432.6	163313	456
457	208849	95443993	21.3776	7.7026	1435.7	164030	457
458	209764	96071912	21.4009	7.7082	1438.9	164748	458
459	210681	96702579	21.4243	7.7138	1442.0	165468	459
460	211600	97336000	21.4476	7.7194	1445.1	166190	460
461	212521	97972181	21.4709	7.7250	1448.3	166914	461
462	213444	98611128	21.4942	7.7306	1451.4	167639	462
463	214369	99252847	21.5174	7.7362	1454.6	168365	463
464	215296	99897344	21.5407	7.7418	1457.7	169093	464
465	216225	100544625	21.5639	7.7473	1460.8	169823	465
466	217156	101194696	21.5870	7.7529	1464.0	170554	466
467	218089	101847563	21.6102	7.7584	1467.1	171287	467
468	219024	102503232	21.6333	7.7639	1470.3	172021	468
469	219961	103161709	21.6564	7.7695	1473.4	172757	469
470	220900	103823000	21.6795	7.7750	1476.5	173494	470
471	221841	104487111	21.7025	7.7805	1479.7	174234	471
472	222784	105154048	21.7256	7.7860	1482.8	174974	472
473	223729	105823817	21.7486	7.7915	1486.0	175716	473
474	224676	106496424	21.7715	7.7970	1489.1	176460	474
475	225625	107171875	21.7945	7.8025	1492.3	177205	475
476	226576	107850176	21.8174	7.8079	1495.4	177952	476
477	227529	108531333	21.8403	7.8134	1498.5	178701	477
478	228484	109215352	21.8632	7.8188	1501.7	179451	478
479	229441	109902239	21.8861	7.8243	1504.8	180203	479

No.	Square	Cube	Square Root	Cube Root	No. = Diam.		No.
					Circum.	Area	
480	230400	110592000	21.9089	7.8297	1508.0	180956	480
481	231361	111284641	21.9317	7.8352	1511.1	181711	481
482	232324	111980168	21.9545	7.8406	1514.3	182467	482
483	233289	112678587	21.9773	7.8460	1517.4	183225	483
484	234256	113379904	22.0000	7.8514	1520.5	183984	484
485	235225	114084125	22.0227	7.8568	1523.7	184745	485
486	236196	114791256	22.0454	7.8622	1526.8	185508	486
487	237169	115501303	22.0681	7.8676	1530.0	186272	487
488	238144	116214272	22.0907	7.8730	1533.1	187038	488
489	239121	116930169	22.1133	7.8784	1536.2	187805	489
490	240100	117649000	22.1359	7.8837	1539.4	188574	490
491	241081	118370771	22.1585	7.8891	1542.5	189345	491
492	242064	119095488	22.1811	7.8944	1545.7	190117	492
493	243049	119823157	22.2036	7.8998	1548.8	190890	493
494	244036	120553784	22.2261	7.9051	1551.9	191665	494
495	245025	121287375	22.2486	7.9105	1555.1	192442	495
496	246016	122023936	22.2711	7.9158	1558.2	193221	496
497	247009	122763473	22.2935	7.9211	1561.4	194000	497
498	248004	123505992	22.3159	7.9264	1564.5	194782	498
499	249001	124251499	22.3383	7.9317	1567.7	195565	499
500	250000	125000000	22.3607	7.9370	1570.8	196350	500
501	251001	125751501	22.3830	7.9423	1573.9	197136	501
502	252004	126506008	22.4054	7.9476	1577.1	197923	502
503	253009	127263527	22.4277	7.9528	1580.2	198713	503
504	254016	128024064	22.4499	7.9581	1583.4	199504	504
505	255025	128787625	22.4722	7.9634	1586.5	200296	505
506	256036	129554216	22.4944	7.9686	1589.7	201090	506
507	257049	130323843	22.5167	7.9739	1592.8	201886	507
508	258064	131096512	22.5389	7.9791	1595.9	202683	508
509	259081	131872229	22.5610	7.9843	1599.1	202482	509
510	260100	132651000	22.5832	7.9896	1602.2	204282	510
511	261121	133432831	22.6053	7.9948	1605.4	205084	511
512	262144	134217728	22.6274	8.0000	1608.5	205887	512
513	263169	135005697	22.6495	8.0052	1611.6	206692	513
514	264196	135796744	22.6716	8.0104	1614.8	207499	514
515	265225	136590875	22.6936	8.0156	1617.9	208307	515
516	266256	137388096	22.7156	8.0208	1621.1	209117	516
517	267289	138188413	22.7376	8.0260	1624.2	209928	517
518	268324	138991832	22.7596	8.0311	1627.3	210741	518
519	269361	139798359	22.7816	8.0363	1630.5	211556	519

No.	Square	Cube	Square Root	Cube Root	No. = Diam. Circum.	No. = Diam. Area	No.
520	270400	140608000	22.8035	8.0415	1633.6	212372	520
521	271441	141420761	22.8254	8.0466	1636.8	213189	521
522	272484	142236648	22.8473	8.0517	1639.9	214008	522
523	273529	143055667	22.8692	8.0569	1643.1	214829	523
524	274576	143877824	22.8910	8.0620	1646.2	215651	524
525	275625	144703125	22.9129	8.0671	1649.3	216475	525
526	276676	145531576	22.9347	8.0723	1652.5	217301	526
527	277729	146363183	22.9565	8.0774	1655.6	218128	527
528	278784	147197952	22.9783	8.0825	1658.8	218956	528
529	279841	148035889	23.0000	8.0876	1661.9	219787	529
530	280900	148877000	23.0217	8.0927	1665.0	220618	530
531	281961	149721291	23.0434	8.0978	1668.2	221452	531
532	283024	150568768	23.0651	8.1028	1671.3	222287	532
533	284089	151419437	23.0868	8.1079	1674.5	223123	533
534	285156	152273304	23.1084	8.1130	1677.6	223961	534
535	286225	153130375	23.1301	8.1180	1680.8	224801	535
536	287296	153990656	23.1517	8.1231	1683.9	225642	536
537	288369	154854153	23.1733	8.1281	1687.0	226484	537
538	289444	155720872	23.1948	8.1332	1690.2	227329	538
539	290521	156590819	23.2164	8.1382	1693.3	228175	539
540	291600	157464000	23.2379	8.1433	1696.5	229022	540
541	292681	158340421	23.2594	8.1483	1699.6	229871	541
542	293764	159220088	23.2809	8.1533	1702.7	230722	542
543	294849	160103007	23.3024	8.1583	1705.9	231574	543
544	295936	160989184	23.3238	8.1633	1709.0	232428	544
545	297025	161878625	23.3452	8.1683	1712.2	233283	545
546	298116	162771336	23.3666	8.1733	1715.3	234140	546
547	299209	163667323	23.3880	8.1783	1718.5	234998	547
548	300304	164566592	23.4094	8.1833	1721.6	235858	548
549	301401	165469149	23.4307	8.1882	1724.7	236720	549
550	302500	166375000	23.4521	8.1932	1727.9	237583	550
551	303601	167284151	23.4734	8.1982	1731.0	238448	551
552	304704	168196608	23.4947	8.2031	1734.2	239314	552
553	305809	169112377	23.5160	8.2081	1737.3	240182	553
554	306916	170031464	23.5372	8.2130	1740.4	241051	554
555	308025	170953875	23.5584	8.2180	1743.6	241922	555
556	309136	171879616	23.5797	8.2229	1746.7	242795	556
557	310249	172808693	23.6008	8.2278	1749.9	243669	557
558	311364	173741112	23.6220	8.2327	1753.0	244545	558
559	312481	174676879	23.6432	8.2377	1756.2	245422	559

No.	Square	Cube	Square Root	Cube Root	No. = Diam.		No.
					Circum.	Area	
560	313600	175616000	23.6643	8.2426	1759.3	246301	560
561	314721	176558481	23.6854	8.2475	1762.4	247181	561
562	315844	177504328	23.7065	8.2524	1765.6	248063	562
563	316969	178453547	23.7276	8.2573	1768.7	248947	563
564	318096	179406144	23.7487	8.2621	1771.9	249832	564
565	319225	180362125	23.7697	8.2670	1775.0	250719	565
566	320356	181321496	23.7908	8.2719	1778.1	251607	566
567	321489	182284263	23.8118	8.2768	1781.3	252497	567
568	322624	183250432	23.8328	8.2816	1784.4	253388	568
569	323761	184220009	23.8537	8.2865	1787.6	254281	569
570	324900	185193000	23.8747	8.2913	1790.7	255176	570
571	326041	186169411	23.8956	8.2962	1793.9	256072	571
572	327184	187149248	23.9165	8.3010	1797.0	256970	572
573	328329	188132517	23.9374	8.3059	1800.1	257869	573
574	329476	189119224	23.9583	8.3107	1803.3	258770	574
575	330625	190109375	23.9792	8.3155	1806.4	259672	575
576	331776	191102976	24.0000	8.3203	1809.6	260576	576
577	332929	192100033	24.0208	8.3251	1812.7	261482	577
578	334084	193100552	24.0416	8.3300	1815.8	262389	578
579	335241	194104539	24.0624	8.3348	1819.0	263298	579
580	336400	195112000	24.0832	8.3396	1822.1	264208	580
581	337561	196122941	24.1039	8.3443	1825.3	265120	581
582	338724	197137368	24.1247	8.3491	1828.4	266033	582
583	339889	198155287	24.1454	8.3539	1831.6	266948	583
584	341056	199176704	24.1661	8.3587	1834.7	267865	584
585	342225	200201625	24.1868	8.3634	1837.8	268783	585
586	343396	201230056	24.2074	8.3682	1841.0	269701	586
587	344569	202262003	24.2281	8.3730	1844.1	270624	587
588	345744	203297472	24.2487	8.3777	1847.3	271547	588
589	346921	204336469	24.2693	8.3825	1850.4	272471	589
590	348100	205379000	24.2899	8.3872	1853.5	273397	590
591	349281	206425071	24.3105	8.3919	1856.7	274325	591
592	350464	207474688	24.3311	8.3967	1859.8	275254	592
593	351649	208527857	24.3516	8.4014	1863.0	276184	593
594	352836	209584584	24.3721	8.4061	1866.1	277117	594
595	354025	210644875	24.3926	8.4108	1869.3	278051	595
596	355216	211708736	24.4131	8.4155	1872.4	278986	596
597	356409	212776173	24.4336	8.4202	1875.5	279923	597
598	357604	213847192	24.4540	8.4249	1878.7	280862	598
599	358801	214921799	24.4745	8.4296	1881.8	281802	599

No.	Square	Cube	Square Root	Cube Root	No. = Diam. Circum.	Area	No.
600	360000	216000000	24.4949	8.4343	1885.0	282743	600
601	361201	217081801	24.5153	8.4390	1888.1	283687	601
602	362404	218167208	24.5357	8.4437	1891.2	284631	602
603	363609	219256227	24.5561	8.4484	1894.4	285578	603
604	364816	220348864	24.5764	8.4530	1897.5	286526	604
605	366025	221445125	24.5967	8.4577	1900.7	287475	605
606	367236	222545016	24.6171	8.4623	1903.8	288426	606
607	368449	223648543	24.6374	8.4670	1907.0	289379	607
608	369664	224755712	24.6577	8.4716	1910.1	290333	608
609	370881	225866529	24.6779	8.4763	1913.2	291289	609
610	372100	226981000	24.6982	8.4809	1916.4	292247	610
611	373321	228099131	24.7184	8.4856	1919.5	293206	611
612	374544	229220928	24.7386	8.4902	1922.7	294166	612
613	375769	230346397	24.7588	8.4948	1925.8	295128	613
614	376996	231475544	24.7790	8.4994	1928.9	296092	614
615	378225	232608375	24.7992	8.5040	1932.1	297057	615
616	379456	233744896	24.8193	8.5086	1935.2	298024	616
617	380689	234885113	24.8395	8.5132	1938.4	298992	617
618	381924	236029032	24.8596	8.5178	1941.5	299962	618
619	383161	237176659	24.8797	8.5224	1944.7	300934	619
620	384400	238328000	24.8998	8.5270	1947.8	301907	620
621	385641	239483061	24.9199	8.5316	1950.9	302882	621
622	386884	240641848	24.9399	8.5462	1954.1	303858	622
623	388129	241804367	24.9600	8.5408	1957.2	304836	623
624	389376	242970624	24.9800	8.5453	1960.4	305815	624
625	390625	244140625	25.0000	8.5499	1963.5	306796	625
626	391876	245314376	25.0200	8.5544	1966.6	307779	626
627	393129	246491883	25.0400	8.5590	1969.8	308763	627
628	394384	247673152	25.0599	8.5635	1972.9	309748	628
629	395641	248858189	25.0799	8.5681	1976.1	310736	629
630	396900	250047000	25.0998	8.5726	1979.2	311725	630
631	398161	251239591	25.1197	8.5772	1982.4	312715	631
632	399424	252435968	25.1396	8.5817	1985.5	313707	632
633	400689	253636137	25.1595	8.5862	1988.6	314700	633
634	401956	254840104	25.1794	8.5907	1991.8	315696	634
635	403225	256047875	25.1992	8.5952	1994.9	316692	635
636	404496	257259456	25.2190	8.5997	1998.1	317690	636
637	405769	258474853	25.2389	8.6043	2001.2	318690	637
638	407044	259694072	25.2587	8.6088	2004.3	319692	638
639	408321	260917119	25.2784	8.6132	2007.5	320695	639

No.	Square	Cube	Square Root	Cube Root	No. = Diam. Circum.	No. = Diam. Area	No.
640	409600	262144000	25.2982	8.6177	2010.6	321699	640
641	410881	263374721	25.3180	8.6222	2013.8	322705	641
642	412164	264609288	25.3377	8.6267	2016.9	323713	642
643	413449	265847707	25.3574	8.6312	2020.0	324722	643
644	414736	267089984	25.3772	8.6357	2023.2	325733	644
645	416025	268336125	25.3969	8.6401	2026.3	326745	645
646	417316	269586136	25.4165	8.6446	2029.5	327759	646
647	418609	270840023	25.4362	8.6490	2032.6	328775	647
648	419904	272097792	25.4558	8.6535	2035.8	329792	648
649	421201	273359449	25.4755	8.6579	2038.9	330810	649
650	422500	274625000	25.4951	8.6624	2042.0	331831	650
651	423801	275894451	25.5147	8.6668	2045.2	332853	651
652	425104	277167808	25.5343	8.6713	2048.3	333876	652
653	426409	278445077	25.5539	8.6757	2051.5	334901	653
654	427716	279726264	25.5734	8.6801	2054.6	335927	654
655	429025	281011375	25.5930	8.6845	2057.7	336955	655
656	430336	282300416	25.6125	8.6890	2060.9	337985	656
657	431649	283593393	25.6320	8.6934	2064.0	339016	657
658	432964	284890312	25.6515	8.6978	2067.2	340049	658
659	434281	286191179	25.6710	8.7022	2070.3	341084	659
660	435600	287496000	25.6905	8.7066	2073.5	342119	660
661	436921	288804781	25.7099	8.7110	2076.6	343157	661
662	438244	290117528	25.7294	8.7154	2079.7	344196	662
663	439569	291434247	25.7488	8.7198	2082.9	345237	663
664	440896	292754944	25.7682	8.7241	2086.0	346279	664
665	442225	294079625	25.7876	8.7285	2089.2	347323	665
666	443556	295408296	25.8070	8.7329	2092.3	348368	666
667	444889	296740963	25.8263	8.7373	2095.4	349415	667
668	446224	298077632	25.8457	8.7416	2098.6	350464	668
669	447561	299418309	25.8650	8.7460	2101.7	351514	669
670	448900	300763000	25.8844	8.7503	2104.9	352565	670
671	450241	302111711	25.9037	8.7547	2108.0	353618	671
672	451584	303464448	25.9230	8.7590	2111.2	354673	672
673	452929	304821217	25.9422	8.7634	2114.3	355730	673
674	454276	306182024	25.9615	8.7677	2117.4	356788	674
675	455625	307546875	25.9808	8.7721	2120.6	357847	675
676	456976	308915776	26.0000	8.7764	2123.7	358908	676
677	458329	310288733	26.0192	8.7807	2126.9	359971	677
678	459684	311665752	26.0384	8.7850	2130.0	361035	678
679	461041	313046839	26.0576	8.7893	2133.1	362101	679

No.	Square	Cube	Square Root	Cube Root	No. = Diam.		No.
					Circum.	Area	
680	462400	314432000	26.0768	8.7937	2136.3	363168	680
681	463761	315821241	26.0960	8.7980	2139.4	364237	681
682	465124	317214568	26.1151	8.8023	2142.6	365308	682
683	466489	318611987	26.1343	8.8066	2145.7	366380	683
684	467856	320013504	26.1534	8.8109	2148.9	367453	684
685	469225	321419125	26.1725	8.8152	2152.0	368528	685
686	470596	322828856	26.1916	8.8194	2155.1	369605	686
687	471969	324242703	26.2107	8.8237	2158.3	370684	687
688	473344	325660672	26.2298	8.8280	2161.4	371764	688
689	474721	327082769	26.2488	8.8323	2164.6	372845	689
690	476100	328509000	26.2679	8.8366	2167.7	373928	690
691	477481	329939371	26.2869	8.8408	2170.8	375013	691
692	478864	331373888	26.3059	8.8451	2174.0	376099	692
693	480249	332812557	26.3249	8.8493	2177.1	377187	693
694	481636	334255384	26.3439	8.8536	2180.3	378276	694
695	483025	335702375	26.3629	8.8578	2183.4	379367	695
696	484416	337153536	26.3818	8.8621	2186.6	380459	696
697	485809	338608873	26.4008	8.8663	2189.7	381554	697
698	487204	340068392	26.4197	8.8706	2192.8	382649	698
699	488601	341532099	26.4386	8.8748	2196.0	383746	699
700	490000	343000000	26.4575	8.8790	2199.1	384845	700
701	491401	344472101	26.4764	8.8833	2202.3	385945	701
702	492804	345948408	26.4953	8.8875	2205.4	387047	702
703	494209	347428927	26.5141	8.8917	2208.5	388151	703
704	495616	348913664	26.5330	8.8959	2211.7	389256	704
705	497025	350402625	26.5518	8.9001	2214.8	390363	705
706	498436	351895816	26.5707	8.9043	2218.0	391471	706
707	499849	353393243	26.5895	8.9085	2221.1	392580	707
708	501264	354894912	26.6083	8.9127	2224.3	393692	708
709	502681	356400829	26.6271	8.9169	2227.4	394805	709
710	504100	357911000	26.6458	8.9211	2230.5	395919	710
711	505521	359425431	26.6646	8.9253	2233.7	397035	711
712	506944	360944128	26.6833	8.9295	2236.8	398153	712
713	508369	362467097	26.7021	8.9337	2240.0	399272	713
714	509796	363994344	26.7208	8.9378	2243.1	400393	714
715	511225	365525875	26.7395	8.9420	2246.2	401515	715
716	512656	367061696	26.7582	8.9462	2249.4	402639	716
717	514089	368601813	26.7769	8.9503	2252.5	403765	717
718	515524	370146232	26.7955	8.9545	2255.7	404892	718
719	516961	371694959	26.8142	8.9587	2258.8	406020	719

No.	Square	Cube	Square Root	Cube Root	No. = Diam. Circum.	No. = Diam. Area	No.
720	518400	373248000	26.8328	8.9628	2261.9	407150	720
721	519841	374805361	26.8514	8.9670	2265.1	408282	721
722	521284	376367048	26.8701	8.9711	2268.2	409416	722
723	522729	377933067	26.8887	8.9752	2271.4	410550	723
724	524176	379503424	26.9072	8.9794	2274.5	411687	724
725	525625	381078125	26.9258	8.9835	2277.7	412825	725
726	527076	382657176	26.9444	8.9876	2280.8	413965	726
727	528529	384240583	26.9629	8.9918	2283.9	415106	727
728	529984	385828352	26.9815	8.9959	2287.1	416248	728
729	531441	387420489	27.0000	9.0000	2290.2	417393	729
730	532900	389017000	27.0185	9.0041	2293.4	418539	730
731	534361	390617891	27.0370	9.0082	2296.5	419686	731
732	535824	392223168	27.0555	9.0123	2299.7	420835	732
733	537289	393832837	27.0740	9.0164	2302.8	421986	733
734	538756	395446904	27.0924	9.0205	2305.9	423138	734
735	540225	397065375	27.1109	9.0246	2309.1	424293	735
736	541696	398688256	27.1293	9.0287	2312.2	425448	736
737	543169	400315553	27.1477	9.0328	2315.4	426604	737
738	544644	401947272	27.1662	9.0369	2318.5	427762	738
739	546121	403583419	27.1846	9.0410	2321.6	428922	739
740	547600	405224000	27.2029	9.0450	2324.8	430084	740
741	549081	406869021	27.2213	9.0491	2327.9	431247	741
742	550564	408518488	27.2397	9.0532	2331.1	432412	742
743	552049	410172407	27.2580	9.0572	2334.2	433578	743
744	553536	411830784	27.2764	9.0613	2337.3	434746	744
745	555025	413493625	27.2947	9.0654	2340.5	435916	745
746	556516	415160936	27.3130	9.0694	2343.6	437087	746
747	558009	416832723	27.3313	9.0735	2346.8	438259	747
748	559504	418508992	27.3496	9.0775	2349.9	439433	748
749	561001	420189749	27.3679	9.0816	2353.1	440609	749
750	562500	421875000	27.3861	9.0856	2356.2	441786	750
751	564001	423564751	27.4044	9.0896	2359.3	442965	751
752	565504	425259008	27.4226	9.0937	2362.5	444146	752
753	567009	426957777	27.4408	9.0977	2365.6	445328	753
754	568516	428661064	27.4591	9.1017	2368.8	446511	754
755	570025	430368875	27.4773	9.1057	2371.9	447697	755
756	571536	432081216	27.4955	9.1098	2375.0	448883	756
757	573049	433798093	27.5136	9.1138	2378.2	450072	757
758	574564	435519512	27.5318	9.1178	2381.3	451262	758
759	576081	437245479	27.5500	9.1218	2384.5	452453	759

No.	Square	Cube	Square Root	Cube Root	No. = Diam. Circum.	Area	No.
760	577600	438976000	27.5681	9.1258	2387.6	453646	760
761	579121	440711081	27.5862	9.1298	2390.8	454841	761
762	580644	442450728	27.6043	9.1338	2393.9	456037	762
763	582169	444194947	27.6225	9.1378	2397.0	457234	763
764	583696	445943744	27.6405	9.1418	2400.2	458434	764
765	585225	447697125	27.6586	9.1458	2403.3	459635	765
766	586756	449455096	27.6767	9.1498	2406.5	460837	766
767	588289	451217663	27.6948	9.1537	2409.6	462042	767
768	589824	452984832	27.7128	9.1577	2412.7	463247	768
769	591361	454756609	27.7308	9.1617	2415.9	464454	769
770	592900	456533000	27.7489	9.1657	2419.0	465663	770
771	594441	458314011	27.7669	9.1696	2422.2	466873	771
772	595984	460099648	27.7849	9.1736	2425.3	468085	772
773	597529	461889917	27.8029	9.1775	2428.5	469298	773
774	599076	463684824	27.8209	9.1815	2431.6	470513	774
775	600625	465484375	27.8388	9.1855	2434.7	471730	775
776	602176	467288576	27.8568	9.1894	2437.9	472948	776
777	603729	469097433	27.8747	9.1933	2441.0	474168	777
778	605284	470910952	27.8927	9.1973	2444.2	475389	778
779	606841	472729139	27.9106	9.2012	2447.3	476612	779
780	608400	474552000	27.9285	9.2052	2450.4	477836	780
781	609961	476379541	27.9464	9.2091	2453.6	479062	781
782	611524	478211768	27.9643	9.2130	2456.7	480290	782
783	613089	480048687	27.9821	9.2170	2459.9	481519	783
784	614656	481890304	28.0000	9.2209	2463.0	482750	784
785	616225	483736625	28.0179	9.2248	2466.2	483982	785
786	617796	485587656	28.0357	9.2287	2469.3	485216	786
787	619369	487443403	28.0535	9.2326	2472.4	486451	787
788	620944	489303872	28.0713	9.2365	2475.6	487688	788
789	622521	491169069	28.0891	9.2404	2478.7	488927	789
790	624100	493039000	28.1069	9.2443	2481.9	490167	790
791	625681	494913671	28.1247	9.2482	2485.0	491409	791
792	627264	496793088	28.1425	9.2521	2488.1	492652	792
793	628849	498677257	28.1603	9.2560	2491.3	493897	793
794	630436	500566184	28.1780	9.2599	2494.4	495143	794
795	632025	502459875	28.1957	9.2638	2497.6	496391	795
796	633616	504358336	28.2135	9.2677	2500.7	497641	796
797	635209	506261573	28.2312	9.2716	2503.8	498892	797
798	636804	508169592	28.2489	9.2754	2507.0	500145	798
799	638401	510082399	28.2666	9.2793	2510.1	501399	799

No.	Square	Cube	Square Root	Cube Root	No. = Diam.		No.
					Circum.	Area	
800	640000	512000000	28.2843	9.2832	2513.3	502655	800
801	641601	513922401	28.3019	9.2870	2516.4	503912	801
802	643204	515849608	28.3196	9.2909	2519.6	505171	802
803	644809	517781627	28.3373	9.2948	2522.7	506432	803
804	646416	519718464	28.3549	9.2986	2525.8	507694	804
805	648025	521660125	28.3725	9.3025	2529.0	508958	805
806	649636	523606616	28.3901	9.3063	2532.1	510223	806
807	651249	525557943	28.4077	9.3102	2535.3	511490	807
808	652864	527514112	28.4253	9.3140	2538.4	512758	808
809	654481	529475129	28.4429	9.3179	2541.5	514028	809
810	656100	531441000	28.4605	9.3217	2544.7	515300	810
811	657721	533411731	28.4781	9.3255	2547.8	516573	811
812	659344	535387328	28.4956	9.3294	2551.0	517848	812
813	660969	537367797	28.5132	9.3332	2554.1	519124	813
814	662596	539353144	28.5307	9.3370	2557.3	520402	814
815	664225	541343375	28.5482	9.3408	2560.4	521681	815
816	665856	543338496	28.5657	9.3447	2563.5	522962	816
817	667489	545338513	28.5832	9.3485	2566.7	524245	817
818	669124	547343432	28.6007	9.3523	2569.8	525529	818
819	670761	549353259	28.6182	9.3561	2573.0	526814	819
820	672400	551368000	28.6356	9.3599	2576.1	528102	820
821	674041	553387661	28.6531	9.3637	2579.2	529391	821
822	675684	555412248	28.6705	9.3675	2582.4	530681	822
823	677329	557441767	28.0880	9.3713	2585.5	531973	823
824	678976	559476224	28.7054	9.3751	2588.7	533267	824
825	680625	561515625	28.7228	9.3789	2591.8	534562	825
826	682276	563559976	28.7402	9.3827	2595.0	535858	826
827	683929	565609283	28.7576	9.3865	2598.1	537157	827
828	685584	567663552	28.7750	9.3902	2601.2	538456	828
829	687241	569722789	28.7924	9.3940	2604.4	539758	829
830	688900	571787000	28.8097	9.3978	2607.5	541061	830
831	690561	573856191	28.8271	9.4016	2610.7	542365	831
832	692224	575930368	28.8444	9.4053	2613.8	543671	832
833	693889	578009537	28.8617	9.4091	2616.9	544979	833
834	695556	580093704	28.8791	9.4129	2620.1	546288	834
835	697225	582182875	28.8964	9.4166	2623.2	547599	835
836	698896	584277056	28.9137	9.4204	2626.4	548912	836
837	700569	586376253	28.9310	9.4241	2629.5	550226	837
838	702244	588480472	28.9482	9.4279	2632.7	551541	838
839	703921	590589719	28.9655	9.4316	2635.8	552858	839

No.	Square	Cube	Square Root	Cube Root	No. = Diam. Circum.	Area	No.
840	705600	592704000	28.9828	9.4354	2638.9	554177	840
841	707281	594823321	29.0000	9.4391	2642.1	555497	841
842	708964	596947688	29.0172	9.4429	2645.2	556819	842
843	710649	599077107	29.0345	9.4466	2648.4	558142	843
844	712336	601211584	29.0517	9.4503	2651.5	559467	844
845	714025	603351125	29.0689	9.4541	2654.6	560794	845
846	715716	605495736	29.0861	9.4578	2657.8	562122	846
847	717409	607645423	29.1033	9.4615	2660.9	563452	847
848	719104	609800192	29.1204	9.4652	2664.1	564783	848
849	720801	611960049	29.1376	9.4690	2667.2	566116	849
850	722500	614125000	29.1548	9.4727	2670.4	567450	850
851	724201	616295051	29.1719	9.4764	2673.5	568786	851
852	725904	618470208	29.1890	9.4801	2676.6	570124	852
853	727609	620650477	29.2062	9.4838	2679.8	571463	853
854	729316	622835864	29.2233	9.4875	2682.9	572803	854
855	731025	625026375	29.2404	9.4912	2686.1	574146	855
856	732736	627222016	29.2575	9.4949	2689.2	575490	856
857	734449	629422793	29.2746	9.4986	2692.3	576835	857
858	736164	631628712	29.2916	9.5023	2695.5	578182	858
859	737881	633839779	29.3087	9.5060	2698.6	579530	859
860	739600	636056000	29.3258	9.5097	2701.8	580880	860
861	741321	638277381	29.3428	9.5134	2704.9	582232	861
862	743044	640503928	29.3598	9.5171	2708.1	583585	862
863	744769	642735647	29.3769	9.5207	2711.2	584940	863
864	746496	644972544	29.3939	9.5244	2714.3	586297	864
865	748225	647214625	29.4109	9.5281	2717.5	587655	865
866	749956	649461896	29.4279	9.5317	2720.6	589014	866
867	751689	651714363	29.4449	9.5354	2723.8	590375	867
868	753424	653972032	29.4618	9.5391	2726.9	591738	868
869	755161	656234909	29.4788	9.5427	2730.0	593102	869
870	756900	658503000	29.4958	9.5464	2733.2	594468	870
871	758641	660776311	29.5127	9.5501	2736.3	595835	871
872	760384	663054848	29.5296	9.5537	2739.5	597204	872
873	762129	665338617	29.5466	9.5574	2742.6	598575	873
874	763876	667627624	29.5635	9.5610	2745.8	599947	874
875	765625	669921875	29.5804	9.5647	2748.9	601320	875
876	767376	672221376	29.5973	9.5683	2752.0	602696	876
877	769129	674526133	29.6142	9.5719	2755.2	604073	877
878	770884	676836152	29.6311	9.5756	2758.3	605451	878
879	772641	679151439	29.6479	9.5792	2761.5	606831	879

No.	Square	Cube	Square Root	Cube Root	No. = Diam. Circum.	No. = Diam. Area	No.
880	774400	681472000	29.6648	9.5828	2764.6	608212	880
881	776161	683797841	29.6816	9.5865	2767.7	609595	881
882	777924	686128968	29.6985	9.5901	2770.9	610980	882
883	779689	688465387	29.7152	9.5937	2774.0	612366	883
884	781456	690807104	29.7321	9.5973	2777.2	613754	884
885	783225	693154125	29.7489	9.6010	2780.3	615143	885
886	784996	695506456	29.7658	9.6046	2783.5	616534	886
887	786769	697864103	29.7825	9.6082	2786.6	617927	887
888	788544	700227072	29.7993	9.6118	2789.7	619321	888
889	790321	702595369	29.8161	9.6154	2792.9	620717	889
890	792100	704969000	29.8329	9.6190	2796.0	622114	890
891	793881	707347971	29.8496	9.6226	2799.2	623513	891
892	795664	709732288	29.8664	9.6262	2802.3	624913	892
893	797449	712121957	29.8831	9.6298	2805.4	626315	893
894	799236	714516984	29.8998	9.6334	2808.6	627718	894
895	801025	716917375	29.9166	9.6370	2811.7	629124	895
896	802816	719323136	29.9333	9.6406	2814.9	630530	896
897	804609	721734273	29.9500	9.6442	2818.0	631938	897
898	806404	724150792	29.9666	9.6477	2821.2	633348	898
899	808201	726572699	29.9833	9.6513	2824.3	634760	899
900	810000	729000000	30.0000	9.6549	2827.4	636173	900
901	811801	731432701	30.0167	9.6585	2830.6	637587	901
902	813604	733870808	30.0333	9.6620	2833.7	639003	902
903	815409	736314327	30.0500	9.6656	2836.9	640421	903
904	817216	738763264	30.0666	9.6692	2840.0	641840	904
905	819025	741217625	30.0832	9.6727	2843.1	643261	905
906	820836	743677416	30.0998	9.6763	2846.3	644683	906
907	822649	746142643	30.1164	9.6799	2849.4	646107	907
908	824464	748613312	30.1330	9.6834	2852.6	647533	908
909	826281	751089429	30.1496	9.6870	2855.7	648960	909
910	828100	753571000	30.1662	9.6905	2858.8	650388	910
911	829921	756058031	30.1828	9.6941	2862.0	651818	911
912	831744	758550528	30.1993	9.6976	2865.1	653250	912
913	833569	761048497	30.2159	9.7012	2868.3	654684	913
914	835396	763551944	30.2324	9.7047	2871.4	656118	914
915	837225	766060875	30.2490	9.7082	2874.6	657555	915
916	839056	768575296	30.2655	9.7118	2877.7	658993	916
917	840889	771095213	30.2820	9.7153	2880.8	660433	917
918	842724	773620632	30.2985	9.7188	2884.0	661874	918
919	844561	776151559	30.3150	9.7224	2887.1	663317	919

No.	Square	Cube	Square Root	Cube Root	No. = Diam. Circum.	No. = Diam. Area	No.
920	846400	778688000	30.3315	9.7259	2890.3	664761	920
921	848241	781229961	30.3480	9.7294	2893.4	666207	921
922	850084	783777448	30.3645	9.7329	2896.5	667654	922
923	851929	786330467	30.3809	9.7364	2899.7	669103	923
924	853776	788889024	30.3974	9.7400	2902.8	670554	924
925	855625	791453125	30.4138	9.7435	2906.0	672006	925
926	857476	794022776	30.4302	9.7470	2909.1	673460	926
927	859329	796597983	30.4467	9.7505	2912.3	674915	927
928	861184	799178752	30.4631	9.7540	2915.4	676372	928
929	863041	801765089	30.4795	9.7575	2918.5	677831	929
930	864900	804357000	30.4959	9.7610	2921.7	679291	930
931	866761	806954491	30.5123	9.7645	2924.8	680752	931
932	868624	809557568	30.5287	9.7680	2928.0	682216	932
933	870489	812166237	30.5450	9.7715	2931.1	683680	933
934	872356	814780504	30.5614	9.7750	2934.2	685147	934
935	874225	817400375	30.5778	9.7785	2937.4	686615	935
936	876096	820025856	30.5941	9.7819	2940.5	688084	936
937	877969	822656953	30.6105	9.7854	2943.7	689555	937
938	879844	825293672	30.6268	9.7889	2946.8	691028	938
939	881721	827936019	30.6431	9.7924	2950.0	692502	939
940	883600	830584000	30.6594	9.7959	2953.1	693978	940
941	885481	833237621	30.6757	9.7993	2956.2	695455	941
942	887364	835896888	30.6920	9.8028	2959.4	696934	942
943	889249	838561807	30.7083	9.8063	2962.5	698415	943
944	891136	841232384	30.7246	9.8097	2965.7	699897	944
945	893025	843908625	30.7409	9.8132	2968.8	701380	945
946	894916	846590536	30.7571	9.8167	2971.9	702865	946
947	896809	849278123	30.7734	9.8201	2975.1	704352	947
948	898704	851971392	30.7896	9.8236	2978.2	705840	948
949	900601	854670349	30.8058	9.8270	2981.4	707330	949
950	902500	857375000	30.8221	9.8305	2984.5	708822	950
951	904401	860085351	30.8383	9.8339	2987.7	710315	951
952	906304	862801408	30.8545	9.8374	2990.8	711809	952
953	908209	865523177	30.8707	9.8408	2993.9	713306	953
954	910116	868250664	30.8869	9.8443	2997.1	714803	954
955	912025	870983875	30.9031	9.8477	3000.2	716303	955
956	913936	873722816	30.9192	9.8511	3003.4	717804	956
957	915849	876467493	30.9354	9.8546	3006.5	719306	957
958	917764	879217912	30.9516	9.8580	3009.6	720810	958
959	919681	881974079	30.9677	9.8614	3012.8	722316	959

No.	Square	Cube	Square Root	Cube Root	No. = Diam.		No.
					Circum.	Area	
960	921600	884736000	30.9839	9.8648	3015.9	723823	960
961	923521	887503681	31.0000	9.8683	3019.1	725332	961
962	925444	890277128	31.0161	9.8717	3022.2	726842	962
963	927369	893056347	31.0322	9.8751	3025.4	728354	963
964	929296	895841344	31.0483	9.8785	3028.5	729867	964
965	931225	898632125	31.0644	9.8819	3031.6	731382	965
966	933156	901428696	31.0805	9.8854	3034.8	732899	966
967	935089	904231063	31.0966	9.8888	3037.9	734417	967
968	937024	907039232	31.1127	9.8922	3041.1	735937	968
969	938961	909853209	31.1288	9.8956	3044.2	737458	969
970	940900	912673000	31.1448	9.8990	3047.3	738981	970
971	942841	915498611	31.1609	9.9024	3050.5	740506	971
972	944784	918330048	31.1769	9.9058	3053.6	742032	972
973	946729	921167317	31.1929	9.9092	3056.8	743559	973
974	948676	924010424	31.2090	9.9126	3059.9	745088	974
975	950625	926859375	31.2250	9.9160	3063.1	746619	975
976	952576	929714176	31.2410	9.9194	3066.2	748151	976
977	954529	932574833	31.2570	9.9227	3069.3	749685	977
978	956484	935441352	31.2730	9.9261	3072.5	751221	978
979	958441	938313739	31.2890	9.9295	3075.6	752758	979
980	960400	941192000	31.3050	9.9329	3078.8	754296	980
981	962361	944076141	31.3209	9.9363	3081.9	755837	981
982	964324	946966168	31.3369	9.9396	3085.0	757378	982
983	966289	949862087	31.3528	9.9430	3088.2	758922	983
984	968256	952763904	31.3688	9.9464	3091.3	760466	984
985	970225	955671625	31.3847	9.9497	3094.5	762013	985
986	972196	958585256	31.4006	9.9531	3097.6	763561	986
987	974169	961504803	31.4166	9.9565	3100.8	765111	987
988	976144	964430272	31.4325	9.9598	3103.9	766662	988
989	978121	967361669	31.4484	9.9632	3107.0	768214	989
990	980100	970299000	31.4643	9.9666	3110.2	769769	990
991	982081	973242271	31.4802	9.9699	3113.3	771325	991
992	984064	976191488	31.4960	9.9733	3116.5	772882	992
993	986049	979146657	31.5119	9.9766	3119.6	774441	993
994	988036	982107784	31.5278	9.9800	3122.7	776002	994
995	990025	985074875	31.5436	9.9833	3125.9	777564	995
996	992016	988047936	31.5595	9.9866	3129.0	779128	996
997	994009	991026973	31.5753	9.9900	3132.2	780693	997
998	996004	994011992	31.5911	9.9933	3135.3	782260	998
999	998001	997002999	31.6070	9.9967	3138.5	783828	999

TABLE 2
NATURAL TRIGONOMETRIC FUNCTIONS

Degrees	Sin	Cos	Tan	Cot	Sec	Csc	
0°00′	.0000	1.0000	.0000	—	1.000	—	90°00′
10	029	000	029	343.8	000	343.8	50
20	058	000	058	171.9	000	171.9	40
30	.0087	1.0000	.0087	114.6	1.000	114.6	30
40	116	.9999	116	85.94	000	85.95	20
50	145	999	145	68.75	000	68.76	10
1°00′	.0175	.9998	.0175	57.29	1.000	57.30	89°00′
10	204	998	204	49.10	000	49.11	50
20	233	997	233	42.96	000	42.98	40
30	.0262	.9997	.0262	38.19	1.000	38.20	30
40	291	996	291	34.37	000	34.38	20
50	320	995	320	31.24	001	31.26	10
2°00′	.0349	.9994	.0349	28.64	1.001	28.65	88°00′
10	378	993	378	26.43	001	26.45	50
20	407	992	407	24.54	001	24.56	40
30	.0436	.9990	.0437	22.90	1.001	22.93	30
40	465	989	466	21.47	001	21.49	20
50	494	988	495	20.21	001	20.23	10
3°00′	.0523	.9986	.0524	19.08	1.001	19.11	87°00′
10	552	985	553	18.07	002	18.10	50
20	581	983	582	17.17	002	17.20	40
30	.0610	.9981	.0612	16.35	1.002	16.38	30
40	640	980	641	15.60	002	15.64	20
50	669	978	670	14.92	002	14.96	10
4°00′	.0698	.9976	.0699	14.30	1.002	14.34	86°00′
10	727	974	729	13.73	003	13.76	50
20	756	971	758	13.20	003	13.23	40
30	.0785	.9969	.0787	12.71	1.003	12.75	30
40	814	967	816	12.25	003	12.29	20
50	843	964	846	11.83	004	11.87	10
5°00′	.0872	.9962	.0875	11.43	1.004	11.47	85°00′
10	901	959	904	11.06	004	11.10	50
20	929	957	934	10.71	004	10.76	40
30	.0958	.9954	.0963	10.39	1.005	10.43	30
40	987	951	992	10.08	005	10.13	20
50	.1016	948	.1022	9.788	005	9.839	10
6°00′	.1045	.9945	.1051	9.514	1.006	9.567	84°00′
10	074	942	080	9.255	006	9.309	50
20	103	939	110	9.010	006	9.065	40
30	.1132	.9936	.1139	8.777	1.006	8.834	30
40	161	932	169	8.556	007	8.614	20
50	190	929	198	8.345	007	8.405	10
7°00′	.1219	.9925	.1228	8.144	1.008	8.206	83°00′
10	248	922	257	7.953	008	8.016	50
20	276	918	287	7.770	008	7.834	40
30	.1305	.9914	.1317	7.596	1.009	7.661	30
40	334	911	346	7.429	009	7.496	20
50	363	907	376	7.269	009	7.337	10
8°00′	.1392	.9903	.1405	7.115	1.010	7.185	82°00′
10	421	899	435	6.968	010	7.040	50
10	449	894	465	6.827	011	6.900	40
30	.1478	.9890	.1495	6.691	1.011	6.765	30
40	507	886	524	6.561	012	6.636	20
50	536	881	554	6.435	012	6.512	10
9°00′	.1564	.9877	.1584	6.314	1.012	6.392	81°00′
	Cos	Sin	Cot	Tan	Csc	Sec	Degrees

Degrees	Sin	Cos	Tan	Cot	Sec	Csc	
9°00'	.1564	.9877	.1584	6.314	1.012	6.392	81°00'
10	593	872	614	197	013	277	50
20	622	868	644	084	013	166	40
30	.1650	.9863	.1673	5.976	1.014	6.059	30
40	679	858	703	871	014	5.955	20
50	708	853	733	769	015	855	10
10°00'	.1736	.9848	.1763	5.671	1.015	5.759	80°00'
10	765	843	793	576	016	665	50
20	794	838	823	485	016	575	40
30	.1822	.9833	.1853	5.396	1.017	5.487	30
40	851	827	883	309	018	403	20
50	880	822	914	226	018	320	10
11°00'	.1908	.9816	.1944	5.145	1.019	5.241	79°00'
10	937	811	974	066	019	164	50
20	965	805	.2004	4.989	020	089	40
30	.1994	.9799	.2035	4.915	1.020	5.016	30
40	.2022	793	065	843	021	4.945	20
50	051	787	095	773	022	876	10
12°00'	.2079	.9781	.2126	4.705	1.022	4.810	78°00'
10	108	775	156	638	023	745	50
20	136	769	186	574	024	682	40
30	.2164	.9763	.2217	4.511	1.024	4.620	30
40	193	757	247	449	025	560	20
50	221	750	278	390	026	502	10
13°00'	.2250	.9744	.2309	4.331	1.026	4.445	77°00'
10	278	737	339	275	027	390	50
20	306	730	370	219	028	336	40
30	.2334	.9724	.2401	4.165	1.028	4.284	30
40	363	717	432	113	029	232	20
50	391	710	462	061	030	182	10
14°00'	.2419	.9703	.2493	4.011	1.031	4.134	76°00'
10	447	696	524	3.962	031	086	50
20	476	689	555	914	032	039	40
30	.2504	.9681	.2586	3.867	1.033	3.994	30
40	532	674	617	821	034	950	20
50	560	667	648	776	034	906	10
15°00'	.2588	.9659	.2679	3.732	1.035	3.864	75°00'
10	616	652	711	689	036	822	50
20	644	644	742	647	037	782	40
30	.2672	.9636	.2773	3.606	1.038	3.742	30
40	700	628	805	566	039	703	20
50	728	621	836	526	039	665	10
16°00'	.2756	.9613	.2867	3.487	1.040	3.628	74°00'
10	784	605	899	450	041	592	50
20	812	596	931	412	042	556	40
30	.2840	.9588	.2962	3.376	1.043	3.521	30
40	868	580	994	340	044	487	20
50	896	572	.3026	305	045	453	10
17°00'	.2924	.9563	.3057	3.271	1.046	3.420	73°00'
10	952	555	089	237	047	388	50
20	979	546	121	204	048	357	40
30	.3007	.9537	.3153	3.172	1.048	3.326	30
40	035	528	185	140	049	295	20
50	062	520	217	108	050	265	10
18°00'	.3090	.9511	.3249	3.078	1.051	3.236	72°00'
	Cos	Sin	Cot	Tan	Csc	Sec	Degrees

Degrees	Sin	Cos	Tan	Cot	Sec	Csc	
18°00′	.3090	.9511	.3249	3.078	1.051	3.236	72°00′
10	118	502	281	047	052	207	50
20	145	492	314	018	053	179	40
30	.3173	.9483	.3346	2.989	1.054	3.152	30
40	201	474	378	960	056	124	20
50	228	465	411	932	057	098	10
19°00′	.3256	.9455	.3443	2.904	1.058	3.072	71°00′
10	283	446	476	877	059	046	50
20	311	436	508	850	060	021	40
30	.3338	.9426	.3541	2.824	1.061	2.996	30
40	365	417	574	798	062	971	20
50	393	407	607	773	063	947	10
20°00′	.3420	.9397	.3640	2.747	1.064	2.924	70°00′
10	448	387	673	723	065	901	50
20	475	377	706	699	066	878	40
30	.3502	.9367	.3739	2.675	1.068	2.855	30
40	529	356	772	651	069	833	20
50	557	346	805	628	070	812	10
21°00′	.3584	.9336	.3839	2.605	1.071	2.790	69°00′
10	611	325	872	583	072	769	50
20	638	315	906	560	074	749	40
30	.3665	.9304	.3939	2.539	1.075	2.729	30
40	692	293	973	517	076	709	20
50	719	283	.4006	496	077	689	10
22°00′	.3746	.9272	.4040	2.475	1.079	2.669	68°00′
10	773	261	074	455	080	650	50
20	800	250	108	434	081	632	40
30	.3827	.9239	.4142	2.414	1.082	2.613	30
40	854	228	176	394	084	595	20
50	881	216	210	375	085	577	10
23°00′	.3907	.9205	.4245	2.356	1.086	2.559	67°00′
10	934	194	279	337	088	542	50
20	961	182	314	318	089	525	40
30	.3987	.9171	.4348	2.300	1.090	2.508	30
40	.4014	159	383	282	092	491	20
50	041	147	417	264	093	475	10
24°00′	.4067	.9135	.4452	2.246	1.095	2.459	66°00′
10	094	124	487	229	096	443	50
20	120	112	522	211	097	427	40
30	.4147	.9100	.4557	2.194	1.099	2.411	30
40	173	088	592	177	100	396	20
50	200	075	628	161	102	381	10
25°00′	.4226	.9063	.4663	2.145	1.103	2.366	65°00′
10	253	051	699	128	105	352	50
20	279	038	734	112	106	337	40
30	.4305	.9026	.4770	2.097	1.108	2.323	30
40	331	013	806	081	109	309	20
50	358	001	841	066	111	295	10
26°00′	.4384	.8988	.4877	2.050	1.113	2.281	64°00′
10	410	975	913	035	114	268	50
20	436	962	950	020	116	254	40
30	.4462	.8949	.4986	2.006	1.117	2.241	30
40	488	936	.5022	1.991	119	228	20
50	514	923	059	977	121	215	10
27°00′	.4540	.8910	.5095	1.963	1.122	2.203	63°00′
	Cos	Sin	Cot	Tan	Csc	Sec	Degrees

Degrees	Sin	Cos	Tan	Cot	Sec	Csc	
27°00′	.4540	.8910	.5095	1.963	1.122	2.203	63°00′
10	566	897	132	949	124	190	50
20	592	884	169	935	126	178	40
30	.4617	.8870	.5206	1.921	1.127	2.166	30
40	643	857	243	907	129	154	20
50	669	843	280	894	131	142	10
28°00′	.4695	.8829	.5317	1.881	1.133	2.130	62°00′
10	720	816	354	868	134	118	50
20	746	802	392	855	136	107	40
30	.4772	.8788	.5430	1.842	1.138	2.096	30
40	797	774	467	829	140	085	20
50	823	760	505	816	142	074	10
29°00′	.4848	.8746	.5543	1.804	1.143	2.063	61°00′
10	874	732	581	792	145	052	50
20	899	718	619	780	147	041	40
30	.4924	.8704	.5658	1.767	1.149	2.031	30
40	950	689	696	756	151	020	20
50	975	675	735	744	153	010	10
30°00′	.5000	.8660	.5774	1.732	1.155	2.000	60°00′
10	025	646	812	720	157	1.990	50
20	050	631	851	709	159	980	40
30	.5075	.8616	.5890	1.698	1.161	1.970	30
40	100	601	930	686	163	961	20
50	125	587	969	675	165	951	10
31°00′	.5150	.8572	.6009	1.664	1.167	1.942	59°00′
10	175	557	048	653	169	932	50
20	200	542	088	643	171	923	40
30	.5225	.8526	.6128	1.632	1.173	1.914	30
40	250	511	168	621	175	905	20
50	275	496	208	611	177	896	10
32°00′	.5299	.8480	.6249	1.600	1.179	1.887	58°00′
10	324	465	289	590	181	878	50
20	348	450	330	580	184	870	40
30	.5373	.8434	.6371	1.570	1.186	1.861	30
40	398	418	412	560	188	853	20
50	422	403	453	550	190	844	10
33°00′	.5446	.8387	.6494	1.540	1.192	1.836	57°00′
10	471	371	536	530	195	828	50
20	495	355	577	520	197	820	40
30	.5519	.8339	.6619	1.511	1.199	1.812	30
40	544	323	661	501	202	804	20
50	568	307	703	1.492	204	796	10
34°00′	.5592	.8290	.6745	1.483	1.206	1.788	56°00′
10	616	274	787	473	209	781	50
20	640	258	830	464	211	773	40
30	.5664	.8241	.6873	1.455	1.213	1.766	30
40	688	225	916	446	216	758	20
50	712	208	959	437	218	751	10
35°00′	.5736	.8192	.7002	1.428	1.221	1.743	55°00′
10	760	175	046	419	223	736	50
20	783	158	089	411	226	729	40
30	.5807	.8141	.7133	1.402	1.228	1.722	30
40	831	124	177	393	231	715	20
50	854	107	221	385	233	708	10
36°00′	.5878	.8090	.7265	1.376	1.236	1.701	54°00′
	Cos	Sin	Cot	Tan	Csc	Sec	Degrees

Degrees	Sin	Cos	Tan	Cot	Sec	Csc	
36°00′	.5878	.8090	.7265	1.376	1.236	1.701	**54°00′**
10	901	073	310	368	239	695	50
20	925	056	355	360	241	688	40
30	.5948	.8039	.7400	1.351	1.244	1.681	30
40	972	021	445	343	247	675	20
50	995	004	490	335	249	668	10
37°00′	.6018	.7986	.7536	1.327	1.252	1.662	**53°00′**
10	041	969	581	319	255	655	50
20	065	951	627	311	258	649	40
30	.6088	.7934	.7673	1.303	1.260	1.643	30
40	111	916	720	295	263	636	20
50	134	898	766	288	266	630	10
38°00′	.6157	.7880	.7813	1.280	1.269	1.624	**52°00′**
10	180	862	860	272	272	618	50
20	202	844	907	265	275	612	40
30	.6225	.7826	.7954	1.257	1.278	1.606	30
40	248	808	.8002	250	281	601	20
50	271	790	050	242	284	595	10
39°00′	.6293	.7771	.8098	1.235	1.287	1.589	**51°00′**
10	316	753	146	228	290	583	50
20	338	735	195	220	293	578	40
30	.6361	.7716	.8243	1.213	1.296	1.572	30
40	383	698	292	206	299	567	20
50	406	679	342	199	302	561	10
40°00′	.6428	.7660	.8391	1.192	1.305	1.556	**50°00′**
10	450	642	441	185	309	550	50
20	472	623	491	178	312	545	40
30	.6494	.7604	.8541	1.171	1.315	1.540	30
40	517	585	591	164	318	535	20
50	539	566	642	157	322	529	10
41°00′	.6561	.7547	.8693	1.150	1.325	1.524	**49°00′**
10	583	528	744	144	328	519	50
20	604	509	796	137	332	514	40
30	.6626	.7490	.8847	1.130	1.335	1.509	30
40	648	470	899	124	339	504	20
50	670	451	952	117	342	499	10
42°00′	.6691	.7431	.9004	1.111	1.346	1.494	**48°00′**
10	713	412	057	104	349	490	50
20	734	392	110	098	353	485	40
30	.6756	.7373	.9163	1.091	1.356	1.480	30
40	777	353	217	085	360	476	20
50	799	333	271	079	364	471	10
43°00′	.6820	.7314	.9325	1.072	1.367	1.466	**47°00′**
10	841	294	380	066	371	462	50
20	862	274	435	060	375	457	40
30	.6884	.7254	.9490	1.054	1.379	1.453	30
40	905	234	545	048	382	448	20
50	926	214	601	042	386	444	10
44°00′	.6947	.7193	.9657	1.036	1.390	1.440	**46°00′**
10	967	173	713	030	394	435	50
20	988	153	770	024	398	431	40
30	.7009	.7133	.9827	1.018	1.402	1.427	30
40	030	112	884	012	406	423	20
50	050	092	942	006	410	418	10
45°00′	.7071	.7071	1.000	1.000	1.414	1.414	**45°00′**
	Cos	Sin	Cot	Tan	Csc	Sec	Degrees

TABLE 3
CONVERTING MILLIMETERS TO INCHES AND DECIMALS

Milli-meters	Inches	Milli-meters	Inches	Milli-meters	Inches	Milli-meters	Inches
1	= .03937	26	= 1.02362	51	= 2.00787	76	= 2.99213
2	= .07874	27	= 1.06299	52	= 2.04724	77	= 3.03150
3	= .11811	28	= 1.10236	53	= 2.08661	78	= 3.07087
4	= .15748	29	= 1.14173	54	= 2.12598	79	= 3.11024
5	= .19685	30	= 1.18110	55	= 2.16535	80	= 3.14961
6	= .23622	31	= 1.22047	56	= 2.20472	81	= 3.18898
7	= .27559	32	= 1.25984	57	= 2.24409	82	= 3.22835
8	= .31496	33	= 1.29921	58	= 2.28346	83	= 3.26772
9	= .35433	34	= 1.33858	59	= 2.32283	84	= 3.30709
10	= .39370	35	= 1.37795	60	= 2.36220	85	= 3.34646
11	= .43307	36	= 1.41732	61	= 2.40157	86	= 3.38583
12	= .47244	37	= 1.45669	62	= 2.44094	87	= 3.42520
13	= .51181	38	= 1.49606	63	= 2.48031	88	= 3.46457
14	= .55118	39	= 1.53543	64	= 2.51969	89	= 3.50394
15	= .59055	40	= 1.57480	65	= 2.55906	90	= 3.54331
16	= .62992	41	= 1.61417	66	= 2.59843	91	= 3.58268
17	= .66929	42	= 1.65354	67	= 2.63780	92	= 3.62204
18	= .70866	43	= 1.69291	68	= 2.67717	93	= 3.66142
19	= .74803	44	= 1.73228	69	= 2.71654	94	= 3.70079
20	= .78740	45	= 1.77165	70	= 2.75591	95	= 3.74016
21	= .82677	46	= 1.81102	71	= 2.79528	96	= 3.77953
22	= .86614	47	= 1.85039	72	= 2.83465	97	= 3.81890
23	= .90551	48	= 1.88976	73	= 2.87402	98	= 3.85827
24	= .94488	49	= 1.92913	74	= 2.91339	99	= 3.89764
25	= .98425	50	= 1.96850	75	= 2.95276	100	= 3.93701

TABLE 4
DECIMAL AND METRIC EQUIVALENTS OF FRACTIONS OF AN INCH

Fraction	$\frac{1}{32}$ds	$\frac{1}{64}$ths	Decimal	Millimeters	Fraction	$\frac{1}{32}$ds	$\frac{1}{64}$ths	Decimal	Millimeters
		1	.015625	0.3969			33	.515625	13.0969
	1	2	.03125	0.7938		17	34	.53125	13.4938
		3	.046875	1.1906			35	.546875	13.8906
$\frac{1}{16}$	2	4	.0625	1.5875	$\frac{9}{16}$	18	36	.5625	14.2875
		5	.078125	1.9844			37	.578125	14.6844
	3	6	.09375	2.3812		19	38	.59375	15.0812
		7	.109375	2.7781			39	.609375	15.4781
$\frac{1}{8}$	4	8	.125	3.1750	$\frac{5}{8}$	20	40	.625	15.8750
		9	.140625	3.5719			41	.640625	16.2719
	5	10	.15625	3.9688		21	42	.65625	16.6688
		11	.171875	4.3656			43	.671875	17.0656
$\frac{3}{16}$	6	12	.1875	4.7625	$\frac{11}{16}$	22	44	.6875	17.4625
		13	.203125	5.1594			45	.703125	17.8594
	7	14	.21875	5.5562		23	46	.71875	18.2562
		15	.234375	5.9531			47	.734375	18.6531
$\frac{1}{4}$	8	16	.25	6.3500	$\frac{3}{4}$	24	48	.75	19.0500
		17	.265625	6.7469			49	.765625	19.4469
	9	18	.28125	7.1438		25	50	.78125	19.8438
		19	.296875	7.5406			51	.796875	20.2406
$\frac{5}{16}$	10	20	.3125	7.9375	$\frac{13}{16}$	26	52	.8125	20.6375
		21	.328125	8.3344			53	.828125	21.0344
	11	22	.34375	8.7312		27	54	.84375	21.4312
		23	.359375	9.1281			55	.859375	21.8281
$\frac{3}{8}$	12	24	.375	9.5250	$\frac{7}{8}$	28	56	.875	22.2250
		25	.390625	9.9219			57	.890625	22.6219
	13	26	.40625	10.3188		29	58	.90625	23.0188
		27	.421875	10.7156			59	.921875	23.4156
$\frac{7}{16}$	14	28	.4375	11.1125	$\frac{15}{16}$	30	60	.9375	23.8125
		29	.453125	11.5094			61	.953125	24.2094
	15	30	.46875	11.9062		31	62	.96875	24.6062
		31	.484375	12.3031			63	.984375	25.0031
$\frac{1}{2}$	16	32	.5	12.7000	1	32	64	1.	25.4000

TABLE 5
CONVERTING DEGREES FAHRENHEIT TO DEGREES CELSIUS

32°F = 0.00°C	78°F = 25.56°C	123°F = 50.56°C	168°F = 75.56°C
33°F = 0.56°C	79°F = 26.11°C	124°F = 51.11°C	169°F = 76.11°C
34°F = 1.11°C	80°F = 26.67°C	125°F = 51.67°C	170°F = 76.67°C
35°F = 1.67°C	81°F = 27.22°C	126°F = 52.22°C	171°F = 77.22°C
36°F = 2.22°C	82°F = 27.78°C	127°F = 52.78°C	172°F = 77.78°C
37°F = 2.78°C	83°F = 28.33°C	128°F = 53.33°C	173°F = 78.33°C
38°F = 3.33°C	84°F = 28.89°C	129°F = 53.89°C	174°F = 78.89°C
39°F = 3.89°C	85°F = 29.45°C	130°F = 54.45°C	175°F = 79.45°C
40°F = 4.45°C	86°F = 30.00°C	131°F = 55.00°C	176°F = 80.00°C
41°F = 5.00°C	87°F = 30.56°C	132°F = 55.56°C	177°F = 80.56°C
42°F = 5.56°C	88°F = 31.11°C	133°F = 56.11°C	178°F = 81.11°C
43°F = 6.11°C	89°F = 31.67°C	134°F = 56.67°C	179°F = 81.67°C
44°F = 6.67°C	90°F = 32.22°C	135°F = 57.22°C	180°F = 82.22°C
45°F = 7.22°C	91°F = 32.78°C	136°F = 57.78°C	181°F = 82.78°C
46°F = 7.78°C	92°F = 33.33°C	137°F = 58.33°C	182°F = 83.33°C
47°F = 8.33°C	93°F = 33.89°C	138°F = 58.89°C	183°F = 83.89°C
48°F = 8.89°C	94°F = 34.45°C	139°F = 59.45°C	184°F = 84.45°C
49°F = 9.45°C	95°F = 35.00°C	140°F = 60.00°C	185°F = 85.00°C
50°F = 10.00°C	96°F = 35.56°C	141°F = 60.56°C	186°F = 85.56°C
51°F = 10.56°C	97°F = 36.11°C	142°F = 61.11°C	187°F = 86.11°C
52°F = 11.11°C	98°F = 36.67°C	143°F = 61.67°C	188°F = 86.67°C
53°F = 11.67°C	99°F = 37.22°C	144°F = 62.22°C	189°F = 87.22°C
54°F = 12.22°C	100°F = 37.78°C	145°F = 62.78°C	190°F = 87.78°C
55°F = 12.78°C	101°F = 38.33°C	146°F = 63.33°C	191°F = 88.33°C
56°F = 13.33°C	102°F = 38.89°C	147°F = 63.89°C	192°F = 88.89°C
57°F = 13.89°C	103°F = 39.45°C	148°F = 64.46°C	193°F = 89.45°C
58°F = 14.45°C	104°F = 40.00°C	149°F = 65.00°C	194°F = 90.00°C
59°F = 15.00°C	105°F = 40.56°C	150°F = 65.56°C	195°F = 90.56°C
60°F = 15.56°C	106°F = 41.11°C	151°F = 66.11°C	196°F = 91.11°C
61°F = 16.11°C	107°F = 41.67°C	152°F = 66.67°C	197°F = 91.67°C
62°F = 16.67°C	108°F = 42.22°C	153°F = 67.22°C	198°F = 92.22°C
63°F = 17.22°C	109°F = 42.78°C	154°F = 67.78°C	199°F = 92.78°C
64°F = 17.78°C	110°F = 43.33°C	155°F = 68.33°C	200°F = 93.33°C
65°F = 18.33°C	111°F = 43.89°C	156°F = 68.89°C	201°F = 93.89°C
66°F = 18.89°C	112°F = 44.45°C	157°F = 69.45°C	202°F = 94.45°C
67°F = 19.45°C	113°F = 45.00°C	158°F = 70.00°C	203°F = 95.00°C
68°F = 20.00°C	114°F = 45.56°C	159°F = 70.56°C	204°F = 95.56°C
69°F = 20.56°C	115°F = 46.11°C	160°F = 71.11°C	205°F = 96.11°C
70°F = 21.11°C	116°F = 46.67°C	161°F = 71.67°C	206°F = 96.67°C
71°F = 21.67°C	117°F = 47.22°C	162°F = 72.22°C	207°F = 97.22°C
72°F = 22.22°C	118°F = 47.78°C	163°F = 72.78°C	208°F = 97.78°C
73°F = 22.78°C	119°F = 48.33°C	164°F = 73.33°C	209°F = 98.33°C
74°F = 23.33°C	120°F = 48.89°C	165°F = 73.89°C	210°F = 98.89°C
75°F = 23.89°C	121°F = 49.45°C	166°F = 74.45°C	211°F = 99.45°C
76°F = 24.45°C	122°F = 50.00°C	167°F = 75.00°C	212°F = 100.00°C
77°F = 25.00°C			

TABLE 6
CONVERTING DEGREES CELSIUS TO DEGREES FAHRENHEIT

0°C = 32.00°F	26°C = 78.80°F	51°C = 123.80°F	76°C = 168.80°F
1°C = 33.80°F	27°C = 80.60°F	52°C = 125.60°F	77°C = 170.60°F
2°C = 35.60°F	28°C = 82.40°F	53°C = 127.40°F	78°C = 172.40°F
3°C = 37.40°F	29°C = 84.20°F	54°C = 129.20°F	79°C = 174.20°F
4°C = 39.20°F	30°C = 86.00°F	55°C = 131.00°F	80°C = 176.00°F
5°C = 41.00°F	31°C = 87.80°F	56°C = 132.80°F	81°C = 177.80°F
6°C = 42.80°F	32°C = 89.60°F	57°C = 134.60°F	82°C = 179.60°F
7°C = 44.60°F	33°C = 91.40°F	58°C = 136.40°F	83°C = 181.40°F
8°C = 46.40°F	34°C = 93.20°F	59°C = 138.20°F	84°C = 183.20°F
9°C = 48.20°F	35°C = 95.00°F	60°C = 140.00°F	85°C = 185.00°F
10°C = 50.00°F	36°C = 96.80°F	61°C = 141.80°F	86°C = 186.80°F
11°C = 51.80°F	37°C = 98.60°F	62°C = 143.60°F	87°C = 188.60°F
12°C = 53.60°F	38°C = 100.40°F	63°C = 145.40°F	88°C = 190.40°F
13°C = 55.40°F	39°C = 102.20°F	64°C = 147.20°F	89°C = 192.20°F
14°C = 57.20°F	40°C = 104.00°F	65°C = 149.00°F	90°C = 194.00°F
15°C = 59.00°F	41°C = 105.80°F	66°C = 150.80°F	91°C = 195.80°F
16°C = 60.80°F	42°C = 107.60°F	67°C = 152.60°F	92°C = 197.60°F
17°C = 62.60°F	43°C = 109.40°F	68°C = 154.40°F	93°C = 199.40°F
18°C = 64.40°F	44°C = 111.20°F	69°C = 156.20°F	94°C = 201.20°F
19°C = 66.20°F	45°C = 113.00°F	70°C = 158.00°F	95°C = 203.00°F
20°C = 68.00°F	46°C = 114.80°F	71°C = 159.80°F	96°C = 204.80°F
21°C = 69.80°F	47°C = 116.60°F	72°C = 161.60°F	97°C = 206.60°F
22°C = 71.60°F	48°C = 118.40°F	73°C = 163.40°F	98°C = 208.40°F
23°C = 73.40°F	49°C = 120.20°F	74°C = 165.20°F	99°C = 210.20°F
24°C = 75.20°F	50°C = 122.00°F	75°C = 167.00°F	100°C = 212.00°F
25°C = 77.00°F			

TABLE 7
WEIGHTS OF MATERIALS

Material	Average Weight in Pounds		Average Weight in Grams per Cm³
	Per Cubic Foot	Per Cubic Inch	
Aluminum	160	0.0924	2.56
Brass	512	0.2960	8.20
Brick, pressed	150		2.40
Brick, common, hard	125		2.00
Cement, American, Rosendale	56		0.90
Cement, Portland	90		1.44
Clay, loose	63		1.01
Coal, broken, loose, Anthracite	54		0.86
Coal, broken, loose, Bituminous	49		0.78
Concrete	154		2.46
Copper	550	0.3184	8.80
Earth, common loam	76		1.22
Earth, packed	95		1.52
Gravel, dry, loose	90 to 106		1.44 to 1.70
Gravel, well shaken	99 to 117		1.58 to 1.87
Gold	1206	0.6975	19.30
Ice	58.7		0.94
Iron, cast	450	0.2600	7.20
Iron, wrought	480	0.2778	7.68
Lead	710	0.4109	11.36
Lime	53		0.85
Masonry, well dressed	165		2.64
Masonry, dry rubble	138		2.21
Mortar, hardened	103		1.65
Nickel	549	0.3177	8.78
Quartz	165		2.64
Sand, dry, loose	90 to 106		1.44 to 1.70
Sand, well shaken	99 to 117		1.58 to 1.87
Silver	657	0.3802	10.51
Snow, freshly fallen	5 to 12		0.08 to 0.19
Snow, wet and compacted	15 to 50		0.24 to 0.80
Steel	490	0.2835	7.84
Stone, gneiss	168		2.67
Stone, granite	170		2.72
Stone, limestone	168		2.69
Stone, marble	168		2.69
Stone, sandstone	151		2.42
Stone, shale	162		2.59
Stone, slate	175		2.80
Tar	62		0.99
Tin	455	0.2632	7.28
Water	62.5		1.00
Wood, dry, ash	38		0.61
Wood, dry, cherry	42		0.67
Wood, dry, chestnut	41		0.66
Wood, dry, elm	35		0.56
Wood, dry, hemlock	25		0.40
Wood, dry, hickory	53		0.85
Wood, dry, lignum vitae	83		1.33
Wood, dry, mahogany	53		0.85
Wood, dry, maple	49		0.78
Wood, dry, oak, white	50		0.80
Wood, dry, oak, other kinds	32 to 45		0.51 to 0.72
Wood, dry, pine, white	25		0.40
Wood, dry, pine, yellow	34 to 45		0.54 to 0.72
Wood, dry, spruce	25		0.40
Wood, dry, sycamore	37		0.59
Wood, dry, walnut, black	38		0.61
Zinc	438	0.2528	7.01

TABLE 8
NATIONAL COARSE-THREAD SERIES

Size	Threads per Inch	Basic Diameters Major Diam. (Inches)	Basic Diameters Minor Diam. (Inches)	Cross-section at Root of Thread (Square Inches)	Tap Drill Size Commercial Size $\frac{3}{4}$ Depth of Thread	Tap Drill Size $\frac{5}{6}$ Depth of Thread
1	64	0.073	0.0538	0.0022	53	0.0561
2	56	.086	.0641	.0031	50	.0667
3	48	.099	.0734	.0041	47	.0764
4	40	.112	.0813	.0050	43	.0849
5	40	.125	.0943	.0067	38	.0979
6	32	.138	.0997	.0075	36	.1042
8	32	.164	.1257	.0120	29	.1302
10	24	.190	.1389	.0145	25	.1449
12	24	.216	.1649	.0206	16	.1709
$\frac{1}{4}$	20	.2500	.1887	.0269	7	.1959
$\frac{5}{16}$	18	.3125	.2443	.0454	F	.2524
$\frac{3}{8}$	16	.3750	.2983	.0678	$\frac{5}{16}$	.3073
$\frac{7}{16}$	14	.4375	.3499	.0933	U	.3602
$\frac{1}{2}$	13	.5000	.4056	.1257	$\frac{27}{64}$	.4167
$\frac{9}{16}$	12	.5625	.4603	.1620	$\frac{31}{64}$	.4723
$\frac{5}{8}$	11	.6250	.5135	.2018	$\frac{17}{32}$	.5266
$\frac{3}{4}$	10	.7500	.6273	.3020	$\frac{21}{32}$	.6417
$\frac{7}{8}$	9	.8750	.7387	.4193	$\frac{49}{64}$	.7547
1	8	1.0000	.8466	.5510	$\frac{7}{8}$	.8647
$1\frac{1}{8}$	7	1.1250	.9497	.6931	$\frac{63}{64}$	.9704
$1\frac{1}{4}$	7	1.2500	1.0747	.8898	$1\frac{7}{64}$	1.0954
$1\frac{3}{8}$	6	1.3750	1.1705	1.0541	$1\frac{7}{32}$	1.1946
$1\frac{1}{2}$	6	1.5000	1.2955	1.2938	$1\frac{11}{32}$	1.3196
$1\frac{3}{4}$	5	1.7500	1.5046	1.7441	$1\frac{9}{16}$	1.5335
2	$4\frac{1}{2}$	2.0000	1.7274	2.3001	$1\frac{25}{32}$	1.7594
$2\frac{1}{4}$	$4\frac{1}{2}$	2.2500	1.9774	3.0212	$2\frac{1}{32}$	2.0094
$2\frac{1}{2}$	4	2.5000	2.1933	3.7161	$2\frac{1}{4}$	2.2294
$2\frac{3}{4}$	4	2.7500	2.4433	4.6194	$2\frac{1}{2}$	2.4794
3	4	3.0000	2.6933	5.6209	$2\frac{23}{32}$	2.7294
$3\frac{1}{4}$	4	3.2500	2.9433	6.7205	$2\frac{31}{32}$	2.9794
$3\frac{1}{2}$	4	3.5000	3.1933	7.9183	$3\frac{7}{32}$	3.2294
$3\frac{3}{4}$	4	3.7500	3.4433	9.2143	$3\frac{15}{32}$	3.4794
4	4	4.0000	3.6933	10.6084	$3\frac{23}{32}$	3.7294

TABLE 9
NATIONAL FINE-THREAD SERIES

Size	Threads per Inch	Basic Diameters		Cross-section at Root of Thread (Square Inches)	Tap Drill Size	
		Major (Inches)	Minor (Inches)		Commercial Size $\frac{3}{4}$ Depth of Thread	$\frac{5}{6}$ Depth of Thread
0	80	0.060	0.0447	0.0015	$\frac{3}{64}$	0.0465
1	72	.073	.0560	.0024	53	.0580
2	64	.086	.0668	.0034	50	.0691
3	56	.099	.0771	.0045	45	.0797
4	48	.112	.0864	.0057	42	.0894
5	44	.125	.0971	.0072	37	.1004
6	40	.138	.1073	.0087	33	.1109
8	36	.164	.1299	.0128	29	.1339
10	32	.190	.1517	.0175	21	.1562
12	28	.216	.1722	.0226	14	.1773
$\frac{1}{4}$	28	.2500	.2062	.0326	3	.2113
$\frac{5}{16}$	24	.3125	.2614	.0524	I	.2674
$\frac{3}{8}$	24	.3750	.3239	.0809	Q	.3299
$\frac{7}{16}$	20	.4375	.3762	.1090	$\frac{25}{64}$	.3834
$\frac{1}{2}$	20	.5000	.4387	.1486	$\frac{29}{64}$	.4459
$\frac{9}{16}$	18	.5625	.4943	.1888	$\frac{33}{64}$	.5024
$\frac{5}{8}$	18	.6250	.5568	.2400	$\frac{37}{64}$	.5649
$\frac{3}{4}$	16	.7500	.6733	.3513	$\frac{11}{16}$	.6823
$\frac{7}{8}$	14	.8750	.7874	.4805	$\frac{13}{16}$	.7977
1	14	1.0000	.8978	.6464	$\frac{15}{16}$	.9227
$1\frac{1}{8}$	12	1.1250	1.0228	.8118	$1\frac{3}{64}$	1.0348
$1\frac{1}{4}$	12	1.2500	1.1478	1.0238	$1\frac{11}{64}$	1.1598
$1\frac{3}{8}$	12	1.3750	1.2728	1.2602	$1\frac{19}{64}$	1.2848
$1\frac{1}{2}$	12	1.5000	1.3978	1.5212	$1\frac{27}{64}$	1.4098

Answers to Self-Tests

Chapter 1
1. 11,576
2. 15,888
3. 479
4. 3,440,596
5. 5,490,000
6. 21,772,089
7. 13
8. 25
9. $170,558,830,458
10. 9,828,898,692 bills

Chapter 2
1. $\frac{7}{8}$
2. $7\frac{1}{4}$
3. $\frac{5}{16}$
4. $1\frac{7}{8}$
5. $\frac{8}{15}$
6. $2\frac{5}{8}$
7. $\frac{2}{3}$
8. $1\frac{2}{3}$
9. 1
10. $\frac{10}{27}$

Chapter 3
1. 0.627
2. 2.102
3. 4.704
4. 5.81
5. 2.4
6. 20.21
7. 10.6
8. 1.838
9. 1.065
10. 73.5

Chapter 4
1. 50%
2. 33.33%
3. 7.5
4. 20%
5. 200
6. $2,640
7. 0.8
8. 1
9. 3%
10. $184.40

Chapter 5
1. 4 to 3
2. 1 to 4
3. 4 to 3
4. 5 to 6
5. 1 to 2
6. 1.08 ohms
7. 1 to 2
8. $60; $40
9. 7 lb
10. $7,800

Chapter 6
1. 1
2. 11
3. 3
4. 15
5. 336
6. $9ab^2$
7. $8x^2y$
8. $6x^2 - 11xy - 10y^2$
9. $x + y$
10. $a = \dfrac{3V}{\pi r^2}$

Chapter 7
1. 80 sq in.
4. 13 in.
7. 10 in.
10. 43.3 sq in.

2. 36 in.
5. 1,600
8. 51°

3. 12 in.
6. 33 in.
9. 21 sq in.

Chapter 8
1. 5.20 in.
4. 18.85 in.
7. 0.59 sq in.
10. 16.02 in.

2. 18 sq in.
5. 78.54 sq in.
8. 4.71 sq in.

3. 3.46 in.
6. 14.14 in.
9. 94.25 sq in.

Chapter 9
1. 160 cu in.
4. 25.62 sq in.
7. 1,089 lb
10. 168 bd ft

2. 63.58 cu in.
5. 113.04 cu in.
8. 180 lb

3. 33.49 cu in.
6. 7.4 cu in.
9. 672 bd ft

Chapter 10
1. 4,800 m
4. 9 mg
7. 16.13 cm^2
10. 37.8°

2. 30 cm^2
5. 63.8 cm
8. 3,540 cm^3

3. 3,663 cm^3
6. 348.4 cm^3
9. 22.7 kg

Chapter 14
1. 3
4. 1.6609
7. 4.444
10. 12.79

2. 2
5. $\overline{2}$.7443
8. 32,360

3. $\overline{2}$
6. 23,390
9. 0.7522

Chapter 15
1. 0.5
4. 24.9″
7. 20.3 mm
10. 713.7 mm^2

2. 50°
5. 19.1″
8. 120°

3. 0.6428
6. 40.6 mm
9. 70.3 mm

Chapter 17
1. 0.5 hp
4. 182.8 hp
7. 5.5 amp
10. 1.1 hp

2. 960 ft-lb
5. 1.2 kw
8. 87%

3. 228.4 hp
6. 0.5 hp
9. 0.86 hp

Chapter 18
1. 0.24 in.
2. 0.4 in.
3. 0.0625 in.
4. 0.375 in.
5. 0.25 in.
6. 0.028 in.
7. 0.336 in.
8. 0.28 in.
9. 3°34'
10. 0°54'

Chapter 19
1. 90 rpm
2. 40 teeth
3. 90 rpm
4. 7,000 rpm
5. 3 rpm
6. 1,312.5 rpm
7. 4.4 in.
8. 583 rpm; 729 rpm; 862 rpm
9. no
10. clockwise

Chapter 20
1. $\frac{1}{13}$-in. pitch
2. $\frac{1}{4}$ in.
3. $\frac{1}{10}$ in.
4. 0.096 in.
5. 0.068 in.
6. 0.185 in.
7. 0.201 or 7 drill
8. 0.573 in.
9. 0.057 in.
10. $\frac{2}{3}$

Chapter 21
1. 65.5 ft per min
2. 1,886 ft per min
3. 1,909 rpm
4. 14,667 ft per min
5. 42 ft per min
6. 9.5 in.
7. 24 ft per min
8. 32 ft per min
9. 1.25 min
10. 0.01 in.

Chapter 22
1. 0.3927 in.
2. 4.5 in.
3. 48
4. 16 pitch
5. 3 in.
6. 0.3927 in.
7. 0.0196 in.
8. 0.1798 in.
9. 4.167 in.
10. 7 in.

Index

VOLUME, CAPACITY, AND WEIGHT EQUIVALENTS

1 gallon = 231 cubic inches
1 cubic foot = $7\frac{1}{2}$ gallons (approximately)
1 bushel = $1\frac{1}{4}$ cubic feet
= 2,150.42 cubic inches
1 cubic foot of water weighs $62\frac{1}{2}$ pounds (approximately)

DRY MEASURE

1 quarter = 2 pints = 67.2 cu in. = 1.101 liters
1 peck = 8 quarts = 537.6 cu in. = 8.81 liters
1 bushel = 4 pecks = 1.244 cu ft = 35.24 liters
1 cord of wood = 128 cu ft = 4 ft × 4 ft × 8 ft = 3.625 m³

FORMULAS

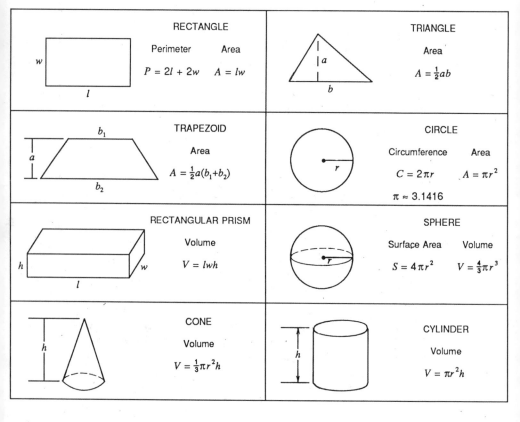

RECTANGLE

Perimeter   Area

$P = 2l + 2w$ $A = lw$

TRIANGLE

Area

$A = \frac{1}{2}ab$

TRAPEZOID

Area

$A = \frac{1}{2}a(b_1 + b_2)$

CIRCLE

Circumference   Area

$C = 2\pi r$ $A = \pi r^2$

$\pi \approx 3.1416$

RECTANGULAR PRISM

Volume

$V = lwh$

SPHERE

Surface Area   Volume

$S = 4\pi r^2$ $V = \frac{4}{3}\pi r^3$

CONE

Volume

$V = \frac{1}{3}\pi r^2 h$

CYLINDER

Volume

$V = \pi r^2 h$